ANNALS OF
THE NEW YORK ACADEMY
OF SCIENCES

Volume 1046

EDITORIAL STAFF

Director, Publishing and New Media
SARAH GREENE

Managing Editor
JUSTINE CULLINAN

Associate Editor
TRUMBULL ROGERS

The New York Academy of Sciences
2 East 63rd Street
New York, New York 10021

THE NEW YORK ACADEMY OF SCIENCES
(Founded in 1817)

BOARD OF GOVERNORS, September 2004 – September 2005

TORSTEN N. WIESEL, *Chairman of the Board*
GERALD D. FISCHBACH, *Vice Chairman*
MICHAEL SCHMERTZLER, *Treasurer*
ELLIS RUBINSTEIN, *Chief Executive Officer* [ex officio]

Honorary Life Governors
WILLIAM T. GOLDEN   JOSHUA LEDERBERG

Governors

KAREN E. BURKE	VIRGINIA W. CORNISH	PETER B. CORR
R. BRIAN FERGUSON	RONALD L. GRAHAM	MARNIE IMHOFF
WENDY EVANS JOSEPH	JACQUELINE LEO	ROBERT W. LUCKY
PAUL MARKS	BRUCE McEWEN	RONAY MENSCHEL
JOHN T. MORGAN	JOHN F. NIBLACK	SANDRA PANEM
PETER RINGROSE		DAVID D. SABATINI

VICTORIA BJORKLUND, *Counsel* [ex officio]   LARRY R. SMITH, *Secretary* [ex officio]

ANNALS OF THE NEW YORK ACADEMY OF SCIENCES
Volume 1046

BIRD HORMONES AND BIRD MIGRATIONS

ANALYZING HORMONES IN DROPPINGS AND EGG YOLKS AND ASSESSING ADAPTATIONS IN LONG-DISTANCE MIGRATION

Edited by Ulf Bauchinger, Wolfgang Goymann, and Susanne Jenni-Eiermann

The New York Academy of Sciences
New York, New York
2005

Copyright © 2005 by the New York Academy of Sciences. All rights reserved. Under the provisions of the United States Copyright Act of 1976, individual readers of the Annals *are permitted to make fair use of the material in them for teaching or research. Permission is granted to quote from the* Annals *provided that the customary acknowledgment is made of the source. Material in the* Annals *may be republished only by permission of the Academy. Address inquiries to the Permissions Department (editorial@nyas.org) at the New York Academy of Sciences.*

Copying fees: *For each copy of an article made beyond the free copying permitted under Section 107 or 108 of the 1976 Copyright Act, a fee should be paid through the Copyright Clearance Center, Inc., 222 Rosewood Drive, Danvers, MA 01923 (www.copyright.com).*

∞ *The paper used in this publication meets the minimum requirements of the American National Standard for Information Sciences—Permanence of Paper for Printed Library Materials, ANSI Z39.48-1984.*

Library of Congress Cataloging-in-Publication Data

Bird hormones and bird migrations: analyzing hormones in droppings and egg yolks and assessing adaptations in long-distance migration / edited by Ulf Bauchinger, Wolfgang Goymann, and Susanne Jenni-Eiermann.
    p. cm. – (Annals of the New York Academy of Sciences ; v. 1046)
  Papers from a technical meeting on the analysis of hormones in droppings and egg yolk of birds, and a workshop on bird migration.
  Includes bibliographical references and index.
  ISBN 1-57331-576-1 (cloth : alk. paper) – ISBN 1-57331-577-X (pbk. : alk. paper)
  1. Birds–Endocrinology. 2. Bird droppings–Analysis. 3. Birds–Migration. I. Bauchinger, Ulf II. Goymann, Wolfgang. III. Jenni-Eiermann, Susanne. IV. Series.
  Q11.N5 vol. 1046
  [QL698]
  500 s–dc22
  [573.4/4
                                      2005010183

GYAT/PCP
Printed in the United States of America
ISBN 1-57331-576-1 (cloth)
ISBN 1-57331-577-X (paper)
ISSN 0077-8923

ANNALS OF THE NEW YORK ACADEMY OF SCIENCES
Volume 1046
June 2005

BIRD HORMONES AND BIRD MIGRATIONS

ANALYZING HORMONES IN DROPPINGS AND EGG YOLKS AND ASSESSING ADAPTATIONS IN LONG-DISTANCE MIGRATION

Editors
ULF BAUCHINGER, WOLFGANG GOYMANN, AND SUSANNE JENNI-EIERMANN

Conference Organizers
Hormone Workshop
SUSANNE JENNI-EIERMANN AND WOLFGANG GOYMANN

Migration Workshop
ULF BAUCHINGER, CHRISTIAAN BOTH, AND THEUNIS PIERSMA

This volume is the result of two workshops entitled **Analysis of Hormones in Droppings of Birds and Optimality in Bird Migration: Are There Specific Adaptations for Long-Distance Migration in Birds? The Search for Adaptive Syndromes**, sponsored by the European Science Foundation, held on October 20–23, 2004 (Hormones Workshop) and September 16–19, 2004 (Migration Workshop), in Seewiesen, Germany.

CONTENTS

Two Workshops in Ornithology: A General Introduction. Dedication to Eberhard Gwinner. *By* ULF BAUCHINGER AND WOLFGANG GOYMANN .. ix

Measurement of Hormones from Droppings and Egg Yolk of Birds

Introduction to the European Science Foundation Technical Meeting: Analysis of Hormones in Droppings and Egg Yolk of Birds. *By* WOLFGANG GOYMANN AND SUSANNE JENNI-EIERMANN 1

Potential Impact of Nutritional Strategy on Noninvasive Measurements of Hormones in Birds. *By* KIRK C. KLASING 5

Measurement of Corticosterone Metabolites in Birds' Droppings: An Analytical Approach. *By* ERICH MÖSTL, SOPHIE RETTENBACHER, AND RUPERT PALME 17

Noninvasive Monitoring of Hormones in Bird Droppings: Physiological Validation, Sampling, Extraction, Sex Differences, and the Influence of Diet on Hormone Metabolite Levels. *By* WOLFGANG GOYMANN 35

Measuring Fecal Glucocorticoid Metabolites in Mammals and Birds: The Importance of Validation. *By* CHADI TOUMA AND RUPERT PALME 54

Measuring Fecal Steroids: Guidelines for Practical Application. *By* RUPERT PALME 75

A Noninvasive Technique to Evaluate Human-Generated Stress in the Black Grouse. *By* MARJANA BALTIC, SUSANNE JENNI-EIERMANN, RAPHAËL ARLETTAZ, AND RUPERT PALME 81

Measuring Corticosterone Metabolites in Droppings of Capercaillies (*Tetrao urogallus*). *By* DOMINIK THIEL, SUSANNE JENNI-EIERMANN, AND RUPERT PALME 96

Noninvasive Measures of Reproductive Function and Disturbance in the Barred Owl, Great Horned Owl, and Northern Spotted Owl. *By* SAMUEL K. WASSER AND KATHLEEN E. HUNT 109

Synthesis of Measuring Steroid Metabolites in Goose Feces. *By* KATHARINA HIRSCHENHAUSER, KURT KOTRSCHAL, AND ERICH MÖSTL 138

Sampling Effort/Frequency Necessary to Infer Individual Acute Stress Responses from Fecal Analysis in Greylag Geese (*Anser anser*). *By* ISABELLA B. R. SCHEIBER, SIMONA KRALJ, AND KURT KOTRSCHAL . 154

Investigating Maternal Hormones in Avian Eggs: Measurement, Manipulation, and Interpretation. *By* TON G. G. GROOTHUIS AND NIKOLAUS VON ENGELHARDT 168

Measuring Steroid Hormones in Avian Eggs. *By* NIKOLAUS VON ENGELHARDT AND TON G. G. GROOTHUIS 181

Corticosterone in Chicken Eggs. *By* S. RETTENBACHER, E. MÖSTL, R. HACKL, AND R. PALME 193

Steroids in Allantoic Waste: An Integrated Measure of Steroid Exposure in *Ovo*. *By* Z MORGAN BENOWITZ-FREDERICKS, ALEXANDER S. KITAYSKY, AND JOHN C. WINGFIELD 204

Optimality in Bird Migration: Adaptations for Long-Distance Migration in Birds? The Search for Adaptive Syndromes

Are There Specific Adaptations for Long-Distance Migration in Birds? The Search for Adaptive Syndromes: Outline of the European Science Foundation Workshop: *By* ULF BAUCHINGER, CHRISTIAAN BOTH, AND THEUNIS PIERSMA 214

Flexible Seasonal Timing and Migratory Behavior: Results from Stonechat Breeding Programs. *By* BARBARA HELM, EBERHARD GWINNER, AND LISA TROST 216

Are Long-Distance Migrants Constrained in Their Evolutionary Response to Environmental Change? Causes of Variation in the Timing of Autumn Migration in a Blackcap (*S. atricapilla*) and in Two Garden Warbler (*Sylvia borin*) Populations . *By* FRANCISCO PULIDO AND MICHAEL WIDMER 228

Spatial Behavior of Medium and Long-Distance Migrants at Stopovers Studied by Radio Tracking. *By* NIKITA CHERNETSOV 242

Ecomorphology of the External Flight Apparatus of Blackcaps (*Sylvia atricapilla*) with Different Migration Behavior. *By* WOLFGANG FIEDLER 253

Melatonin and Nocturnal Migration. *By* LEONIDA FUSANI AND EBERHARD GWINNER 264

Phenotypic Flexibility of Skeletal Muscles during Long-Distance Migration of Garden Warblers: Muscle Changes Are Differentially Related to Body Mass. *By* ULF BAUCHINGER AND HERBERT BIEBACH 271

Is There a "Migratory Syndrome" Common to All Migrant Birds? *By* THEUNIS PIERSMA, JAVIER PÉREZ-TRIS, HENRIK MOURITSEN, ULF BAUCHINGER, AND FRANZ BAIRLEIN 282

Index of Contributors 295

Financial assistance was received from:

- EUROPEAN SCIENCE FOUNDATION
- NATIONAL SCIENCE FOUNDATION

> The New York Academy of Sciences believes it has a responsibility to provide an open forum for discussion of scientific questions. The positions taken by the participants in the reported conferences are their own and not necessarily those of the Academy. The Academy has no intent to influence legislation by providing such forums.

EBO GWINNER DURING ONE OF HIS LAST FIELD TRIPS IN SPRING 2004.
(PHOTOGRAPH COURTESY OF DIETER SCHMIDL.)

Two Workshops in Ornithology: A General Introduction

Dedication to Eberhard Gwinner

ULF BAUCHINGER[a] AND WOLFGANG GOYMANN[b]

[a]*University of Munich (LMU), Department of Biology II, Section of Functional Morphology, 82152 Martinsried, Germany*

[b]*Max Planck Institute for Ornithology, D-82346 Andechs, Germany*

In October 2004 and in January 2005, two workshops in ornithology were held at the Max Planck Institute for Ornithology. The first workshop was a technical meeting within the E-BIRD network entitled Adaptation and Constraints in Avian Reproduction: Integrating Ecology and Endocrinology. This workshop analyzed hormones in droppings and egg yolk of birds. The theme of the second workshop asked the question, "Are there specific adaptations for long-distance migration in birds? The search for adaptive syndromes," and was held within the framework of the European Science Foundation (ESF) program entitled Optimality in Bird Migration.

Both meetings took place during very difficult times for the Max Planck Institute for Ornithology, as the director of the Department of Biological Rhythms and Behaviour, Eberhard Gwinner, unexpectedly passed away on September 7, 2004, after a short but severe illness. Ebo Gwinner was the founding director of the Max Planck Research Centre for Ornithology in Andechs and Seewiesen, which is now the Max Planck Institute for Ornithology. Thanks to his outstanding scientific reputation, his tremendous capacity for work, and his diplomatic skillfulness and persistence, Ebo Gwinner had gained an international reputation for the Institute. Beyond this, in the recent difficult days of restructuring and closure of many Max Planck Institutes, he had successfully managed to maintain and even expand the Max Planck Institute for Ornithology as a stable unit.

We, the organizers of both meetings, had the pleasure to work under Ebo Gwinner's protective wings while we were still graduate students at the Max Planck Institute. His passion for science and birds was clearly evident. Hands-on work with birds, both in the lab and in the field, and caring for the animals—even on Christmas Eve or during weekends off—were a matter of course for him and gave us an immediate feeling of his sense of responsibility.

Ebo encouraged his students to discuss their data critically, and he gave them the opportunity for frequent scientific exchange. He organized an outstanding series of seminars, bringing renowned speakers from all over the world to the Institute. More-

---

Address for correspondence: Ulf Bauchinger, Ph.D., University of Munich (LMU), Department of Biology II, Section of Functional Morphology, Grosshadernder Str. 2, 82152 Martinsried, Germany. Voice: +4989218074132; fax: +498921809974134.
 Bauchinger@lmu.de

over, he encouraged numerous international guests, including Ph.D. students, postdocs, and established scientists, to stay and do research at the Institute. Many scientific collaborations and friendships have developed from this.

Ebo was always open-minded about initiating new areas of research, even if risky and new techniques had to be established. His support and encouragement could be relied upon at (almost!) any time. He frequently said "No!" in the beginning, but after a second and third thought the "No!" eventually turned into a supportive "Yes!" Ebo's modesty, thoughtfulness, and liberal spirit made it easy for his students to do creative work within a family-like atmosphere at the Institute.

Today, after his untimely death, it is still very important for us to have experienced Ebo as a person and mentor. On many occasions, we benefited from his immense human warmth and friendship. He enjoyed dancing at parties, and although he avoided being the focus of attention, he loved to talk about birds, nature, music, sports, and everything else during social get-togethers. Appreciating a fine vintage, he could very well be a quiet listener at a crackling campfire. We will never forget his *joie de vivre,* his warm personality, and his creative scientific work; all that will keep on living vividly in our memories.

The three large and complex themes of the two workshops—bird migration, endogenous rhythms, and endocrinology—cover some of the main research topics of Ebo Gwinner and his department at the Max Planck Institute. It was Ebo who initially encouraged us to organize these meetings. During the preparation of both meetings we received his unlimited support. Ebo was looking forward to his own active participation, which unfortunately was no longer granted to him. On behalf of all participants, we would like to dedicate, in sincere gratitude, the proceedings of both workshops to him.

Introduction to the European Science Foundation Technical Meeting

Analysis of Hormones in Droppings and Egg Yolk of Birds

WOLFGANG GOYMANN[a] AND SUSANNE JENNI-EIERMANN[b]

[a]*Max Planck Institute for Ornithology, Department of Biological Rhythms and Behaviour, D-82346 Andechs, Germany*

[b]*Swiss Ornithological Institute, CH-6204 Sempach, Switzerland*

In 2002, a workshop entitled "Adaptations and Constraints in Avian Reproduction: Integrating Ecology and Endocrinology" took place in Wageningen, the Netherlands. The aim of the workshop was to build a network between scientists of different disciplines to exchange their knowledge and methods. This initiative resulted in the formation of the E-Bird network with the same name: "Adaptation and Constraints in Avian Reproduction: Integrating Ecology and Endocrinology." Within this E-Bird framework we organized this technical meeting on the analysis of hormones from bird droppings and egg yolk, supported both by the European Science Foundation and the U.S.-based National Science Foundation (ESF and NSF).

Starting in the late 1970s and early 1980s with pioneering studies in birds[1,2] and mammals,[3] noninvasive methods to measure steroid hormone levels in feces have been developed for an increasing number of species.[4] In birds, the term "fecal analysis" is a bit problematic, because feces and urine are excreted together in the form of droppings and in general, a complete separation of the fecal and urine fraction is not possible.[5] Hence, it may be more appropriate to speak of excreta or droppings.

Increasing interest in noninvasive methods to monitor hormonal profiles in birds and other vertebrates, combined with common misconceptions about the ease of applying such methods, triggered the idea to organize this technical meeting. The intention was to discuss this method among a small group of experts and come up with some guidelines for its proper application.[6] During the organizational phase of the meeting it became clear that the analysis of hormones from egg yolk is accompanied by very similar methodological issues. Hence, we decided to extend the scope of the meeting and include the analysis of hormones from egg yolk. Although this meeting was held with a focus on birds, most contributions are of equal importance for studies in mammals and other vertebrates.

Noninvasive methods to measure hormones from droppings have the major advantage that samples can be collected repeatedly with no handling stress. Hence, re-

Address for correspondence: Dr. Wolfgang Goymann, Max Planck Institute for Ornithology, Von-der-Tann-Str. 7, D-82346 Andechs, Germany. Voice: +49-8152-373-119; fax: +49-8152-373-133.

goymann@orn.mpg.de

searchers of various disciplines, such as conservation, physiology, animal behavior, veterinary medicine, and reproductive and zoo biology, are interested in applying this method. However, one major problem is that the hormones are extensively metabolized before excretion and that the metabolites may differ between species, sexes, and to a minor degree, even between individuals.[4,6–8] Thus, before applying the method, each investigator needs proof that the immunoreactivity of the assay used truly measures metabolites of the hormone in question. The articles in this volume extensively discuss the biochemical aspects of the method, stress the importance of analytical and physiological validations,[4,6–8] and provide some good examples for applications.[9–11]

Klasing[5] provides an overview of the nutritional strategies and digestive tracts of birds, and discusses the potential impact of diet and nutritional strategies on the composition of feces and hormone metabolites. Goymann,[7] Möstl et al.,[8] and Touma and Palme[4] stress the importance of physiological validations and provide examples of why these validations are so crucial. Möstl et al.[8] focus on the biochemical aspects of validations, whereas Goymann[7] uses the example of the European stonechat (*Saxicola torquata rubicola*) to explain why physiological validations are crucial for the interpretation of the data. He demonstrates sex differences in the formation and excretion of metabolites, and describes the impact of different diets on metabolite levels in birds. Further studies on the impact of diet on hormone metabolite levels seem especially important in birds, because their diet may change on a seasonal basis to a much larger degree than in most mammals. Furthermore, the digestive organs of birds may go through major changes during different life-history stages, as some of the contributions of the second workshop on bird migration suggest. Touma and Palme[4] provide an overview for mammalian and avian species, for which physiologically validated methods to measure steroid metabolites noninvasively are so far available, and again stress the importance of analytical, physiological, and biological validations. Palme[6] summarizes these previous chapters and offers guidelines on what is important to consider when conducting noninvasive hormone research. Baltic et al.,[9] Thiel et al.,[10] and Wasser and Hunt[11] provide examples for extensive biochemical and physiological validations to measure corticosteroids in black grouse (*Tetrao tetrix*), capercaillies (*Tetrao urogallus*), and three North American owl species. Hirschenhauser et al.[12] provide an overview of noninvasive hormone research in greylag geese (*Anser anser*), probably the bird species for which the most extensive data regarding noninvasive hormone research are available so far. Finally, Scheiber et al.[13] explore how many samples from an individual are necessary to measure an acute stress response in greylag geese. Although the focus of all these contributions was on birds, most of the issues and problems discussed apply equally well to mammals and other vertebrates.

The analysis of hormones from egg yolk was pioneered by Hubert Schwabl in the early 1990s,[14] and since then has become a extensively used tool for the study of maternal effects. In two complementary contributions, Groothuis and von Engelhardt[15] and von Engelhardt and Groothuis[16] critically evaluate the surge in studies of avian egg hormones. The first study focuses on conceptual issues, including the biological interpretation of the data,[15] whereas the second study focuses on methodology, that is, issues regarding the actual measurement of steroid hormones from egg yolk.[16] So far, most studies on yolk steroids have concentrated on androgens or sex-steroids. For the first time, Rettenbacher et al.[17] demonstrate if, how, and when maternal adrenocortical activity is reflected by corticosterone concentrations in the egg of domestic chickens. Finally, Benowitz-Fredericks et al.[18] close the circle by combining the analysis of hormones in droppings and egg yolk of birds by measuring "allantoic waste," a green, sticky, and pasty substance in the eggshell that includes several components: allantois, extra embryonic membranes, blood vessels, and sometimes a "true" fecal/urine sample excreted by the embryo during hatching. This sample may potentially be used to study the combined effects of maternal and endogenous hormone levels *in ovo*.

We are grateful to the European Science Foundation, the National Science Foundation, the late Ebo Gwinner, and the Max Planck Society for generously funding this meeting.

Without the help of several people, the organization of this meeting would have been much more difficult and certainly less fun. We particularly thank Raimund Barth, Helga Gwinner, Barbara Helm, Anke Hundrisser, Christina Muck, Michael Raess, Christian Völk, and the administrative staff of the Max Planck Institute for Ornithology, all of whom helped to make this meeting a success. Very special thanks go to Nicole Hoiss for her thoughtful and professional assistance during the coordination and organization of the meeting, as well as her patience and dedication while editing and handling correspondence for all manuscripts for this issue of the *Annals* of the New York Academy of Sciences. Finally, we thank the people in the editorial office of the *Annals,* in particular Linda Hotchkiss Mehta, Ralph W. Brown, and Trumbull Rogers, for very smooth, friendly, and professional cooperation. We trust that they will now view the bird droppings on the windshields of their cars in a different light.

REFERENCES

1. CZEKALA, N.M. & B.L. LASLEY. 1977. A technical note on sex determination in monomorphic birds using fecal steroid analysis. Int. Zoo Yearb. **17:** 209–211.
2. BERCOVITZ, A.B. *et al.* 1982. Noninvasive assessment of seasonal hormone profile in captive bald eagles (*Haliaeetus leucocephalus*). Zoo Biol. **1:** 111–117.
3. MÖSTL, E. *et al.* 1984. Pregnancy diagnosis in cows and heifers by determination of oestradiol-17α in faeces. Br. Vet. J. **140:** 287–291.
4. TOUMA, C. & R. PALME. 2005. Measuring fecal glucocorticoid metabolites in mammals and birds: the importance of validation. Ann. N.Y. Acad. Sci. **1046:** 54–74.
5. KLASING, K.C. 2005. Potential impact of nutritional strategy on noninvasive measurements of hormones in birds. Ann. N.Y. Acad. Sci. **1046:** 5–16.
6. PALME, R. 2005. Measuring fecal steroids: guidelines for a practical application. Ann. N.Y. Acad. Sci. **1046:** 75–80.
7. GOYMANN, W. 2005. Noninvasive monitoring of hormones in bird droppings: biological validations, sampling, extraction, sex differences, and the influence of diet on hormone metabolite levels. Ann. N.Y. Acad. Sci. **1046:** 35–53.
8. MÖSTL, E., S. RETTENBACHER & R. PALME. 2005. Measurement of corticosterone metabolites in birds' droppings: an analytical approach. Ann. N.Y. Acad. Sci. **1046:** 17–34.
9. BALTIC, M. *et al.* 2005. A noninvasive technique to evaluate human-generated stress in the black grouse. Ann. N.Y. Acad. Sci. **1046:** 96–108.
10. THIEL, D., S. JENNI-EIERMANN & R. PALME. 2005. Measuring corticosterone metabolites in droppings of Capercaillies (*Tetrao urogallus*). Ann. N.Y. Acad. Sci. **1046:** 96–108.
11. WASSER, S.K. & K. HUNT. 2005. Noninvasive measures of reproductive function and disturbance in the barred owl, great horned owl, and northern spotted owl. Ann. N.Y. Acad. Sci. **1046:** 109–137.

12. HIRSCHENHAUSER, K., K. KOTRSCHAL & E. MÖSTL. 2005. Synthesis of measuring steroid metabolites in goose feces. Ann. N.Y. Acad. Sci. **1046:** 138–153.
13. SCHEIBER, I., S. KRALJ & K. KOTRSCHAL. 2005. Sampling effort/frequency necessary to infer individual acute stress responses from fecal analysis in greylag geese (*Anser anser*). Ann. N.Y. Acad. Sci. **1046:** 154–167.
14. SCHWABL, H. 1993. Yolk is a source of maternal testosterone for developing birds. Proc. Natl. Acad. Sci. U.S.A. **90:** 11446–11450.
15. GROOTHUIS, T.G. & N. VON ENGELHARDT. 2005. Investigating maternal hormones in avian eggs: measurement, manipulation, and interpretation. Ann. N.Y. Acad. Sci. **1046:** 168–180.
16. VON ENGELHARDT, N. & T.G.G. GROOTHUIS. 2005. Measuring steroid hormones in avian eggs. Ann. N.Y. Acad. Sci. **1046:** 181–192.
17. RETTENBACHER, S. *et al.* 2005. Corticosterone in chicken eggs. Ann. N.Y. Acad. Sci. **1046:** 193–203.
18. BENOWITZ-FREDERICKS, Z.M., A.S. KITAYSKY & J.C. WINGFIELD. 2005. Steroids in allantoic waste—an integrated measure of steroid exposure *in ovo*. Ann. N.Y. Acad. Sci. **1046:** 204–213.

Potential Impact of Nutritional Strategy on Noninvasive Measurements of Hormones in Birds

KIRK C. KLASING

Department of Animal Science, University of California, Davis, California 95616, USA

ABSTRACT: The dietary preferences, gastrointestinal anatomy, digestive physiology, biochemical capabilities, and commensal microflora of a bird are collectively known as its nutritional strategy. Measurement of hormones in droppings requires an appreciation of an animal's nutritional strategy in order to optimize collection protocols, validate techniques, interpret results, and minimize variability and artifacts. Foods of animal origin, nectar, and seeds are highly digestible by relatively simple digestive tracts and result in low rates of feces production. Most frugivorous species also have simple digestive tracts, and they digest the fruit's simple sugars and proteins, but not the fiber in its pulp. Consequently, retention time of food in the digestive tract is short, and their droppings are voluminous. Herbivorous species possess enlarged ceca that house microorganisms that aid in the digestion of fibrous components of their food. Part of the digesta enters the ceca and is subjected to lengthy microbial fermentation. The rest is excluded and quickly passes through the rectum, and is quickly defecated. For measurement of hormones in droppings it appears prudent to collect only rectal feces and to avoid cecal feces. One-third of the avian families are omnivorous and consume a wide variety of foods. Their digestive strategies are highly variable and change with diet, as does the amount and composition of feces and the rate of passage.

KEYWORDS: nutrition; intestines; anatomy; fiber; excretion; droppings

INTRODUCTION

The dietary preferences, gastrointestinal anatomy, digestive physiology, biochemical capabilities, and commensal microflora of a bird are collectively known as its "nutritional strategy." Nutritional strategies are diverse and permit specialization on foods ranging from nectar to grass to other vertebrates. The nutritional strategy determines the excretion rate, metabolism, and resorption of food components and endogenous losses, including hormones. The large variation in nutritional strategies is an impediment of interspecies comparisons of hormone levels in droppings. Seasonal variations in nutritional strategies may, in some cases, affect intraspecies comparisons made over time, and local variations in available foods may affect

Address for correspondence: Kirk C. Klasing, Department of Animal Science, University of California, 1 Shields Ave., Davis, CA 95616. Voice: 530-752-1901; fax: 530-752-0175.
kcklasing@ucdavis.edu

comparisons between populations. In birds, there is an added complication due to the simultaneous voiding of excreta originating from the intestinal tract with excreta originating from the urinary tract. To avoid confusion, the term "feces" will be used to refer to excreta of intestinal origin, and the term "droppings" will be used to refer to the combined fecal and urinary excreta. An appreciation of an animal's nutritional strategy is helpful in designing excreta collection protocols, validating techniques, interpreting results, and minimizing variability and artifacts. This article provides a general overview of the nutritional strategies of birds and focuses on those processes most likely to affect the excretion of endogenous compounds, using steroid hormones as an example.

NUTRITIONAL STRATEGIES

The specific gastrointestinal anatomy, digestive physiology, and metabolic capabilities of each species have coevolved with the types of foods consumed so that species eating similar food types often have similar nutritional strategies.[1–3] The rate of passage of digesta, microbial metabolism of endogenous compounds, and dilution of endogenous compounds by undigested food components often can be predicted by the dietary preferences of a species (TABLE 1). The primary factor affecting digestive anatomy and fecal composition is the amount of plant fiber in the diet. Consequently, species that consume high-fiber foods such as grasses, buds, shoots, and leaves have very different nutritional strategies than species that consume animal matter or readily digestible plant components, such as nectar, fruits, and seeds.

Dietary Preferences

Across the more than 9000 species of birds, there is an almost continuous distribution of dietary patterns and, consequently, digestive-tract morphologies, from the simplest, which belong to nectarivores and frugivores, to the most elaborate, which belong to terrestrial herbivores.[1–3] However, the largest number of species are toward the center of this distribution, including insectivores, granivores, and omnivores. The types of foods consumed by a species can have considerable seasonal variability, especially in temperate climates, and the nutritional strategy adjusts accordingly. For example, the mixture of foods consumed by many omnivores is often higher in insects during the summer and higher in foods of plant origin during the winter. This fact has been detailed in rock ptarmigan, in which the winter shift to high-fiber foods is matched by increased cecal size, microbial fermentation, and retention time of the digesta.[4] Some species make even more extreme annual shifts. A common dietary pattern is a diet of insects during the summer and a complete switch to fruits during the winter. The change in digestive anatomy and physiology after such switches can be dramatic.[5–7]

Digestive Anatomy

Avian digestive tracts have more organs than their mammalian counterparts, and there is greater interorgan cooperation. Storage, enzymatic digestion, and mechanical digestion are divided among the crop, proventriculus, and gizzard, respectively. When ceca are present, they are paired. Except for ostriches, the "large intestine" of

TABLE 1. Overview of nutritional strategies and their implications for measurement of hormones in droppings

Consumption category	Nutritional strategy[a]	Retention time[b] (min)	Microbial activity	Implications for hormone measurements
Nectarivores	1	30–50	Very low	Droppings are predominantly urine. No cecal feces.
Carnivores	1	360–600	Very low, unless ceca are present	Cecal feces are likely to have different hormone concentrations and metabolites than found in rectal feces
Piscivores	1	360–780	Very low, unless ceca are present	Cecal feces are likely to have different hormone concentrations and metabolites than found in rectal feces.
Insectivores	1	30–90	Very low	No cecal feces.
Granivores	1	40–100	Very low	No cecal feces.
Frugivores	1	15–60	Very low	Hormones diluted by high amount of undigested fiber.
Herbivores				
Flying	2, 3, or 4	50–300	Very high	Hormones diluted by high amount of undigested fiber; cecal feces are likely to have different hormone concentrations and metabolites than found in rectal feces.
Terrestrial	2 or 4	300–1440	Very high	
Omnivores	1, 2, or 3	Highly variable	Highly variable	Cecal feces are likely to have different hormone concentrations and metabolites than found in rectal feces.

[a]See FIGURE 1 for description of nutritional strategy.
[b]From Karasov[6] and Klasing.[3]

birds is considerably smaller than the "small intestine." Consequently, the intestine distal to the ileum is usually referred to as the rectum and not the large intestine. Digestive morphology is extremely variable among species, and ranges from the very simple, almost tube-like tracts of frugivores and nectarivores to the very complex, highly convoluted tracts of herbivores. Although the variability in digestive anatomy is large, there are several rough relationships with adult body size.[8] Among passerines, the volume of digesta that can be held in the intestines is directly related to body weight (allometric constant of 1). To accommodate this proportional increase in volume, the length and diameter scale to body mass$^{0.33}$. The total surface area of the small intestine scales at about body mass$^{0.72}$, which is similar to the scaling of the metabolic rate. The average time that the digesta are retained in the GI tract (retention time) increases only slightly with increasing body weight; body mass.$^{0.21}$ Thus, as species get larger, they maintain digestive efficiency by increasing surface area in the tract and only slightly increasing the time that they subject food to digestive processes.

STRATEGIES AND ADAPTATIONS

Distantly related birds eating similar diets often have very similar nutritional strategies and can extract similar nutritional value from their diet.[3,6] At the morphological level, all birds have their digestive tracts built from the same units in the same order, but wide variations occur in design, depending on the type of diet typically consumed. In general, faunivores and granivores can digest their food with the enzymes they are capable of producing (autoenzymatic digestion) and possess gastrointestinal (GI) tracts with relatively small ceca and rectums, but large proventriculus, gizzard, and small intestines. Conversely, herbivorous species require symbiotic microflora to aid in digestion (alloenzymatic digestion) and have tracts with large ceca. More subtle functional adaptations are required to accommodate the physical and nutritional characteristics of the food. Adaptations found in various birds include the following: the capacity to egest the exoskeleton of arthropods, the bones of vertebrate prey, or seeds of fruits; the ability to concentrate dietary lipids in the proventriculus; the capacity to sort fermentable from refractory cell-wall components; the ability to consume feces originating from the ceca to obtain microbial protein, energy, and vitamins; and the ability to modulate the rate of passage of digesta to match the type of feedstuff consumed.

FIGURE 1 summarizes four typical nutritional strategies. The majority of avian species, including almost all passerines, use their capacity of flight to procure the most nutritious and digestible food items, which they digest autoenzymatically (Strategy 1). The simple sugars of nectar are easily absorbed, and the digestive tract of nectarivores is very simple and short. However, the water content of this food is very high, which results in droppings that contain very low amounts of feces but high amounts of urine. Foods of animal origin, such as vertebrates, insects, and mollusks, are highly digestible. Faunivorous species usually have a large proventriculus and gizzard to accommodate digestion of the high level of protein and fat in these foods. Seeds also are highly digestible by granivores, which possess large gizzards for mechanically reducing the structure of the seeds. The very high digestibility of foods of animal origin, nectar, and seeds results in a low rate of feces production. Fruits con-

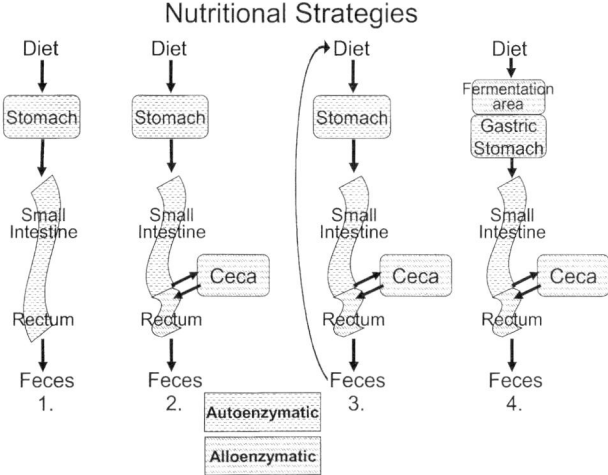

FIGURE 1. Nutritional strategies of birds. (Adapted from Klasing.[3]) Digestion mediated by enzymes produced by the bird (autoenzymatic digestion) is dominant in all strategies. Areas of the digestive tract housing large numbers of microbes permit alloenzymatic digestion in Strategies 2 through 4.

tain relatively high amounts of fiber that cannot be digested autoenzymatically. Most frugivorous species autoenzymatically digest simple sugars and proteins, but digestion of the fiber-containing components of the pulp is often completely lacking. Frugivores compensate for the nutrient-dilute composition of fruits by consuming large amounts and moving them through the digestive tract quickly.[9,10] This "skimming" strategy also precludes the digestion of seeds in the fruit. Consequently, the droppings of frugivores are voluminous relative to the droppings of faunivores or granivores, and the retention time of food in the digestive tract is short (TABLE 1).

Many species possess enlarged ceca that house microorganisms that aid in the digestion of fibrous components of their food (Strategy 2, FIG. 1). Not all of the undigested components of food that leaves the ileum move into the ceca. This is particularly well studied in Anseriformes and Galliformes, where the anatomy of their ileal–cecal–rectal junction permits sieving of digesta so that only small particulates and wate-soluble compounds enter the ceca.[11–13] Digesta that enters the ceca is retained and subjected to microbial fermentation, whereas the digesta that is excluded quickly passes through the rectum and is defecated (rectal feces). Consequently, the retention times of these two fractions of digesta are considerably different. In addition, the composition of the feces that arises from the ceca (cecal feces) is very different from the rectal feces. Cecal feces are enriched in bacteria and have higher levels of protein, fat, and vitamins.[4,12] As a result, the cecal feces may be consumed by birds of some species (Strategy 3, FIG. 1). To the extent that hormones enter the ceca, the resulting microbial metabolism and coprophagy would likely affect their form and recoverability in droppings. To measure hormones in droppings, it appears prudent to collect only rectal feces and to avoid cecal feces. In many Anseriformes and Galliformes, rectal feces can be distinguished from cecal feces by color and texture. Rectal feces, which are excreted frequently throughout the day, often have green-colored streaks when fresh due to the presence of unmetabolized biliverdin. Cecal feces are excreted less frequently, and microbial metabolism of biliverdin likely eliminates its green color, resulting in light brown to black coloration. Cecal feces are high in ammonia, putrefactive products from microbial metabolism, volatile fatty acids, and hydrophilic dietary fibers that impart a uniform viscous texture and a distinctive odor.[14,15] In chickens and turkeys, cecal evacuation occurs mainly in the early morning hours.[15]

One-third of the avian families are omnivorous, and they consume a wide variety of foods, including both animals and plants. The digestive strategy of omnivores is highly variable, and Strategy 1, 2, or 3 may be used. Most omnivores switch between strategies as the availability of food items change. When readily digestible foods are available, they use Strategy 1, but when high-fiber plant materials are consumed, Strategies 2 or 3 may be used. Consequently, the amount and composition of feces, the rate of passage, and likelihood of coprophagy is dependent on the foods.

A few species have an area for microbial fermentation in their crop, esophagus, and anterior proventriculus (Strategy 4, FIG. 1). This is taken to the extreme in the hoatzin, but also occurs in mousebirds and kakapo. The digestive tract posterior to this fermentation area is relatively simple and resembles that of omnivores.

CHARACTERISTICS OF AVIAN DIGESTION THAT AFFECT MEASUREMENT OF ENDOGENOUS LOSSES

Birds have several aspects of their digestive anatomy and physiology that differentiate them from mammals and that have the potential to affect the measurements of fecal metabolites. These include bidirectional flow of digesta between many of the digestive organs, varied biliary anatomy, the presence of paired ceca in many species, and, most importantly, the excretion of feces and the urine in a single dropping. As with mammals, other important effects on hormone measurement include (1) the relative amount of hormones excreted via bile, urine, and as a component of the inevitable loss of endogenous compounds across the single-cell intestinal epithelium; (2) the extent of hormone resorption via the intrahepatic circulation; and (3) the extent and type of microbial metabolism.

Bile Secretion

In mammals, the bile is a predominant pathway for steroid excretion; presumably, this is also the case in Aves. Conjugated steroid hormones, bile salts, cholesterol, and phospholipids are secreted into the bile canaliculi and collected by the bile ducts. Because steroid hormone excretion patterns follow bile secretion patterns, it is useful to know the kinetics of bile release and its enterohepatic circulation. This is especially important in birds, because very little is known about the hepatic metabolism and enterohepatic circulation of steroid hormones themselves.[16–18] Among a limited sampling of species, chenodeoxycholic acid is the predominant bile acid, except among faunivores, in which cholic and allocholic acids are dominant.[19,20] β-Phocacholic acid is a major component in the bile of ducks, geese, and flamingos, and 16-hydroxychenodeoxycholic acid is the primary bile acid of several species of herons.[21,22] Bile acids are usually conjugated to taurine in chickens and turkeys, but some species also conjugate with glycine.[23] Depending on the species, the duct from the right side of the liver may enlarge into, or branch into, a gall bladder.[19,24] However, a gall bladder is absent in many species (e.g., ostrich, hummingbirds, and many species of passerines, doves, and parrots). In species that possess a gall bladder, bile is stored and concentrated until it is released in large quantities at the time of consumption of a meal. In species lacking a gall bladder, storage within the bile duct is limited, and bile flows relatively continuously into the intestines. In some species (e.g., chickens), the gall bladder stores bile only from the right side of the liver, and bile produced by the left side drains directly, and somewhat continuously, into the intestines. In other species (e.g., ducks), bile produced in the left side of the liver can reach the gall bladder via a common sinus. Chickens,[25] pigeons,[26] and presumably all birds possess an efficient enterohepatic circulation that recovers bile acids and other sterols by active transport in distal areas of the small intestine.[27] Resorbed bile acids enter the portal vein and are returned to the liver where they are secreted again. The efficiency of bile acid absorption is approximately 90%.[28]

In chickens, turkeys, geese, petrels, and buzzards, bile is released into the small intestine and then reverse peristalsis sends some of it back into the gizzard where it can help emulsify lipids in the food.[3,19] It is unknown how widespread this phenomenon occurs across species. However, refluxing of bile is indicated by a characteristic green staining of the gizzard lining. Gulls, cormorants, boobies, gannets, petrels, and penguins store partially digested food, especially the lipid fraction, in their proventriculus for long periods of time, sometimes several days. This storage pool contains variable amounts of bile,[29] and presumably would contain steroid hormones and markedly change the kinetics of the birds' excretion.

Diversion of Digesta into the Ceca

In many species, paired ceca are present and provide a place where intestinal contents may be diverted for extended periods of time. The ceca originate from the rectum at a location immediately distal to its junction with the small intestine. There is tremendous variation in the size and function of ceca.[30] Ceca are absent or vestigial in many species, including in the families Sphenisciformes, Procellariiformes, Pelecaniformes, Ciconiiformes, Falconiformes, Columbiformes, Psittaciformes, Apodiformes, Coliiformes, Piciformes, and Passeriformes. Ceca are especially large and voluminous in Strigiformes, Caprimulgiformes, Galliformes, Gruiformes, Struthioniformes, Rheiformes, Apterygiformes, Tinamiformes and some Anseriformes, Charadriiformes, and Cuculiformes. They are moderate in size in Casuariiformes, Gaviiformes, Trogoniformes and some Podicipediformes, Anseriformes, Charadriiformes, Cuculiformes, and Coraciiformes.

The ceca of faunivorous birds serve as a site for water absorption and microbial assistance with nitrogen detoxification. The ceca of herbivores and omnivores serve as a site for microbial fermentation of undigested food components and endogenous compounds. Small particles and water-miscible compounds in the digesta are diverted into the ceca, whereas large particles of digesta are excluded by the sieve-like arrangement of the ileal–cecal–rectal junction.[13] These large particles move quickly into the rectum to form the rectal feces. In many species, urine from the cloaca also can flow into the ceca by reverse peristalsis. Consequently, it is likely that a fraction of hormones in bile and urine are shunted into the ceca where it would be subjected to extended periods of microbial metabolism. In spotted owls, cecal feces contain higher levels of immunoreactive corticosterone than rectal feces.[31]

Microbial Action

The number and type of microbes in the gastrointestinal tract is dependent on the nutritional strategy. Microbes are located predominantly in the ceca and rectum, although some species, including emu and geese, have robust populations in their posterior ileum. In chicken, bacteria, especially anaerobes, dominate the microflora with more than 10^{11}/g of digesta in the ceca.[14] The microflora of the gastrointestinal tract is nourished by undigested remnants of food as well as endogenous compounds and possess many enzymatic activities that modify bile acids and steroid hormones, including dehydratases, dehydrogenases, isomerases epimerases, and deconjugation via amide hydrolase.[32–34] *Bacteroides, Eubacterium, Coliforms, Clostridia, Bifidobacteria, Pseudomonas,* and *Streptococcus* species are very active at metabolizing steroid hormones. All of these species are present in the intestines of chickens, and the number is modulated by diet.[14,35] Avian species that have a high rate of passage of digesta and that lack ceca are likely to have the lowest numbers of microflora and the least modification of hormones in the digestive tract. Among species that possess ceca, hormones in the fecal excreta are exposed to fewer microbes for less time and are likely to be metabolized less than hormones in the cecal excreta.

Within an individual, a variety of factors affect the microbial ecology in the digestive tract. Age, diet, stress, and infection influence the types, numbers, and locations of microbes. Young chicks have relatively small ceca, and little microbial modification of bile occurs compared with that in adults.[36] Among dietary factors, fiber and secondary plant compounds are especially active at changing microbial ecology in the cecum.[3,15] In humans and rats, the changes in microflora induced by dietary fiber increases fecal estrogen excretion.[33,37,38]

Cloacal Function

In mammals, the percentage of steroid hormone excreted in the urine versus the feces is extremely variable, depending on the species and the hormone.[39] In birds, the cloaca serves as the storage area for the feces and the urine, and as the exit of the reproductive tract.[40] The rectum empties into the cloaca at the coprodeum, which has a much larger diameter than the rectum. The ureters empty into the cloaca in the urodeum. Because birds lack a urinary bladder, the cloaca serves as the primary storage site for urine before excretion.[1] In some species, the kidney produces dilute urine, and reverse peristalsis transports urine from the cloaca through the rectum and even into the ceca, if present, to recover additional water.[41] This process is most likely to occur when drinking water is not readily available. Retrograde flow of urine to the ceca is also a strategy to fertilize the microflora with an excellent nitrogen source.[42] Species that practice retrograde flow of urine likely have droppings with unabsorbed components of the urine distributed within the feces, and presumably hormones of urinary origin would be intermixed with hormones of biliary origin. In other species, the kidney is responsible for most water resorption, and retrograde flow is minimal. The mucosal fold that separates the coprodeum from the urodeum can prevent the movement of feces into the urodeum. During defecation, these folds can be everted through the vent to prevent microbial contamination of the ureters and the reproductive tract. During evacuation, the urine leaves first and then the feces. In some species, urine in droppings can be sampled separately from the feces, and hormone concentrations can be determined on either the fecal or urinary component.[43] In ostriches and some seabird species, the cloaca is expandable and serves as a bladder so that large amounts of urine can be stored.[1,44] Diet has a large effect on the amount of moisture in droppings, and extensive research has been conducted on this topic recently.[45]

Retention Time

Measurements of hormone levels in blood represent a single time point and can be highly variable due to episodic release patterns. A theoretical advantage of measuring hormones in the droppings is that it integrates hormone levels over the time during which the feces were produced. The length of time that it takes to produce the fecal component of droppings can be estimated from the rate of passage of the food through the gastrointestinal tract.[46] This interval is highly variable and ranges from less than 30 min to a few days, depending on the nutritional strategy of the animal (TABLE 1). The mean retention time is the average amount of time that all of the components of food spend in the gastrointestinal tract. Food tends to clear the digestive tract in an exponential fashion so that the time required to clear 98% of a food is equal to approximately four times the time required to clear 50%.[46] Frugivores consuming nutrient-dilute fruits pass digesta through their tract very quickly with little opportunity for microbial metabolism. Carnivores and piscivores have relatively long retention times, but digesta spends much of this time in the stomach where microbial numbers are very low. The retention time in omnivores is highly variable, depending on the foods consumed and with a wide divergence between the digesta that enters the ceca and the digesta that moves directly through the rectum. In chickens, digesta that moves into the ceca has a residence time that is at least three times as long as components directly excreted.

Diet-dependent changes in the rate of passage can occur rapidly, and differences can be observed between meals of different composition.[7,47,48] Even different components of the same meal may move at greatly different rates through the digestive tract; for example, the retention of soluble fibers in the ceca, or grit in the gizzard. However, the effect of different components of food on the retention time is highly dependent on the species. For example, fruit seeds pass through the digestive tract of small frugivores in less than 10 min, but may be retained by ratites for several weeks. Pectin, hemicellulose, and other plant cell wall components pass through the digestive tract of a frugivore in less than 30 min, but are retained and fermented in the ceca for 12 h or more in many omnivores and herbivores. In herbivorous geese, the retention time, as indicated by defecation rate, markedly changes seasonally as the fiber content of the food changes. When the nutritional quality of their food is poor, they consume large quantities and adopt a skimming strategy somewhat like frugivores. The rate of passage and the defecation rate can be very high when the food quality is poor. For example, barnacle geese feeding on grass defecate every 3 to 5 min, with each dropping averaging 5.5 cm in length.[49] Presumably the high mass of undigested food and the short interval during which endogenous losses accumulate in the droppings of these geese would result in low concentrations of hormones relative to that in the droppings in geese eating more nutritious diets.

ACKNOWLEDGMENT

The financial support of the National Science Foundation (IBN-0212587) is greatly appreciated.

REFERENCES

1. KING, A.S. & J. MCLELLAND. 1984. Birds: Their Structure and Function. Bailliere Tindall. London.
2. ZISWILER, V. & D.S. FARNER. 1972. Digestion and the digestive system. In Avian Biology, Vol. II. D.S. Farner, J.R. King & K.C. Parkes, Eds.: 343–430. Academic Press. New York.
3. 3. KLASING, K.C. 1998. Comp. Avian Nutr. CAB International. New York.
4. GASAWAY, W.C. 1976. Seasonal variation in diet, volatile fatty acid production and size of the cecum of rock ptarmigan. Comp. Biochem. Physiol. **53A**: 109–114.
5. LEVEY, D.J. & W.H. KARASOV. 1989. Digestive responses of temperate birds switched to fruit or insect diets. Auk **106**: 675–686.
6. KARASOV, W.H. 1990. Digestion in birds: chemical and physiological determinants and ecological implications. Studies Avian Biol. **13**: 391–415.
7. AFIK, D. & W.H. KARASOV. 1995. The trade-offs between digestion rate and efficiency in warblers and their ecological implications. Ecology **76**: 2247–2257.
8. RICKLEFS, R.E. 1996. Morphometry of the digestive tracts of some Passerine birds. Condor **98**: 279–292.
9. MARTINEZ DEL RIO, C.M. & W.H. KARASOV. 1990. Digestion strategies in nectar-eating and fruit-eating birds and the sugar composition of plant rewards. Am. Nat. **136**: 618–637.
10. LEVEY, D.J. & G.E. DUKE. 1992. How do frugivores process fruit–gastrointestinal transit and glucose absorption in cedar waxwings (*Bombycilla-cedrorum*). Auk **109**: 722–730.
11. DUKE, G.E. 1989. Relationship of cecal and colonic motility to diet, habitat, and cecal anatomy in several avian species. J. Exp. Zool. **3**(Suppl.): 38–47.
12. BJORNHAG, G. 1989. Transport of water and food particles through the avian ceca and colon. J. Exp. Zool. **3**(Suppl.): 32–37.
13. MAHDI, A.H. & J. MCLELLAND. 1988. The arrangement of the muscle at the ileo-caecorectal junction of the domestic duck (*Anas platyrhynchos*) and the presence of anatomical sphincters. J. Anat. **161**: 133–142.
14. TERADA, A. *et al.* 1994. Effects of dietary supplementation with lactosucrose (4G-beta-D-galactosylsucrose) on cecal flora, cecal metabolites, and performance in broiler chickens. Poult. Sci. **73**: 1663–1672.
15. ZUBAIR, A.K., C.W. FORSBERG & S. LEESON. 1996. Effect of dietary fat, fiber, and monensin on cecal activity in turkeys. Poult. Sci. **75**: 891–899.
16. HAWKINS, R.A. & P.D. TAYLOR. 1967. The metabolism of tritiated estradiol in the laying hen. Can. J. Biochem. **45**: 523–530.
17. WELLS, J.W. 1971. The metabolism of progesterone in the laying hen (*Gallus domesticus*). Comp. Biochem. Physiol. A **40**: 61–70.
18. HELTON, E.D. & W.N. HOLMES. 1973. The distribution and metabolism of labeled corticosteroids in the duck (*Anas platyrhynchos*). J. Endocrinol. **56**: 361–385.
19. STURKIE, P.D. 1976. Avian Physiology. Springer-Verlag. New York.
20. ELKIN, R.G., K.V. WOOD & L.R. HAGEY. 1990. Biliary bile acid profiles of domestic fowl as determined by high performance liquid chromatography and fast atom bombardment mass spectrometry. Comp. Biochem. Physiol. B Comp. Biochem. **96**: 157–161.

21. HAGEY, L.R. et al. 2002. A novel primary bile acid in the Shoebill stork and herons and its phylogenetic significance. J. Lipid. Res. **43:** 685–690.
22. HAGEY, L.R. et al. 1990. b-Phocacholic acid in bile. Condor **92:** 593–597.
23. LINDSAY, O.B. & B.E. MARCH. 1967. Intestinal absorption of bile salts in the cockerel. Poult. Sci. **46:** 164–168.
24. CROMPTON, D.W.T. & C. NESHEIM. 1972. A note on the biliary system of the domestic duck and a method for collecting bile. J. Exp. Biol. **56:** 545–550.
25. GUI, X. et al. 2000. Neurotensin elevates hepatic bile acid secretion in chickens by a mechanism requiring an intact enterohepatic circulation. Comp. Biochem. Physiol. C Comp. Pharmacol. **127:** 61–70.
26. SPITTELL, D., L.K. VONGROVEN & M.T. SUBBIAH. 1976. Concentration changes of bile acids in sequential segments of pigeon intestine and their relation to bile acid absorption. Biochim. Biophys. Acta **441:** 32–7.
27. COLETO, R., J. BOLUFER & C.M. VAZQUEZ. 1998. Taurocholate transport by brush border membrane vesicles from different regions of chicken intestine. Poult. Sci. **77:** 594–599.
28. HURWITZ, S. et al. 1973. Absorption and secretion of fatty acids and bile acids in the intestine of the laying fowl. J. Nutr. **103:** 543–547.
29. PLACE, A.R. 1992. Bile is essential for lipid assimilation in leach storm petrel, oceanodroma-leucorhoa. Am. J. Physiol. **263:** R389–R399.
30. CLENCH, M.H. & J.R. MATHIAS. 1995. The avian cecum—a review. Wilson Bull. **107:** 93–121.
31. TEMPEL, D.J. & R.J. GUTIERREZ. 2004. Factors related to fecal corticosterone levels in California spotted owls: implications for assessing chronic stress. Conserv. Biol. **18:** 538–547.
32. KNARREBORG, A. et al. 2002. Quantitative determination of bile salt hydrolase activity in bacteria isolated from the small intestine of chickens. Appl. Environ. Microbiol. **68:** 6425–6428.
33. EYSSEN, H. 1973. Role of the gut microflora in metabolism of lipids and sterols. Proc. Nutr. Soc. **32:** 59–63.
34. MACDONALD, I.A. et al. 1983. Degradation of steroids in the human gut. J. Lipid Res. **24:** 675–700.
35. KNARREBORG, A. et al. 2002. Effects of dietary fat source and subtherapeutic levels of antibiotic on the bacterial community in the ileum of broiler chickens at various ages. Appl. Environ. Microbiol. **68:** 5918–5924.
36. BURCZAK, J.D., J.L. MCNAUGHTON & T.F. KELLOGG. 1980. Cholesterol metabolism in poultry *Gallus-domesticus* fecal neutral sterol and bile acid excretion. Comp. Biochem. Physiol. B **66:** 385–390.
37. GORBACH, S.L. & B.R. GOLDIN. 1987. Diet and the excretion and enterohepatic cycling of estrogens. Prev. Med. **16:** 525–531.
38. KENDALL, M.E. & L.A. COHEN. 1992. Effect of dietary fiber on mammary tumorigenesis, estrogen metabolism, and lipid excretion in female rats. In Vivo **6:** 239–245.
39. SCHWARZENBERGER, F. et al. 1996. Fecal steroid analysis for non-invasive monitoring of reproductive status in farm, wild and zoo animals. Anim. Reprod. Sci. **42:** 515–526.
40. PROCTOR, N.S. & P.J. LYNCH. 1993. Manual of Ornithology. Yale Univ. Press. New Haven, CT.
41. DUKE, G.E. 1986. Alimentary canal: anatomy, regulation of feeding, and motility. *In* Avian Physiology. P.D. Sturkie, Ed. Springer-Verlag. New York.
42. KARASAWA, Y. & M. MAEDA. 1994. Role of caeca in the nitrogen nutrition of the chicken fed on a moderate protein diet or a low protein diet plus urea. Br. Poult. Sci. **35:** 383–391.
43. WASSER, S.K. et al. 1997. Noninvasive physiological measures of disturbance in the northern spotted owl. Conserv. Biol. **11:** 1019–1022.
44. DUKE, G.E., A.A. DEGEN & J.K. REYNHOUT. 1995. Movement of urine in the lower colon and cloaca of Ostriches. Condor **97:** 165–173.
45. FRANCESCH, M. & J. BRUFAU. 2004. Nutritional factors affecting excreta moisture and quality. World's Poult. Sci. **60:** 64–74.
46. WARNER, A.C.I. 1981. Rate of passage of digesta through the gut of mammals and birds. Nutr. Abstr. Rev. B **51:** 789–820.

47. HERD, R.M. & T.J. DAWSON. 1984. Fiber digestion in the emu, *Dromaius novaehollandiae*, a large bird with a simple gut and high rates of passage. Physiol. Zool. **57:** 70–84.
48. WASHBURN, K.W. 1991. Efficiency of feed utilization and rate of feed passage through the digestive system. Poult. Sci. **70:** 447–452.
49. PROP, J. & T. VULINK. 1992. Digestion by barnacle geese in the annual cycle—the interplay between retention time and food quality. Funct. Ecol. **6:** 180–189.

Measurement of Corticosterone Metabolites in Birds' Droppings: An Analytical Approach

ERICH MÖSTL, SOPHIE RETTENBACHER, AND RUPERT PALME

Institute of Biochemistry, Department of Natural Sciences, University of Veterinary Medicine, A-1210 Vienna, Austria

ABSTRACT: Fecal steroid analyses are becoming increasingly popular among both field and laboratory scientists. The benefits associated with sampling procedures that do not require restraint, anesthesia, and blood collection include less risk to subject and investigator, as well as the potential to obtain endocrine profiles that are not influenced by the sampling procedure itself. In the feces, a species-specific pattern of metabolites is present, because glucocorticoids are extensively metabolized. Therefore, selection of adequate extraction procedures and immunoassays for measuring the relevant metabolites is a serious issue. In this review, emphasis is placed on the establishment and analytical validation of methods to measure glucocorticoid metabolites for a noninvasive evaluation of adrenocortical activity in droppings of birds.

KEYWORDS: radioimmunoassay; enzyme immunoassay; cortisol; corticosterone; glucocorticoids; feces; urine; extraction; analytical validation; conservation biology; animal welfare

INTRODUCTION

In vertebrates, the front-line hormones to overcome stressful situations are the glucocorticoids (GCs) and catecholamines. Their increased secretion enhances adaptive physiological responses of an animal.[1-4] The main GCs produced by the adrenal glands are cortisol and corticosterone, the latter dominating in birds.[5,6] Their quantification in blood samples provides valuable information about an animal's endocrine status and can be used as a parameter of adrenocortical activity. Thus, disturbances are assessable,[2,5] although plasma corticosterone concentrations in birds are also subjected to diurnal and annual rhythms,[2,7,8] and corticosterone is also involved in the induction of ovulation in hens.[9]

However, blood sampling itself is critical, as disturbances of the animals will increase the glucocorticoid concentration within minutes,[10] possibly affecting the results. Therefore, in investigations concerning animal welfare, biology, or veterinary medicine, there is increasing interest in measuring glucocorticoid metabolites (GCMs) noninvasively in feces or droppings. This method is feedback free, as samples can be collected without fixation of the animal. Because feces are a complex

---

Address for correspondence: Erich Möstl, Institute of Biochemistry, Department of Natural Sciences, University of Veterinary Medicine, Vienna, Veterinärplatz 1, A-1210 Vienna, Austria. Voice: 43 1 250 77 4102; fax: 43 1 250 77 4190.
  erich.moestl@vu-wien.ac.at

Ann. N.Y. Acad. Sci. 1046: 17–34 (2005). © 2005 New York Academy of Sciences.
doi: 10.1196/annals.1343.004

matrix, various other factors may influence the determination of GCM levels. A rigorous validation of the whole analytical procedure is necessary to produce reliable results. Questions regarding collection, storage, extraction, and analysis of fecal samples are reviewed in the following. Emphasis is placed on the establishment and analytical validation of such noninvasive methods for assessing adrenocortical activity in droppings of birds.

CORTICOSTERONE METABOLITES IN DROPPINGS OF BIRDS

GCs are extensively metabolized in various organs, mainly in the liver (for a review, see Ref. 11) and are excreted via the bile into the gut and via the kidney into the urine, mainly conjugated as sulfates or glucuronides. Those products are more water-soluble than the parent steroids.[12] A certain portion of the metabolites is reabsorbed from the gut and again transported to the liver (enterohepatic circulation[6,13]). In the gut, microbial enzymes play an additional role in the conversion of the metabolites. In general, the metabolism includes 5α or 5β reduction, hydroxylations, or reductions of functional groups or side-chain cleavage (C-17,20-lyase) in the case of 17α-hydroxylated metabolites (FIG. 1; see also Refs. 5 and 14).

The best way to investigate the metabolism and excretion pattern of GCs are studies using radiolabeled ($^{14}C/^{3}H$) hormones,[6] although this approach is not possible in every species due to economic or welfare restrictions. If ^{3}H-labeled GCs are used, the radioactive metabolites excreted do not necessarily represent the amounts of formed metabolites because some of the tritium may be lost during metabolism or exchanged for unlabeled hydrogen. These effects do not take place in the case of

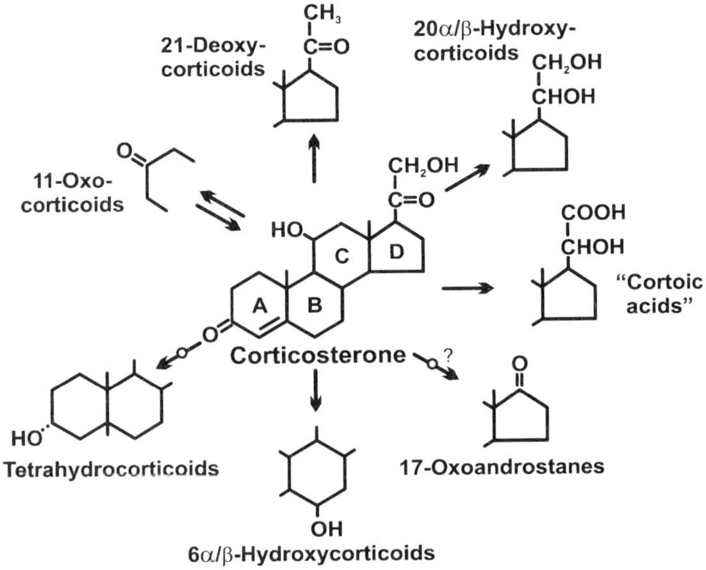

FIGURE 1. Possible pathways of corticosterone metabolism.

[14]C-labeled GCs, because the ring system of steroids is quite stable. However, ^{14}C-GCs are very expensive, and especially ^{14}C-corticosterone requires a custom synthesis. Because the specific activity of ^{3}H-GC is quite high, the additional mass of these radioactive hormones will not substancially increase the total concentration (radioactive and nonradioactive) in the peripheral blood, and therefore plasma levels of the GCs will remain within the physiological range.

In birds, only a few such studies have been conducted so far[15–19] (for a review, see Ref. 6). Following the administration of GCs, a two-peaked excretion curve of radioactivity was found, reflecting urinary (first peak) and fecal (second peak, corresponding to the gut passage time[26]) excretion.[17]

All radiometabolism studies performed so far (for a review, see Ref. 6) demonstrated pronounced species and sometimes even sex differences concerning the formed metabolites.[2,6,16–19] Cortisol or corticosterone is present in fecal samples only in trace amounts, if at all.[6,16–20] In animals with severe diarrhea, however, the situation may be different because there is less time for bacterial metabolism, and albumin or even blood will pass the intestinal mucosa. Until now, definitive characterization of the GC levels in bird droppings is only tentative. In chicken, Rettenbacher et al.[17] found that the dominating ^{3}H-labeled metabolites elute in reversed-phase high-performance liquid chromatography (HPLC) after estrone sulfate and before cortisol, which were used as markers.[16–19] They probably resemble conjugated or polar unconjugated (or at least tetrahydroxylated) metabolites. In all species investigated, enzymatic hydrolysis of these ^{3}H-labeled metabolites did not yield large amounts of diethyl ether–extractable radioactivity.[6,18,19] Analyses using mass spectrometry (MS) should be performed to further characterize these metabolites.

COLLECTION AND STORAGE OF FECAL SAMPLES

Sampling and storing is an important issue for steroid analysis. Any problems arising during this process usually cannot be compensated for by analytical skills afterwards. Fecal GCMs are reported to be further metabolized by bacterial enzymes after defecation, depending on environmental conditions.[5,6,21,22] Therefore, feezing samples immediately after defecation is recommended.[2,6,21,23] Fecal GCMs can be treated with heat or by adding acids or alcohol because the steroids are not affected by these procedures (FIGS. 2 and 3). However, storing preserved samples for longer periods may be critical and must be carefully evaluated (for a review, see Touma and Palme,[2] Palme,[22] Millspaugh and Washburn,[21] and Hunt and Wasser[24]).

EXTRACTION

In mammals, fecal steroids have been reported to be not evenly distributed within samples.[21,25] Therefore, homogenizing the feces before analysis is recommended. The sample homogeneity is very important for measuring fecal metabolites, and a certain amount of feces is necessary to represent a random sample. In birds, droppings usually consist of a urinary and a fecal portion, which in some species cannot be separated because the two excreta are already mixed in the cloaca.[26] This fact must be considered because the pattern and amounts of GCMs may differ in the two components of the droppings.

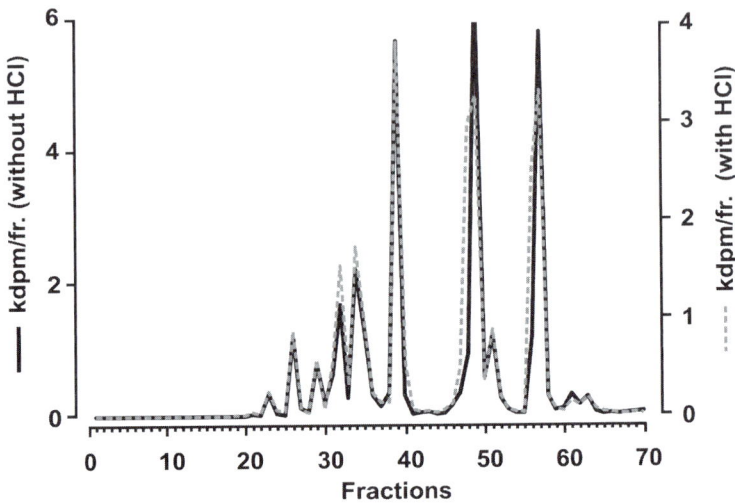

FIGURE 2. Straight phase high-performance liquid chromatographic separation (for conditions, see Palme and Möstl[20]) of ^{14}C-cortisol metabolites derived from an radiometabolism experiment in sheep (Palme et al.[25]) immediately after extraction (*straight line*) and after incubation of the extract with concentrated hydrochloric acid for 18 h at 80°C (Möstl et al.[27]).

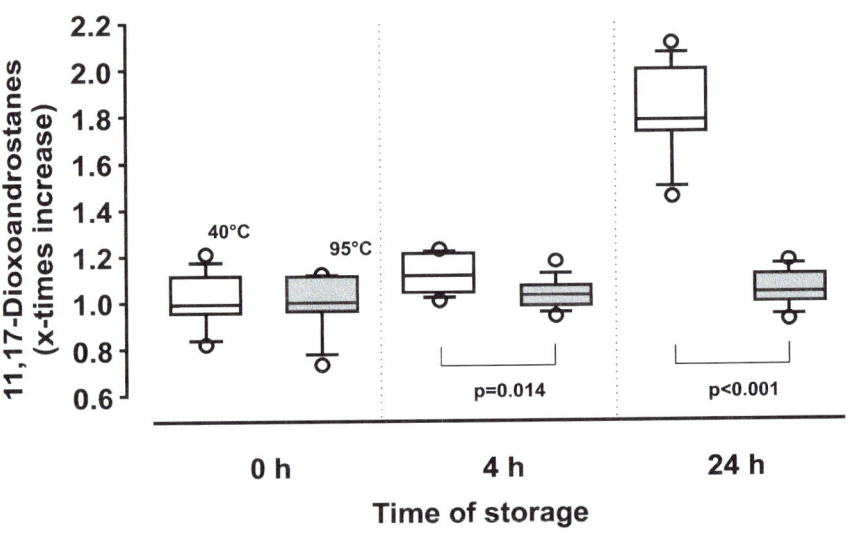

FIGURE 3. Box plot of immunoreactive 11,17-dioxoandrostanes (percent increase) in feces of cows. Samples were defrosted at −40°C and incubated at room temperature (*open boxes*) or at 95°C (*filled boxes*) for 4 or 24 h, respectively (for details, see Möstl et al.[27]).

Deep-freezing fecal samples does not destroy bacterial enzymes, which metabolize steroids. If these enzymes are not inactivated before storing the samples (drying, alcohol, or heat), metabolism can continue after thawing. Defrosting the samples by heating (95°C) destroys the bacterial enzymes[27] (FIG. 3). Another possibility is to keep intervals short between defrosting and the addition of organic solvent.

Attention must be paid to avoid carryover effects during weighting of the samples. This is especially important if there are large differences in sample concentrations. In addition, one must address possible health risks caused by parasites, bacteria, or virus particles, because sometimes samples from animals with unknown health status or even sick animals are investigated, and some pathogenic microorganisms, viruses, or prions have zoonotic potential. Some research groups homogenize lyophilized samples before extraction. The powder causes additional risks because particles may be inhaled.

In general, extraction procedures are developed to concentrate the substances of interest and at the same time exclude possible interfering compounds. GCMs of various polarities are present in the feces. Polarity describes the degree to which a compound (steroid) is soluble in water. Additional hydroxyl groups render the steroid more polar (hydrophilic). A separation of hydrophobic steroids from aqueous media using organic solvents (petroleum or diethyl ether) is relatively simple, but dissolves many other lipids also.[28]

Because extraction procedures are time-consuming and costly, assays, which measure an aliquot of the sample directly with no purification, are used to determine steroid hormone levels in plasma samples. Stanczyk et al.[29] assessed the reliability of many different commercially available (estradiol and testosterone) kits. Large differences in the obtained results were found between these direct immunoassays. The authors claim that a more thorough analytical validation of the assays is necessary with respect to sensitivity and specificity because matrix influences can play a substantial role. Similar effects must be expected when measuring GC levels in droppings, especially because fecal samples represent a much more complex matrix than plasma samples.

Another way to extract and clean up the substances of interest is minicolumns, filled with material for reversed-phase chromatography (e.g., Sep-Pak C_{18} cartridges from Waters, Milford, MA). Because these columns are too expensive for use in routine analysis, it is recommended that they be used only before HPLC separation of the GCMs.[30,31]

Because radiometabolism studies are not available for most avian species, the extraction procedures must cover a broad polarity range (potentially conjugated and unconjugated steroids); otherwise, substances of interest may be excluded.[17] It should be kept in mind that, for example, diethylether is too apolar to extract tetra- or pentahydroxylated steroids (conjugated steroids are also not extractable using this solvent), and therefore more polar solvents must be used. The more hydrophilic, conjugated steroids (sulfates or glucuronides) can be extracted into organic solvents by forcing the extraction by saturating the water phase with salt (NaCl, ammonium sulfate) and lowering the pH.[28]

Although boiling procedures have been described for extraction of GCMs (e.g., Wasser et al.[15]), most authors now use mixtures of methanol or ethanol with water to dissolve the steroids from the feces, which is more a suspension than an extraction.[6,15,22] After shaking (three times for 10 s, using a hand vortex or 30 min with a multivortex), the vials are centrifuged. We recommend not evaporating the alcoholic extract because redissolving the material in assay buffer can cause problems. Instead, an aliquot of the supernatant is used directly in the immunoassay. Because a higher percentage of alcohol will interfere with the steroid–antibody binding, a further dilution step with assay buffer prior to analysis is performed.

Determination of Recovery Rate

As long as the chemical structures of the immunoreactive metabolites are unknown or the metabolites are not available in radiolabeled forms derived from radiometabolism experiments, a reliable determination of recoveries of GCMs is not possible. Recoveries reported in published studies were based mostly on the extraction of radiolabeled cortisol or corticosterone added to the sample just before processing. Results obtained by Palme *et al.*[32] emphasized that a higher proportion of radioactivity could be extracted from feces after ^{14}C-labeled progesterone was added *in vitro* than with fecal samples containing metabolites of ^{14}C-labeled progesterone injected *in vivo*. This is due to the fact that first, the metabolites are of different polarity, and second, complex interactions between sample matrix and steroids, which can affect extraction efficiency, are less pronounced with steroids added *in vitro*. As a result, the recoveries reported in the literature probably do not reflect the true recovery of the metabolites in the feces, but are over- (or under-) estimates, depending upon the steroid added, the metabolites measured by the immunoassay, and the investigated species. Therefore, recovery testing based on naturally metabolized, radiolabeled steroids infused into the animals should be favored.[6,20,27] The influence of different methanol/water mixtures on the recovery of GCMs in mammals is described in a review by Palme *et al.*,[6] but similar studies are lacking in birds.

TEST SYSTEMS FOR MEASURING GCMS

To analyze fecal GCM levels, two different procedures can be used. Traditionally, complex mixtures of steroid hormone metabolites are analyzed after extraction and derivatization using gas chromatography–mass spectrometry (MS). HPLC–MS is now becoming increasingly popular,[33] as derivatization procedures are not necessary. HPLC using direct UV detection can be applied only if a 4-ene-3-oxo structure is present, as is the case with cortisol, cortisone, or corticosterone. The reduced metabolites cannot be measured with such detectors.

The second approach involves immunoassays. Since 1970, there has been a very rapid increase in publications concerning steroid immunoassays.[34] Immunoassays are much cheaper than methods employing MS, and they allow measurement of many samples within a short time, but they are less specific (see below).

IMMUNOASSAYS

For analysis of steroids, competitive immunoassays are mainly used.[35] This means that the label and the steroid to be measured compete for an antibody binding site. Radioimmunoassays (RIAs) or enzyme immunoassays (EIAs) are mainly used.

RIAs certainly have their merits, such as high precision, robustness, and a long-lasting tradition in performance. Various antibodies and labels for cortisol and corticosterone are commercially available, but because the two hormones are not present in the feces of mammals and birds,[6] these assays rely on cross-reactions with the excreted metabolites. If commercial RIAs are used, cross-reactions with GCMs should be carefully analyzed because the manufacturers mostly give only the data relevant for plasma analysis (see below). Assays especially designed for measuring GCMs are mainly EIAs because radioactive labels for the metabolites are not commercially available, and a custom synthesis or biosynthesis is too expensive.

Antibody Production (Immunogen and Immunization)

To establish an immunoassay, an antibody is required. Therefore, animals must be immunized. Because steroids are too small to act as an immunogen for themselves, they must be linked to a macromolecule—for example, a protein—to be an immunogen. However, most of the steroids do not contain a functional group, which can be linked directly to a macromolecule, and therefore a carboxyl group, for example, must be added. To add these functional groups, one can use a variety of reactive steroid derivatives, including chloroformates, hemicuccinates, carboxymethyloximes, and thioether alkanoic acids.[36] The molecule between the steroid and the protein also acts as a spacer (four to six carbon atoms' distance between steroid and protein seemed to be best suited). The carboxyl group is then used to link the steroid to a protein, for example, by using a mixed anhydride or the carbodiamide reaction. The chemistry of the formation of steroid–protein conjugates was reviewed by Kellie *et al.*[37] If an antibody is raised against this steroid–protein conjugate, the

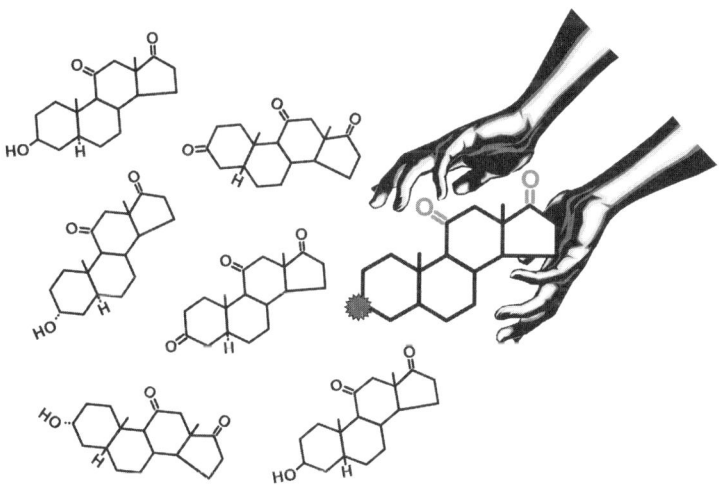

FIGURE 4. Scheme of various 11,17-oxo-androstanes (5α/5β-androstanes, having a 3-oxo or 3α/3β-hydroxyl group), which may show cross-reactions with an antibody raised against 11-oxoetiocholanolone-3-hemisuccinate:bovine serum albumin (e.g., Palme and Möstl[20]). The palpating hands symbolize the antibody, which recognizes only a certain part of the steroid.

functional group in the steroid molecule used for linking to the protein is masked and will not act as an immunological discriminant of the antibody formed (FIG. 4).

For measuring steroid hormones in blood samples, various efforts have been made to increase the specificity of antibodies.[38] For example, many authors described procedures of coupling the steroid at positions remote from functional groups, for example, C-6, C-7, or C-11α to get more specific antisera, using as their basis the fact that potentially cross-reacting steroids are not present in the samples. However, for analysis of fecal GCM levels, the use of highly specific cortisol and corticosterone antibodies must be avoided because those substances are not present in the fecal samples.[6] Therefore, antibodies that are successfully used for measuring fecal GCMs are group specific (e.g., the corticosterone RIA from ICN Biomedicals (Costa Mesa, CA).[15,32]

Because the fecal GCMs vary substantially between species, it may be too time-consuming and costly to raise antibodies against the main metabolites in each species, as this would require establishing a new assay for each species. Unlike in blood samples, in which a highly specific antibody is desired to get a specific assay, most assays for measuring GCMs use antibodies that are more or less group specific (FIG. 4), as due to the masked coupling position, a group of metabolites are recognized by the antibody.

Those metabolites can be measured using the same assay, because the antibodies show mostly sufficient cross-reaction with steroids, which differ from each other only at positions close or at the position, where the hapten was linked to the carrier molecule. For example, several steroids with two oxo-groups at positions C-11 and C-17 in common [11,17-dioxoandrostanes (11,17-DOAs)] are recognized by an antibody raised against 5β-androstane-3α-ol-11,17-dione-3-HS:BSA (bovine serum albumin;[20] FIG. 4).

Most of the antibodies used for steroid analysis are raised in rabbits because a successful immunization of an animal results in enough material for many assays (working dilutions of antibodies using double-antibody techniques are often greater than 1:10,000). To get higher titers, adjuvants must be applied for immunization, and immunization protocols may take several months to obtain suitable antibodies.[39,40]

Labeled Steroids

In radioimmunoassays, tritiated or iodinated steroids are used, and a highly specific activity of the label increases the sensitivity of the assay system. Since the early 1970s, an increase in sensitivity of steroid immunoassays has been achieved by introducing a so-called heterology. This means that the labels do not use the same steroid, site, or bridge as the hapten used for raising the antibody (steroid heterology, bridge heterology, or site heterology;[41] Fig. 5).

For EIAs, mainly horseradish peroxidase (HRPO) or alkaline phosphatase (AP) is used. In competitive immunoassays, mostly the label and not the antibody (immunosorbent), is enzyme linked. Therefore, these assays should be called EIAs and not enzyme-linked immunosorbent assays because the antibody is not labeled. Another way to label the steroids is to link them to biotin. In both cases, the same biochemical procedures can be used to link the steroid derivates with a carboxyl group (COOH) to a protein (e.g., using the ε-amino group of lysine). Biotin derivates with an amino group plus a spacer [Biotin-PEO-LC-Amine = biotinyl-3,6,9-trioxaundecanedi-

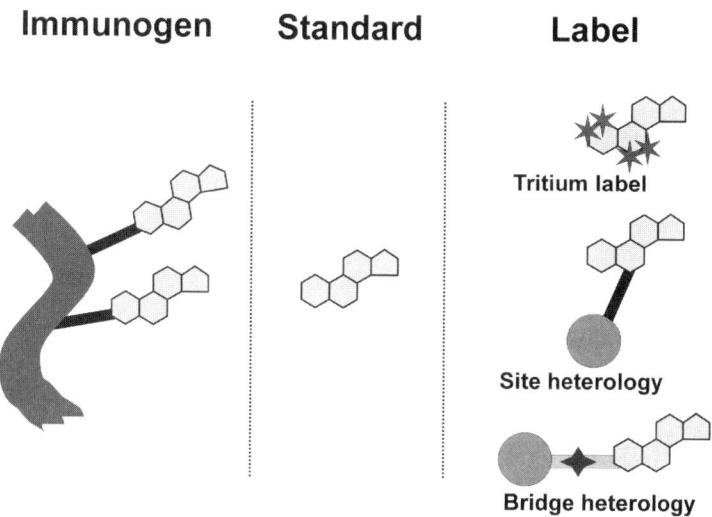

FIGURE 5. Steroid linked to a protein carrier to be used as immunogen for the antibody production, standard, and various labels (tritium label with four tritium atoms at positions C1, C2, C6, and C7 scheme of a potential label with site or bridge heterology).

amine;[46] Biotin-PEO-Amine = biotinyl-3,6-dioxaoctanediamine, 5-(biotinamido)-pentylamine[20]] are commercially available from, for example, Pierce (Rockford, IL).

Biotin is tightly bound by avidin or streptavidin. Streptavidin linked to enzymes like HRPO or AP is commercially available and is used to detect the biotin label. The use of biotin gives a higher specific activity of the system than the direct linking of the steroid to enzymes, as is used in direct labeling. The best ratio of enzyme to steroid achievable by direct linking is 1 to 1, but in many cases more than 1 mole of steroid is bound per mole of enzyme. On the other hand, using streptavidin–enzyme conjugates gives a ratio of two to three enzymes per mole of streptavidin, which results in a higher sensitivity of these assays.

Second-Antibody Technique

In many EIAs, a double-antibody technique is used, which means that the antisteroid antibody is not directly bound to the polystyrol surface of a microtiter plate, but a so-called coating antibody. That antibody was raised in another species as the antisteroid antibody and is directed, for example, against rabbit immunoglobulin G. The coating antibodies used in EIAs are purified mainly by affinity chromatography to give a high coating efficiency for the specific antisera.[42]

ANALYTICAL VALIDATION OF AN IMMUNOASSAY

An established immunoassay must be analytically validated for each given species under investigation. This includes certain criteria, such as accuracy, specificity (cross-reactions), sensitivity, precision, and parallelism of the dose–response relationship for the standard and the unknown.[a]

Accuracy

The absolute concentrations (true values) of steroid hormone metabolites are difficult to evaluate because the matrix, cross-reacting substances, and the purity of the standards used may influence the results. Whitehead[43] stated that the method used for analysis influences the results, and that a particular result obtained by a particular method may not be true, but may give a correct value. For measuring GC metabolites in feces of mammals or droppings of birds, various laboratories are using different assays and extraction procedures. Usually, the assays consider only some of the metabolites and ignore others. Therefore, concentrations published may vary between laboratories, hindering a direct comparison of the results between different laboratories. In cattle and geese, the influence of using different antibodies tomeasure GC metabolites is shown by Morrow *et al.*[44] and Frigerio *et al.*[45] Because there are different metabolites in feces, the question of which metabolite should be measured is important. For example, in fecal samples of sheep, 27 different metabolites were detected using HPLC–MS.[46] Radiometabolism investigations showed similar results also in various other species (for a review, see Palme *et al.*[6]). So there is a broad range of substances that can be measured as parameters of GC production, and the results differ depending on the metabolite(s) measured. Standardization with respect to the results measured by different laboratories should be one goal.

Taken together, independent of a "true" value, a "good" assay should fulfill some criteria,[47] like to provide appropriate specificity and sensitivity, to be precise, to be robust, and to have a "working range" adequate for the study and to be compatible with the environment.

SPECIFICITY AND CROSS-REACTIONS

If we define specificity according to the fact that the substance measured is substantially unique and identical to the standard used, this definition is quite rigorous.[48] As in most assays for GCMs, there is more than one immunoreactive substance measured, and their chemical structures are not characterized. Thus, a specificity test in such a strict sense as described by Cekan[48] cannot be given in most cases.

Parallelism of diluted samples with a standard curve is sometimes given as a parameter of specificity of an immunoassay. However, as has been already described by Ekins,[49] some cross-reacting compounds will yield dilution curves that give acceptable parallelism results with standards, and parallelism is therefore not a marker of specificity, but a proof of a dose–response relationship.

---

[a]To convert steroid concentrations given in nanograms per gram to nanomoles per kilogram (SI unit), one multiplies the original value by a factor of 1000 divided by the molecular weight of the standard of the immunoassay used. For example, the molecular weights of cortisol, corticosterone, and 11-oxoetiocholanolone are 362, 346, and 304, respectively.

Because the GCMs in the feces are a complex mixture of different, structurally closely related steroids, testing cross-reactions is important for the analytical validation of an immunoassay.[34] This provides information, which substances, other than the standard, react with the antibody. Tests should include steroids, which are expected to occur in the samples (GCMs) of the species investigated and not only considering the steroid hormones present in the blood. For comparisons between laboratories, we must consider that incubation times have an influence on the concentrations measured because cross-reactions of an assay system are different, depending on whether an assay is in the equilibrium or not. The cross-reactions indicate to which degree a certain steroid reacts in the immunoassay. However, they do not reveal which substances are really measured because other metabolites not evaluated for their reactivity may be present in the sample (especially in the feces). Therefore, a recommended procedure for assessing the presence of cross-reacting substances is the analysis of samples derived from radiometabolism studies. After extraction, an HPLC separation of the metabolites is performed, and the individual fractions of the eluent are collected. An aliquot of each fraction is used for measuring radioactivity, and another aliquot for the immunoreactivity in the respective assay(s). If the immunoreactive substances coelute (at least partly) with radioactive peaks, this is an indicator that an assay can detect metabolites of the parent steroid. Radioactive peaks, occurring without accompanying immunoreactivity, demonstrate that those metabolites are not measured by the assay system. On the other hand, immunoreactive peaks without coeluting radioactivity are an indicator for cross-reacting substances not originating from the substance injected.[5,6,22,31,50] These so-called HPLC immunograms also characterize the immunoreactive substances present in the droppings, but care must be given to possible exclusion of steroid metabolites during extraction, cleanup procedure, or redissolving of the extract in the mobile phase for chromatography. Therefore, it is also advisable to calculate the total amount of immunoreactive metabolites measured (area under the curve) with and without chromatography to check whether significant amounts of immunoreactivity were lost during the whole chromatographic procedure.

Cortisol and corticosterone are both present in plasma of some vertebrates, and Teskey-Gerstl et al.[30] showed that different fecal metabolites are formed originating from these two hormones. The authors injected ^{14}C-labeled cortisol and ^{3}H-labeled corticosterone separately in European hares (three animals each). Without such radiometabolism investigations, it is an open question whether the immunoreactive GCMs measured in feces originate from cortisol or corticosterone. For example, a commercially available corticosterone RIA (ICN Biomedicals) is widely used,[15,23,24] but its cross-reactions with the fecal steroid metabolites (5α/5β-reduced steroids) are still unknown. This should be tested, especially to elucidate if the ICN antibody measures cortisol or corticosterone metabolites (both GCs differ only at position C-17; cortisol having a hydroxy group, which is lacking in the case of corticosterone).[6] That may be especially important in animals with cortisol as the main GC in the blood, because in those species the adrenal gland is also capable of secreting corticosterone and both hormones may have different biological functions (the latter acting mainly within the brain). Because this corticosterone antibody shows cross-reactions of less than 1% with cortisol, it probably reacts to an even lesser extent with the reduced cortisol metabolites. Although bacteria can produce many different steroid-transforming enzymes, the presence of a 17α-dehydroxylase was not reported.[14] Thus, the corticosterone RIA may measure metabolites reflecting blood corticosterone and not cortisol levels. If there is no proof that the substance measured is identical to the standard, it is advisable to use the term "immunoreactive." For example, an antibody raised against 11-oxoetiocholanolone (linked at position C-3 of the molecule) will show cross-reactions with 11,17-dioxoandrostanes (FIG. 4). This description will be more correct than labeling the measured substances as 11-oxoetiocholanolone. In the case of cortisol or corticosterone immunoassays, the standard is not present in the feces,[2,6,18,19] and the substances measured should be labeled as cortisol/corticosterone metabolites. However, some reduced GCMs may have a biological activity[51–53] that is neither known nor proven for the fecal GCMs. Therefore, the term "fecal glucocorticoids" should also be avoided.

PRECISION

As parameters of precision control, pool samples must be analyzed in each assay and the coefficient of variation (CV) of their measured concentrations calculated within and between assays (intra- and interassay CV). To detect a potential bias within an assay, these samples should be distributed among samples from the experiment. To monitor the precision more thoroughly, two pool samples with different concentrations (high- and low-level pool) can be used.

SENSITIVITY AND BLANK VALUES

The sensitivity of an assay is defined as the smallest value that can be reliably discriminated from zero values with a 95% probability (two standard deviations from the signal given by the zero blanks).[34,54] Therefore, the precision influences the sensitivity of a test system, and a higher precision will also lower the assay sensitivity. Immunoassays usually have higher variations at both ends of the standard curve and lowest in the middle. In addition, blank effects have more consequences in the lower concentrations, whereas interferences concerning nonspecific binding are more pronounced in the higher concentrations (low optical densities). If possible, the dilution of samples should be performed in such a way that most of the values measured are in the range of the optimal precision of the assay, which is evaluated by a so-called precision profile, which shows the variation of pool samples at various dilutions.

As mentioned earlier, blank values (nonspecific interferences) may be a problem in immunoassays, especially when the concentrations of the metabolites to be measured are relatively low. In this case, interferences from the sample matrix or from the organic solvents used for extraction may cause problems. Because samples are normally not available without GCMs, it may be advisable to perform a suppression test using synthetic GC to check if the assay can detect the lower amount of GC produced by the adrenal glands.[2] The dexamethasone or flumethasone metabolites do not cross-react in immunoassays for corticosterone metabolites, but the adrenal production of GCs is reduced in response to the injection of the synthetic substances. Testing blank extracts (extraction procedures done without feces) will give some information concerning blank values.

The use of highly sensitive assays can reduce a possible blank problem because such assays permit a dilution of the sample extract that is much higher than those of less sensitive assays, and the interfering substances may be decreased to a level where their effect is negligible.

ASSAY SELECTION

Selection of an appropriate assay plays an important part in fecal analysis. There are various immunoassays described, measuring GCMs in fecal samples. Some authors (e.g., Kotrschal et al.,[55] Goymann et al.,[56] and Wasser et al.[15]) used corticosterone assays, which show some cross-reactions with the GCMs in feces. Another approach is the use of assays especially designed to measure groups of fecal metabolites, as for example described by Palme and Möstl[20] and Möstl et al.[46] for ruminants. In birds, a tetrahydrocorticosterone EIA (5β-pregnane-3α,11β,21-triol-20-one, a metabolite of corticosterone) was established to measure the corticosterone production in Wilson's storm petrels.[57] The same assay was also used successfully in Adélie penguins.[58]

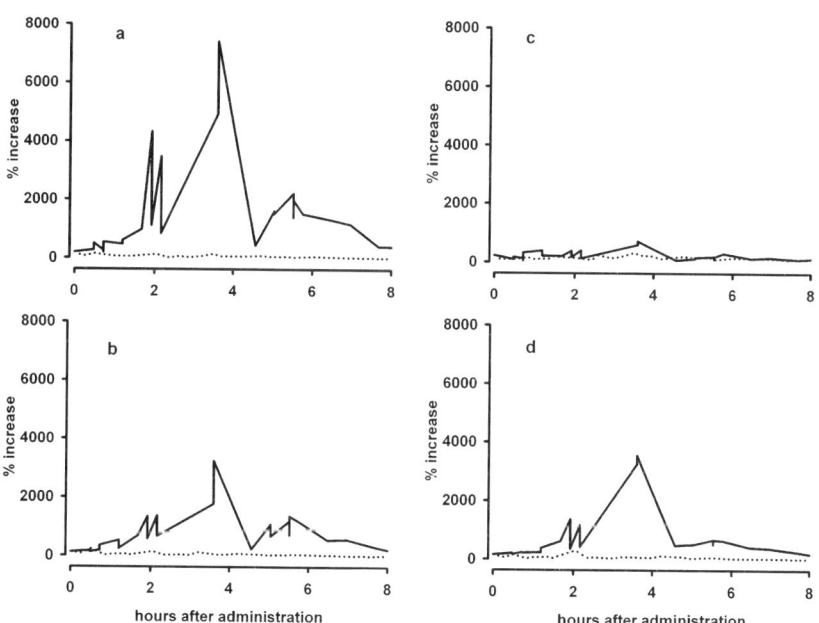

FIGURE 6. Immunoreactive corticosterone metabolites in feces of a gander after ACTH injection. All samples were stored at –20°C after collection and analyzed after methanolic extraction using four different assays: (**a**) 11β-hydroxyetiocholanolone EIA;[45] (**b**) tetrahydrocorticosterone EIA;[57] (**c**) corticosterone EIA;[20] (**d**) 11-oxoetiocholanolone EIA.[46] *Dotted line* represents the basal values measured during the same time on the day before ACTH administration.

On the basis of analytical validations (e.g., HPLC immunograms) alone, it is not possible to select the best-suited EIA for measuring adrenocortical activity in a species. To achieve that, it is important to perform a physiological and biological validation as well (for details, see Touma and Palme[2] and Goymann[50]), which should prove that changes in activity of the hypothalamic–pituitary–adrenal (HPA) axes are reflected in fecal GCM concentrations measured by the respective assay. For example, a higher increase of GCM concentrations following an ACTH challenge is thought to indicate a better-suited immunoassay.

To show this, data from a recently performed ACTH test in domestic geese are given. Because these birds have a long gut passage time, we also expected we would be able to discriminate the immunoreactive metabolites excreted via the urine (first peak) and the feces (second peak). ACTH (0.25 mg; Synacthen, Ciba Geigy, Switzerland) was administered (i.v.) to four geese (two males and two females). and all droppings were collected for the next 10 h. For comparison, all samples were collected from the same animals during the same time on the day before the stimulation test. We measured various immunoreactive metabolites using assays with antibodies produced against corticosterone,[20] 11-oxoetiocholanolone,[46] tetrahydrocorticosterone,[57] and 11β-hydroxyetiocholanolone.[45] All assays showed an increase in immunoreactive substances measured in the animals after ACTH injection (FIG. 6). As expected, two immunoreactive peaks occurred, representing urinary and fecal metabolites. The baseline-to-peak ratio was highest using the 11β-hydroxyetiocholanolone EIA. Therefore, this assay showed the highest reactivity, and thus may detect smaller changes in GC production in this species than, for example, the corticosterone assay used. These results back up the findings of Frigerio et al.[45] There it was shown that adverse weather conditions increased the concentration of the immunoreactive substances measured with the 11β-hydroxyetiocholanolone assay, but not

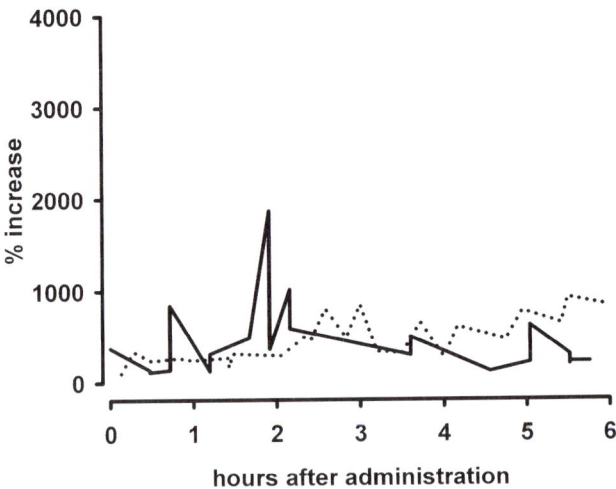

FIGURE 7. Immunoreactive 20-oxopregnanes in droppings of the same gander after ACTH injection. *Dotted line* represents the values measured during the same time on the day before ACTH administration.

those measured with the corticosterone EIA, revealing different biological sensitivities of the assay systems. There is no general rule regarding which assay should be used in birds because the pattern of formed metabolites is species specific. For example, Goymann et al.[16] tested the 11-oxoetiocholanolone EIA[20] in stonechats, but did not get satisfactory results. In general, if small changes in GC production must be measured, a careful selection of the assay systems available is advisable. However, a measured increase of immunoreactive metabolites after a stressful event or ACTH injection is not conclusive proof that an assay measures fecal GCMs. The goose samples mentioned earlier were also analyzed using an assay for 20-oxopregnanes,[59] and the results show an increase in these metabolites also. FIGURE 7 shows the data of the same male individual as shown before (Fig. 6). The increase was even more pronounced in the females. This may be explained by the fact that progesterone is the precursor of GCs and also increases after ACTH injection.[60] In addition, these immunoreactive metabolites were excreted predominantly via the urine. In this case, the only way to definitively prove that the metabolites measured with a certain EIA are derived from plasma GC would be an HPLC immunogram of samples from a respective radiometabolism study.

CONCLUSION

It is mandatory to evaluate the immunoassays used for measuring fecal GCMs for each species under investigation. As highlighted earlier in this article, an immunoassay must be carefully validated analytically. Particularly, it should be demonstrated that the antibody used cross-reacts with metabolites derived from the GC present in the blood. In addition, a physiological and biological validation is necessary. A clear alteration of fecal GCMs in relation to basal levels is thought to be indicative of a better-suited immunoassay for measuring adrenocortical activity. However, depending on the actual experiment performed, stability of the fecal GCMs measured with an immunoassay also plays a crucial role. Because this was reported to vary substantially between assays, these may also be criteria for the selection of an immunoassay. Only appropriate assays fulfilling these criteria will allow investigators to also monitor low degrees of disturbances, and thus provide a high biological sensitivity.

ACKNOWLEDGMENTS

We thank A. Kuchar and M. Fischer for their excellent technical assistance. The goose ACTH experiment was performed at the Konrad Lorenz Reseach Station, Grünau, Germany, with the help of this team and C. Youssif. The helpful comments of C. Touma on an earlier version of the manuscript are also acknowledged.

REFERENCES

1. SAPOLSKY, R.M., L.M. ROMERO & A.U. MUNCK. 2000. How do glucocorticoids influence stress responses? Integrating permissive, suppressive, stimulatory and preparative actions. Endocr. Rev. **21:** 55–89.

2. TOUMA, C. & R. PALME. 2005. Measuring fecal glucocorticoid metabolites in mammals and birds: the importance of validation. Ann. N.Y. Acad. Sci. **1046**: 54–74.
3. WINGFIELD, J.C. & M. RAMENOFSKY. 1999. Hormones and the behavioural ecology of stress. *In* Stress Physiology in Animals. P.H.M. Balm, Ed.: 1–51. Sheffield Academic Press.
4. DE KLOET, E.R., J. GROOTENDORST, A.M. KARSSEN & M.S. OITZL. 2002. Gene x environment interaction and cognitive performance: animal studies on the role of corticosterone. Neurobiol. Learn. Mem. **78**: 570–577.
5. MÖSTL, E. & R. PALME. 2002. Hormones as indicators of stress. Domest. Anim. Endocrinol. **23**: 67–74.
6. PALME, R. *et al.* 2005. Stress hormones in mammals and birds: comparative aspects regarding metabolism, excretion and noninvasive measurement in fecal samples. Ann. N.Y. Acad. Sci. **1040**: 162–171.
7. ETCHES, R.J. 1979. Plasma concentrations of progesterone and corticosterone during the ovulatory cycle of the hen (*Gallus domesticus*). Poult. Sci. **58**: 211–216.
8. ROMERO, L.M. 2002. Seasonal changes in plasma glucocorticoid concentrations in free-living vertebrates. Gen. Comp. Endocrinol. **128**: 1–24.
9. ETCHES, R.J. & F. CROZE. 1983. Plasma concentrations of LH, progesterone, and corticosterone during ACTH- and corticosterone-induced ovulation in the hen (*Gallus domesticus*). Gen. Comp. Endocrinol. **50**: 359–365.
10. BEUVING, G. & G.M.A. VONDER. 1978. Effect of stressing factors on corticosterone levels in the plasma of laying hens. Gen. Comp. Endocrinol. **35**: 153–159.
11. BROWNIE, A.C. 1992. The metabolism of adrenal cortical steroids. *In* The Adrenal Gland. V.H.T James, 2nd., Ed.: 209–224. Raven Press. New York.
12. TAYLOR, W. 1971. The excretion of steroid hormone metabolites in bile and feces. Vitam. Horm. **29**: 201–285.
13. LINDNER, H.R. 1972. Enterohepatic circulation and patterns of urinary excretion of cortisol metabolites in the ewe. J. Endocrinol. **52**: xix–xx.
14. MACDONALD, I.A. *et al.* 1983. Degradation of steroids in the human gut. J. Lipid Res. **24**: 675–700.
15. WASSER, S.K. *et al.* 2000. A generalized fecal glucocorticoid assay for use in a diverse array of nondomestic mammalian and avian species. Gen. Comp. Endocrinol. **120**: 260–275.
16. GOYMANN, W., E. MÖSTL & E. GWINNER. 2002. Corticosterone metabolites can be measured non-invasively in excreta of European stonechats, *Saxicola torquata rubicola*. Auk **119**: 1167–1173.
17. RETTENBACHER, S. *et al.* 2004. Measurement of corticosterone metabolites in chicken droppings. Br. Poult. Sci. **45**: 704–711.
18. BALTIC, M. *et al.* 2005. A noninvasive technique to evaluate human-generated stress in the black grouse. Ann. N.Y. Acad. Sci. **1046**: 81–95.
19. THIEL, D., S. JENNI-EIERMANN & R. PALME. 2005. Measuring corticosterone metabolites in droppings of Capercaillies (*Tetrao urogallus*). Ann. N.Y. Acad. Sci. **1046**: 96–108.
20. PALME, R. & E. MÖSTL. 1997. Measurement of cortisol metabolites in faeces of sheep as a parameter of cortisol concentration in blood. Int. J. Mammal. Biol. **62**(Suppl. II): 192–197.
21. MILLSPAUGH, J.J. & B.E. WASHBURN. 2004. Use of fecal glucocorticoid metabolite measures in conservation biology research: considerations for application and interpretation. Gen. Comp. Endocrinol. **138**: 189–199.
22. PALME, R. 2005. Measuring fecal steroids: guidelines for the practical application. Ann. N.Y. Acad. Sci. **1046**: 75–08.
23. KHAN, M.Z., J. ALTMANN, S.S. ISANI & J. YU. 2002. A matter of time: evaluating the storage of fecal samples for steroid analysis. Gen. Comp. Endocrinol. **128**: 57–64.
24. HUNT, K. & S.K. WASSER. 2003. Effect of long-term preservation methods on fecal glucocorticoid concentrations of grizzly bear and African elephant. Physiol. Biochem. Zool. **76**: 918–928.
25. PALME, R. *et al.* 1996. Excretion of infused ^{14}C-steroid hormones via faeces and urine in domestic livestock. Anim. Reprod. Sci. **43**: 43–63.

26. KLASING, K.C. 2005. Potential impact of nutritional strategy on noninvasive measurements of hormones in birds. Ann. N.Y. Acad. Sci. **1046:** 5–16.
27. MÖSTL, E. et al. 1999. Measurement of glucocorticoid metabolite concentrations in faeces of domestic livestock. J. Vet. Med. A. **46:** 621–632.
28. MAKIN, H.L.J. 1975. Methods of steroid analysis. I. Group estimations and separation techniques. In Biochemistry of Steroid Hormones. H.L.J. Makin, Ed.: 185–210. Blackwell Scientific Publishing, Oxford.
29. STANCZYK, F.Z. et al. 2003. Limitations of direct estradiol and testosterone immunoassay kits. Steroids **68:** 1173–1178.
30. TESKEY-GERSTL, A., E. BAMBERG, T. STEINECK & R. PALME. 2000. Excretion of corticosteroids in urine and faeces of hares (Lepus europaeus). J. Comp. Physiol. B **170:** 163–168.
31. SCHATZ, S. & R. PALME. 2001. Measurement of faecal cortisol metabolites in cats and dogs: a noninvasive method for evaluating adrenocortical function. Vet. Res. Commun. **25:** 271–287.
32. PALME, R. et al. 1997. Faecal metabolites of infused ^{14}C-progesterone in domestic livestock. Reprod. Dom. Anim. **32:** 199–206.
33. MIKSIK, I. 1999. Separation and identification of corticosterone metabolites by liquid chromatography–electrospray ionization mass spectrometry. J. Chromatogr. B Biomed. Sci. Appl. **726:** 59–69.
34. ABRAHAM, G.E. 1975. Characterisation of anti-steroid antisera. In Steroid Immunoassay. E.H.D. Cameron, S.G. Hillier & K. Griffith, Eds.: 67–86. Alpha Omega. Cardiff, Wales.
35. KELLIE, A.E. 1975. Methods of steroid analysis. II. Competitive binding. In Biochemistry of Steroid Hormones. H.L.J. Makin, Ed: 211–226. Blackwell Scientific Publishing. Oxford.
36. KOHEN, F., S. BAUMINGER & H.R. LINDNER, 1975. Preparation of antigenic steroid-protein conjugates. In Steroid Immunoassay. E.H.D. Cameron, S.G. Hillier & K. Griffith, Eds: 11–32. Alpha Omega. Cardiff, Wales.
37. KELLIE, A. E., K.V. LICHMAN & SAMARAJEEWA, 1975. Chemistry of steroid-protein conjugate formation In Steroid Immunoassay. E.H.D. Cameron, S.G. Hillier & K. Griffith, Eds.: 33–60. Alpha Omega. Cardiff, Wales.
38. HILL, M. et al. 1999. Elimination of cross-reactivity by addition of an excess of cross-reactant for radioimmunoassay of 17a-hydroxypregnenolone .Steroids **64:** 341–355.
39. NIESCHLAG, E., H.K. KLEY & G.H. USADEL. 1975. Production of antisera in rabbits. In Steroid Immunoassay. E.H.D. Cameron, S.G. Hillier & K. Griffith, Eds.: 87–96. Alpha Omega. Cardiff, Wales.
40. NECHANSKY, S. et al. 1980. Effect of immunization of male rabbits against androstenedione. Experientia **36:** 1131–1132.
41. VAN WEEMEN, B.K.. Discussion. In Steroid Immunoassay. E.H.D. Cameron, S.G. Hillier & K. Griffith, Eds.: 199. Alpha Omega. Cardiff, Wales.
42. PRAKASHA, B.S., H.H.D. MEYER & D.F.M. VAN DE WIELB, 1988. Sensitive enzyme immunoassay of progesterone in skim milk using second-antibody technique. Anim. Reprod. Sci. **16:** 225–235.
43. WHITEHEAD, T.P. 1981. Introduction: principles of quality control in clinical chemistry. In Quality Control in Clinical Endocrinology. D.W. Wilson, S.J. Gaskell &K.W. Kemp, Eds: 3–5. Alpha Omega. Cardiff, Wales.
44. MORROW, C.J., E.S. KULVER, G.A. VERKERK & L.R. MATTHEWS. 2002. Fecal glucocorticoid metabolites as a measure of adrenal activity in dairy cattle. Gen. Comp. Endocrinol. **126:** 229–241.
45. FRIGERIO, D., J. DITTAMI, E. MÖSTL & K. KOTRSCHAL. 2004. Excreted corticosterone metabolites co-vary with air temperature and air pressure in male greylag geese (Anser anser). Gen. Comp. Endocrinol. **137:** 29–36.
46. MÖSTL, E. et al. 2002. Measurement of cortisol metabolites in faeces of ruminants. Vet. Res. Commun. **26:** 127–139.
47. MIDDLE, J.G. 1995. The quality assessment of steroid hormone assays. In Steroid Analysis. H.L.J. Makin, D.B. Gower & D.N. Kirk, Eds.: 647–696. Blackie Academic & Professional.

48. CEKAN, S.Z. 1979. On the assessment of validity of steroid radioimmunoassays. J. Steroid Biochem. **11:** 135–141.
49. EKINS, R.P. 1975. Discussion. *In* Steroid Immunoassay. E.H.D. Cameron, S.G. Hillier & K. Griffith, Eds.: 58. Alpha Omega. Cardiff, Wales.
50. GOYMANN, W. 2005. Noninvasive monitoring of hormones in bird droppings:physiological validations, sampling, extraction, sex differences, and the influence of diet on hormone metabolite levels. Ann. N.Y. Acad. Sci. **1046:** 35–53.
51. PANIN, L.E. et al. 2002. Interaction of tetrahydrocortisol–apolipoprotein A-I complex with eukaryotic DNA and with single-stranded oligonucleotides. Mol. Biol. (Mosk.) (In Russian). 36: 103–105.
52. PENLAND S.N. & A.L. MORROW. 2004. 3a,5b-reduced cortisol exhibits antagonist properties on cerebral cortical GABA(A) receptors. Eur. J. Pharmacol. **506:** 129–132.
53. SUDO N. et al. 2004. Postnatal microbial colonization programs the hypothalamic-pituitary-adrenal system for stress response in mice. J. Physiol. **558:** 263–275
54. DEHNHARD, M. et al. 2003. Measurement of plasma corticosterone and fecal glucocorticoid metabolites in the chicken (*Gallus domesticus*), the great cormorant (*Phalacrocorax carbo*), and the goshawk (*Accipiter gentilis*). Gen. Comp. Endocrinol. **131:** 345–352.
55. KOTRSCHAL, K., K. HIRSCHENHAUSER & E. MÖSTL. 1998. The relationship between social stress and dominance is seasonal in greylag geese. Anim. Behav. **55:** 171–176.
56. GOYMANN, W. et al. 1999. Noninvasive fecal monitoring of glucocorticoids in spotted hyenas, *Crocuta crocuta*. Gen. Comp. Endocrinol. **114:** 340–348.
57. QUILLFELDT, P. & E. MÖSTL. 2003.Resource allocation in Wilson's storm-petrels (*Oceanites oceanicus*) determined by measurement of glucocorticoid excretion. Acta Ethol. **5:** 115–122.
58. NAKAGAWA, S., E. MÖSTL & J.R. WAAS. 2003. Validation of an enzyme immunoassay to measure faecal glucocorticoid metabolites from Adélie penguins (*Pygoscelis adeliae*): a noninvasive tool for the measurement of stress? Polar Biol. **26:** 491–493.
59. SCHWARZENBERGER, F. 1996. Faecal steroid analysis for non-invasive monitoring of reproductive status in farm, wild and zoo animals. Anim. Reprod. Sci. **42:** 515–526.
60. BAGE, R. et al. 2000, Effect of ACTH-challenge on progesterone and cortisol levels in ovariectomised repeat breeder heifers. Anim. Reprod. Sci. **63:** 65–76.

Noninvasive Monitoring of Hormones in Bird Droppings

Physiological Validation, Sampling, Extraction, Sex Differences, and the Influence of Diet on Hormone Metabolite Levels

WOLFGANG GOYMANN

Max Planck Institute for Ornithology, Department of Biological Rhythms and Behaviour, D-82346 Andechs, Germany

ABSTRACT: During the past several years, the noninvasive measurement of steroid metabolites from mammalian feces and bird droppings has become more and more popular. With an increasing acceptance of the method, investigators may become less aware of the need to validate their assays. It is shown why such validations are essential for each new species investigated and various ways to physiologically validate such noninvasive methods are described. Using the European stonechat (*Saxicola torquata rubicola*) as a model, it is explained why a validated method to measure androgen metabolites in males does not necessarily work in females. In addition the difficulties that may be neglected owing to the superficial ease of sampling and processing of excreta are investigated. Various issues that may arise during sampling, storage, and extraction of excreta are addressed. Finally, results suggesting that experimental manipulations of the diet may affect hormone metabolite levels in European stonechats are presented. So far, only a few studies have investigated the impact of diet on hormone metabolite levels, and these are the first data to report such an impact in birds. More studies are urgently needed to learn more about differences between the sexes, individuals, and populations and the impact of diet and energy metabolism on hormone metabolites.

KEYWORDS: noninvasive; feces; excreta; hormone metabolites; birds; diet; radioimmunoassay (RIA); enzyme immunoassay (EIA)

INTRODUCTION

Traditionally, sex steroids and corticosteroids are measured from blood samples that provide a measure of circulating plasma steroid concentrations. However, this method has some drawbacks, because animals must be caught and handled, which is not always feasible or desired. Particularly in small birds, the practicability of hor-

---

Address for correspondence: Dr. Wolfgang Goymann, Max Planck Institute for Ornithology, Von-der-Tann-Str. 7, D-82346 Andechs, Germany. Voice: +49-8152-373-119; fax: +49-8152-373-133.

goymann@orn.mpg.de

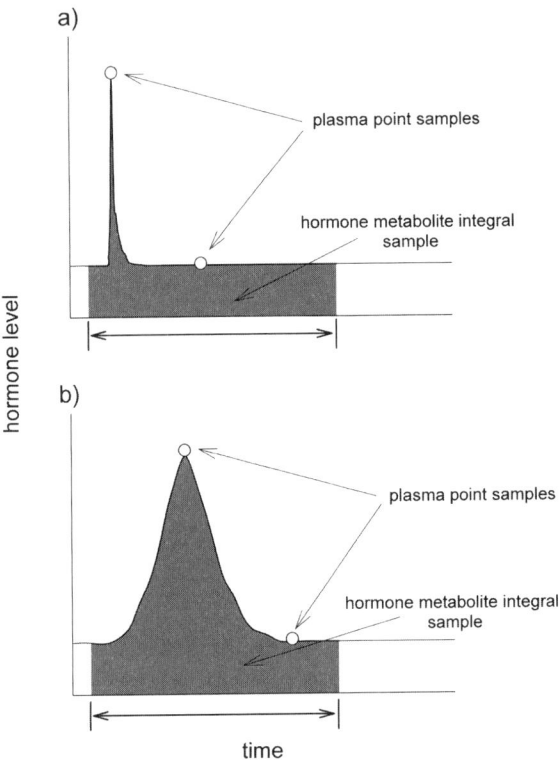

FIGURE 1. Illustration of the difference between hormone measurements from plasma samples and excreta. Plasma samples are point samples (*open dots*) indicating the hormonal status of an individual at a certain point in time. Excreta contain the pooled amount of excreted hormones allocated to one dropping. They represent an integral measure of hormone metabolites (*gray shading*). (**a**) A brief elevation of hormone levels that is unlikely to be traceable in excreta, that is, the small additional area under the peak does not add much to the total gray area. (**b**) A longer-term elevation of hormone levels that is likely to be traceable in excreta, that is, the additional large area under the peak adds significantly to the total gray area. The *arrow* on the time scale indicates the time interval of integration of the excreta sample.

mone studies is often limited by the frequency and volume of blood samples that can be obtained. In addition, handling causes stress,[1] which may influence the measured hormone values. Thus, blood samples must be collected immediately after capture to measure baseline hormone levels. Plasma hormone levels reflect the hormonal status of an individual at a certain point in time (FIG. 1a). If the pattern of hormone release has a strong pulsatory component, a single blood sample may not be sufficient to determine the endocrine status of an individual.

In addition to traditional blood sampling techniques, noninvasive methods, such as measuring steroid metabolites in mammalian urine and feces or droppings of birds, are desirable. In this article, I refer to samples used for these noninvasive techniques as "excreta," because in birds, the urine and the fecal fraction of the excreta are often voided together as droppings, rendering it difficult to distinguish between urine and feces. (For a detailed discussion of this topic, see Ref. 2). When excreta are used for hormone analyses, sampling does not interfere much with behavior, and multiple samples can be obtained from one individual. Steroid metabolites measured in excreta represent pooled fractions of excreted hormones, providing an integrated measure of steroid levels over a longer period of time (gray shading in FIG. 1a and 1b). The length of the integral depends on the defecation frequency of the study organism and ranges from a few minutes or a few hours in birds or ruminants to 1 to 2 d in large carnivores.[3] Depending on the question asked, an integrated measure of steroid levels over a longer period of time may be more desirable than a plasma point measure. In cattle, fecal concentrations of cortisol metabolites reflect the total amount of produced cortisol better than blood concentrations.[4] Conversely, short-term changes in hormone levels (FIG. 1) might not always be detectable in excreta, especially when the defecation rate of the study organism is low.

Unlike blood sampling, the collection of excreta does not require special skills. Typically, samples can be collected with great ease, even in the field, and sampling usually does not conflict with animal welfare considerations and does not require special permits. Because of these benefits and the progress in assay development, the noninvasive approach has caused increasing interest by researchers who work in very different fields, from behavioral ecology to conservation and reproductive biology. However, all participants at the technical meeting agreed that the measurement of steroid metabolites in excreta includes some pitfalls (see also Refs. 5–9) and that investigators should be aware of those pitfalls before applying these techniques. The participants of the tecnical meeting therefore decided to publish guidelines on good practice in noninvasive hormone research.[7]

It cannot be emphasized enough that the validity of noninvasive hormone measurements relies on the assumption that the concentration of hormone metabolites in feces or droppings reflects the circulating levels of the actual hormone (or at least the free fraction of circulating hormones; see Ref. 8 for more details) integrated over a certain period of time. In plasma, the hormone (i.e., corticosterone) is measured. In excreta, however, the circulating hormone itself is typically no longer present at all or present only in minor amounts (see FIG. 2 and Refs. 3, 10, and 11). What is measured in excreta are metabolites of the original hormone. Hence, in this article, hormone levels measured in excreta are referred to as hormone metabolites. For each hormone, typically several metabolites are present in the excreta. Their exact identity is not known and could only be determined with a significant analytical effort. Because antibodies for these metabolites usually are not available, researchers use commercial or custom-made antibodies for the original hormone or similar compounds, hoping that these antibodies cross-react with one or several of the hormone metabolites. However, any *a priori* assumption that an antibody for the original hormone (or similar compounds) detects the metabolites of the hormone also is wrong. Thus, a crucial step in the validation of a noninvasive assay is the proof that metabolites of the hormone are indeed detected (see also Refs. 5, 7–9). This is not a trivial step, because species, or even the sexes, may excrete different metabolites that may or may not cross-react with the antibody used. As a consequence, a method suitable for detecting steroid metabolites in one species or sex does not necessarily work in others, even if the species are closely related. For example, Baltic *et al.*[12] and Thiel

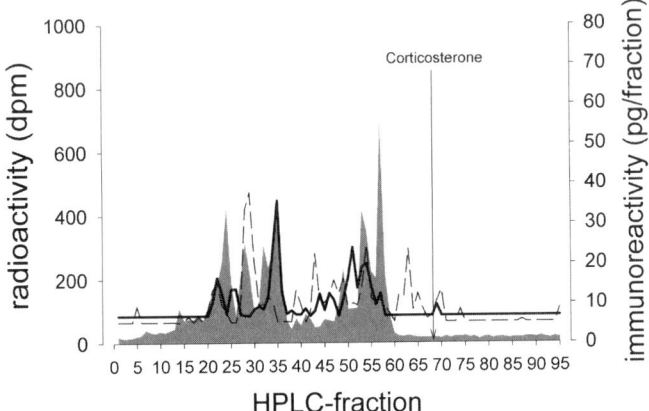

FIGURE 2. HPLC profile of corticosteroid metabolites in a male European stonechat. There are 7 large radioactive peaks indicating the major metabolites of corticosterone (gray shading). Both an EIA (*dashed line*),[11] and an RIA (*solid line*) for corticosterone cross-react with at least some of the major metabolites of corticosterone, indicating that both assays may be suitable to detect corticosterone metabolites in stonechat droppings. Note that none of the radioactive or immunoreactive peaks corresponds with the elution profile of corticosterone. (Modified from Goymann *et al.*[11])

et al.[13] had to test several different antibodies to find a suitable one for corticosterone metabolites in black grouse (*Tetrao tetrao*) and capercaillies *(T. urogallus)*. I had to try four different antibodies for corticosteroids, all of which worked in several other species, to find a suitable assay that was sensitive enough to detect cortisol metabolites in feces of spotted hyenas (*Crocuta crocuta*); I tested five different androgen antibodies to validate an assay for the detection of testosterone metabolites in feces of spotted hyenas, but none gave satisfying results (Goymann, unpublished data). Recently, Dloniak *et al.* developed an assay using another androgen antibody, which finally appears to give satisfying results.[14]

Like conventional assays for hormones in plasma, each assay for steroid metabolites in excreta requires analytical validation steps to demonstrate parallelism, accuracy, and precision. These analytical validations ensure that the assay system works properly and that there are no substances present in the extracts disturbing the binding properties of the antibodies used (so-called matrix effects). Good manuals, such as the one by Chard,[15] provide excellent descriptions on such assay validations. For the reasons previously mentioned, noninvasive hormone assays require an additional validational step, a so-called physiological or biological validation not required when measuring hormones in blood samples. This additional validation is necessary because hormone metabolites rather than the actual hormone are measured; hence it must be proved that the antibody is indeed capable of detecting the metabolites of the hormone in question. Radioinfusion studies, in which a radiolabeled hormone is injected into the animal and its metabolic fate is investigated, are very useful and elegant (but not always feasible). A physiological validation could include a correlation of plasma hormone and excreted hormone metabolite levels, although such correlations may rarely exist because of the very different nature of the measurements (see the following sections for more details). It is better to conduct pharmacological stimulations or suppressions of circulating hormone levels with the expectation of finding the respective changes in excreted hormone metabolite levels. In the following sections, I will describe these three kinds of physiological validations in further detail.

RADIOINFUSION STUDIES

Radioinfusion studies are very useful, because they provide the researcher with a lot of information. They can be used to measure the time lag between secretion and excretion of a hormone,[10,11,16,17] they give information about the proportional allocation of hormone metabolites to urine and feces,[16–20] and they allow the determination of whether the antibody really cross-reacts with metabolites of the hormone in question.[10,11,16–22] The disadvantage of radioinfusion studies is that they typically require special permits and equipment regarding animal care facilities and the generation of radioactive waste. This is relatively easy to accomplish when working with small songbirds, but I have due respect for investigators who have done this with large mammals such as horses, cattle,[17] and elephants.[19] In addition, animal welfare considerations need to be taken into account, especially when working with endangered species.

Radioinfusion studies are performed by injecting a small amount of radiolabeled target hormone into the circulation of the animal and subsequently sampling the excreta. After extraction, steroid metabolites, including the metabolites of the infused radiolabeled hormone, can be separated using liquid chromatographic methods such as high-performance liquid chromatography (HPLC).[10] FIGURE 2 shows the steroid metabolites of one dropping sample of a European stonechat (*Saxicola torquata rubicola*); these metabolites were separated into 95 different fractions. Then a subsample of each of these HPLC separated fractions was counted in a liquid scintillation counter to identify those fractions that contain radioactivity (FIG. 2). These radioactive fractions contained the metabolites of the injected radiolabeled hormone. In all cases investigated so far, the original hormone was no longer present, but was completely metabolized into other compounds (see also Refs. 3, 12, and 13). In the specific case shown in FIGURE 2, seven major radioactive peaks appeared, indicating fractions with remnants of metabolized ^{3}H-labeled corticosterone.

Next, another subsample of each HPLC-fraction is measured with a radioimmunoassay (RIA), enzyme immunoassay (EIA), or fluorescence immunoassay (FIA) to establish the cross-reactivity of each HPLC-fraction with the antibody of the hormone in question. If the cross-reactivity of the antibody matches at least in part the elution profile of radioactive metabolites, there is good evidence that the assay system is able to detect metabolites of the hormone in question (FIG. 2). In the case of the stonechat, both a custom-made corticosterone EIA (FIG. 2)[11] and a commercially available corticosterone antibody (FIG. 2; B3-163, Esoterix Endocrinology, Calabasas Hills, CA) significantly cross-reacted with major metabolites of corticosterone. The original hormone corticosterone was completely metabolized, as there was neither a radioactive nor an immunoreactive peak in the fraction in which corticosterone elutes in the HPLC system (FIG. 2).

CORRELATIONS BETWEEN PLASMA HORMONE AND HORMONE METABOLITE LEVELS

At first sight, an obvious and seemingly straightforward possibility to validate a noninvasive hormone assay is a simple correlation with blood values. In fact, referees often ask for this kind of validation. However, because of the very different nature of circulating hormones and their excreted metabolites and because of the time lag between secretion and excretion, such correlations often may be difficult. As mentioned earlier, plasma is a point-in-time sample, and excreta represent interval samples (FIG. 1). Thus, short-term changes in plasma hormone levels may be muted in excreta samples (FIG. 1a). Hence, the absence of a good correlation of hormone levels between plasma and excreta does not disprove the validity of noninvasive methods (see also Ref. 23), it just demonstrates the different nature of the respective measures. Correlations between plasma hormone and excreted metabolite concentrations are possible only when blood levels are constant over longer periods of time. This is usually the case for estrogens and progestins during the estrous cycle or during pregnancy. In these cases, however, it is usually sufficient to correlate behavioral or morphological signs of estrus with excreted hormone metabolite levels. Hence, plasma levels are not required for a validation.[24–26] In addition, seasonal hormone cycles of males can be used.[10,27] It is usually much more powerful to conduct hormonal stimulation or suppression experiments, especially for androgens and corticosteroids.

PHARMACOLOGICAL STIMULATIONS OR INHIBITIONS OF HORMONE SECRETION

The strongest methods to physiologically validate noninvasive methods are pharmacological stimulations or inhibitions of steroid hormone release. These methods typically involve the administration of high doses or long-acting versions of hormones such as corticotrophin releasing hormone (CRH) or adrenocorticotrophic hormone (ACTH) to stimulate the adrenal gland to produce corticosteroids and gonadotrophic releasing hormone (GnRH) or luteinizing hormone (LH) to stimulate the production of sex steroids. Noninvasive methods also have been validated by administering the steroid hormone in question (i.e., corticosterone).[28] When doing so, it is important to administer physiological doses, otherwise pharmacological rather than physiological effects may be measured. In my opinion, the administration of hormones such as CRH and ACTH or GnRH and LH is more elegant, because these hormones stimulate the natural production of steroids in the original glands, thus generating a physiological increase of the steroid hormone in question.

Goymann *et al.* conducted an ACTH challenge of seven male and one female European stonechat to stimulate the production of corticosterone and measure corticosterone metabolites.[11] Initially, we tested three EIAs with two custom-made antibodies against corticosterone (also used in geese[29]), 11-oxoetiocholanolone from the laboratory of Erich Möstl,[30] and a commercial corticosterone antibody from ICN Biomedicals[31] that is frequently used to measure corticosteroids in mammalian feces noninvasively[32] In contrast to the 11-oxoetiocholanolone antibody, both corticosterone antibodies gave satisfying results (see Ref. 11 and W. Goymann,

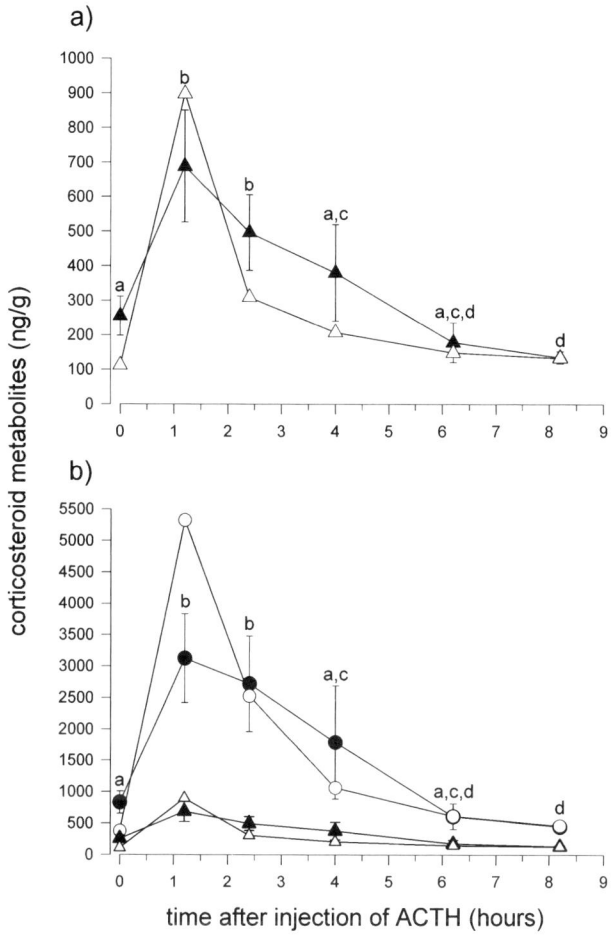

FIGURE 3. Excreted corticosterone metabolites in 1 female (*open symbols*) and 7 male (*closed symbols*) European stonechats after injection with ACTH. Measured (**a**) with a custom-made corticosterone EIA (see Ref. 11) and (**b**) with an RIA with a commercial corticosterone antibody. The EIA resulted in a 2.7-fold (**a**) increase of corticosterone metabolites from baseline levels, and the RIA resulted in a 3.8-fold (**b**) increase. (Modified from Goymann *et al.*[11] For illustrative purposes, the curve of (**a**) is graphed in (**b**) again, using a different scale.

unpublished data). Data from the EIA with the custom-made corticosterone antibody are shown in FIG. 3. Recently, I tested a fourth antibody against corticosterone (B3-163, Esoterix Endocrinology, Calabasas Hills, CA) that I normally use in my laboratory as an RIA to measure plasma corticosterone levels. The HPLC profile of this assay suggests that the antibody detects corticosterone metabolites (FIG. 2). When I measured the extracted samples of the ACTH challenge experiment with this corticosterone RIA, it had a higher cross-reactivity with corticosterone metabolites (FIG. 3b) than the previously used custom-made corticosterone EIA. Also, the magnitude of the response (the signal-to-noise ratio) was higher in the corticosterone RIA than in the corticosterone EIA (3.8- vs. 2.7-fold increase; FIG. 3b). Hence, for future investigations I will use the corticosterone RIA (with the commercial antibody) rather than the previously used corticosterone EIA (with the custom-made antibody). I should point out that the differences between the two assays are caused by different binding properties of the respective antibodies and not by the use of RIA or EIA as a method. Another recent article (by Frigerio *et al.*) demonstrates how much more information about the adrenocortical activity can be obtained with an optimized assay.[33]

The example of the two different corticosterone assays (EIA and RIA) for stonechats shows that it is basically impossible to do comparative studies of absolute hormone metabolite levels if different assays are used. In the case of the stonechat, and using the same samples, the EIA detected a mean maximum corticosteroid metabolite concentration of 690 ng/g (FIG. 3a), whereas the RIA detected a mean maximum levels of 3125 ng/g (FIG. 3b). How can this be? Both antibodies showed different cross-reactivities with different corticosterone metabolites (FIG. 2), and obviously the antibody used in the RIA detected these metabolites to a higher degree. Thus, within-species comparisons of absolute hormone metabolite concentrations can be performed only if the same assay system is used. Between-species comparisons of absolute metabolite levels are problematic, because the species may differ in the kind of metabolites that they produce. Thus, any difference between species may simply be caused by different metabolites that are or are not detectable with the respective extraction procedure and assay system. Relative measures (i.e., the ratio between maximum and baseline hormone levels) may be a bit less problematic, but caution is required here as well. For example, the maximum-to-baseline ratio of the corticosterone RIA is considerably higher (3.8; FIG. 3b) than that of the corticosterone EIA (2.7; FIG. 3a). I will address some of these issues (i.e., sampling and processing of excreta, sex differences in the excretion of metabolites, and the impact of diet and metabolic rate) in more detail in the following sections. Other important points, such as nutritional strategies, time-lag differences between production and excretion of steroid metabolites, and diurnal fluctuations will not be covered, but are elaborated elsewhere.[2,16,34]

SAMPLING AND PROCESSING OF FECES OR BIRD DROPPINGS

Hormone metabolite measures from excreta that were not collected fresh may be erroneous. (For a detailed discussion of this topic, see Refs. 6, 13, and 23.) Hence, ideally, measurements should be taken only from excreta that have been collected fresh (i.e., are not older than a couple of hours).

In larger mammals or birds, it is often inconvenient to process the whole sample. Several studies have shown that hormone metabolites are not equally distributed within the excreta,[19,35] and thus the whole sample should be taken and homogenized before taking a smaller subsample for the analysis. In birds, excreta typically consist of a fecal and a urine part that may or may not be separable, and the allocation of metabolites to urine and feces has not been studied in detail (see also Refs. 2, 12, and 36). Some researchers prefer to take only the fecal part for analysis,[29,37–39] whereas others take the whole sample,[10,11,21] assuming that a true separation is not possible,

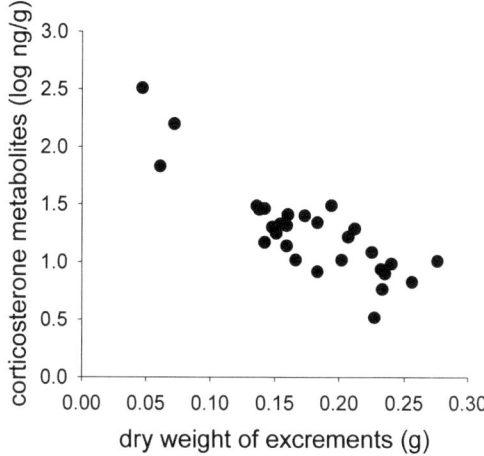

FIGURE 4. Relationships of dropping dry mass and corticosterone metabolite levels in European stonechats. The Pearson correlation coefficient of this correlation is $r_p = -0.849$. Because the y variable is nonindependent of the x variable, the null model is the average r_p generated from an appropriate Monte Carlo simulation.[41] The mean r_p of 256 Monte Carlo simulations for this case was -0.851, and 95% of simulations yeilded r_p-values between -0.114 and -0.964. Thus, the relationship between dropping dry mass and corticosterone metabolite levels does not differ from what would be expected by chance.

because the urine and fecal part of the excreta are already mixed inside the cloaca and when voided on the ground (see also Refs. 2, 12, and 13).

In small birds, an additional difficulty arises because of the small size of each dropping. For example, Millspaugh and Washburn[6] and Tempel and Gutierrez[40] found that very small samples (<0.02 g) resulted in proportionally higher concentrations of hormone metabolites than larger samples. In my studies with stonechats, I found a similar effect (FIG. 4). Millspaugh and Washburn[6] suggested as one possible explanation, that the extraction efficiency may differ between small and large samples. This suggestion can and should be tested. But there is another, more likely explanation for a correlation between sample mass and hormone concentration: the correlation in the studies of Millspaugh and Washburn,[6] Tempel and Guttierez,[40] and Goymann (FIG. 4) may be spurious. Spurious correlations are those caused solely by data transformations. As such, they do not reflect meaningful properties of the underlying data. Any correlation between the concentration (expressed as ng/g fecal mass) and fecal mass (in g) is based on nonindependent variables as each concentration value (representing the y-axis) is calculated, including the corresponding fecal mass value (representing the x-axis; see also FIG. 4). Correlating these variables violates the important statistical assumption that the two variables in question should be independent. When hormone metabolite concentrations are correlated with fecal mass, the variables are nonindependent. For correlations of such nonindependent variables, the null-hypothesis is not that there is no correlation. (For a detailed discussion of this topic, see Ref. 41.) In such cases of statistical nonindependence, the statistical significance (which is equal to the deviation from the null hypothesis) of a correlation coefficient r can be determined through Monte Carlo simulations.[41] In such a simulation, a set of random x and y variables is generated, which is then used to calculate the derived variable y/x by dividing each y through the respective x. Then, a correlation coefficient r is generated by correlating x with y/x. Any correlation coefficient between x and y/x is, by definition, spurious, because both x and y variables were generated randomly and independently from one another. Only the derived variable y/x is nonindependent from x. Repeating this simulation procedure n times with new random sets of x and y variables generates n spurious correlation coefficients, r_n. The distribution of these spurious correlation coefficients, r_n, can be then used to calculate the 95% confidence interval for the expected correlation coefficients that are based on pure chance. This calculation will allow us to estimate whether the observed correlation in the real data is spurious or meaningful.[41]

An example may help to illustrate this concept. To estimate whether the Pearson correlation coefficient, r_p, of 0.849 obtained with the data from FIGURE 4 significantly deviates from an expected spurious correlation coefficient, I ran 256 Monte Carlo simulations, generating 256 sets of random x and y variables with the same sample size ($n = 29$) and distribution pattern (mean and standard deviation) as the data from FIGURE 4. Then, I calculated for each of the 256 random sets of x and y the derived variable y/x, thus simulating a situation in which excreta mass is used as a reference standard for the concentration of excreted metabolites. All 256 sets of x variables were correlated with the respective 256 sets of y/x values. This resulted in 256 different Pearson correlation coefficients, r_p, that were, by definition, all spurious. From these 256 r_p values, I determined the mean and the 2.5 and 97.5 percentiles (95% confidence interval): 0.851, –0.114, and –0.964, respectively. Thus, given the distribution of the particular dataset of FIGURE 4 and a significance level of $\alpha = 5\%$, any correlation coefficient between –0.114 and 0.964 would not deviate from random expectation. The correlation coefficient of the real data ($r_p = 0.849$, see FIG. 4) was almost identical with the expected mean of the spurious correlation coefficients, indicating that the relationship between corticosterone metabolite levels and excreta mass was indeed spurious.

Issues regarding spurious correlations can be avoided when a constant amount of droppings are used for analysis, even in small bird species. However, then either the smallest sample determines the amount to be extracted or many small samples may have to be excluded from or pooled for the analysis. For laboratory studies, it sometimes may be useful to consider alternative reference standards. For example, in our experiments with captive stonechats, I collect all droppings voided over a period of 2 h. This allows me to express the amount of hormone with reference to the time (i.e., picogram of hormone metabolites voided per hour). Using this measure, for example, allows me to control for the potential impact of different diets on hormone metabolite concentrations (see the following discussion). The drawback of this approach is that it is clearly limited, because it is necessary to collect all droppings over a defined period of time. Further studies are required to investigate whether the potential benefits justify the increased effort in data collection.

Extraction of Excreta Samples

Before extracting hormones from plasma, it is good practice to incubate the plasma with a small amount of radiolabeled hormone.[42] This is done to estimate the extraction efficiency (recovery) and to adjust the final concentration of the hormone for losses caused by the extraction procedure. It may be tempting, and in fact it has been suggested,[5] to conduct similar controls for the extraction of hormone metabolites. However, because the exact nature and chemical properties of hormone metabolites are normally unknown, controlling for extraction efficiency is difficult. The extraction properties of the original hormone might differ substantially from those of the metabolites of this hormone. Thus, rather than controlling for extraction efficiency, the addition of a labeled tracer hormone for the determination of the recoveries may introduce additional error. Unless the exact chemical nature of the metabolite is known, it may be better to refrain from adding recoveries and instead try to use extraction procedures that are as simple as possible.[30] By doing so, the amount and variability caused by extraction losses can be minimized.

Storage of Excreta Samples

Several studies have shown that sample storage is a critical issue.[13,43–46] However, all of these studies have been performed in mammals, and studies in birds are still lacking. Because of the large differences between species in the sensitivity of feces or droppings to storage conditions, such studies in birds are urgently needed.[8,9] Still, the best way to store feces or droppings appears to be immediate freezing; in some cases, freeze-drying, oven-drying, or sun-drying of the samples may work as well.[45]

SEX DIFFERENCES IN HORMONE METABOLITE EXCRETION

Touma *et al.*[16] found sex differences in the excretion of corticosteroid metabolites in mice; Rettenbacher *et al.*[36] and Baltic *et al.*[12] reported sex differences in corticosteroid metabolism in domestic chicken and black grouse, respectively. In European stonechats, there appears to be a similar pattern regarding androgens. A male and a female stonechat excreted different metabolites of 3H-testosterone (FIG. 5; for detailed methods, see Ref. 10). Although a testosterone RIA detected biologically relevant metabolites of androgens in male stonechats (FIGS. 5a and 6),[10] this did not appear to be the case in females (FIGS. 5b and 6). In males, there were two major radioactive peaks indicating metabolites of 3H-testosterone. These peaks were found in the same fractions that also showed strong immunoreactivity, indicating that the assay detected true metabolites of testosterone (FIG. 5a). In females, there were three major radioactive peaks indicating metabolites of 3H-testosterone, but the immunoreactivity is distributed more or less evenly in all fractions and only the first radioactive peak coincides with some elevation in the immunoreactivity (FIG. 5b). Hence, it appears as if the testosterone antibody that successfully detects testosterone metabolites in males is far less efficient in females. This may be caused by differences in metabolism, but there is more to this story: In March, plasma concentrations of testosterone in male European stonechats peak with a mean of 1.0 ng/mL (E. Gwinner, unpublished data). At the same time of the year, excreted testosterone metabolites of males peak at around 200 ng/g (FIG. 6). These results and additional data from GnRH-challenge experiments[10] suggest that our testosterone RIA reliably measures testosterone metabolite patterns in male stonechats. For female

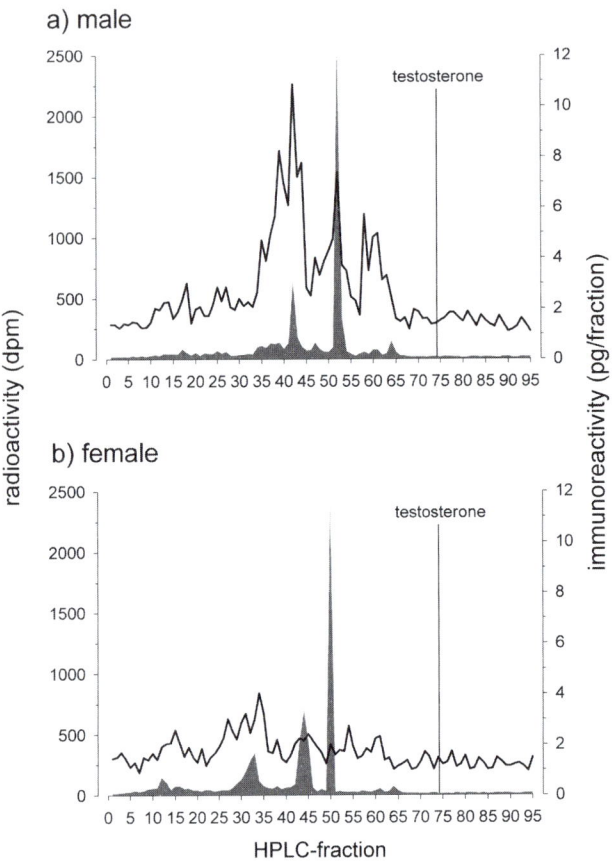

FIGURE 5. HPLC profile of testosterone metabolites in (**a**) a male and (**b**) a female stonechat. There are 2 large radioactive peaks in the male and 3 in the female (*gray shadings*) indicating that the major metabolites of ^{3}H-testosterone are different in males and females. The radioimmunoassay for testosterone[10] clearly cross-reacts with the major testosterone metabolites in males, but only partially so in females (*solid lines*). Note that none of the radioactive or immunoreactive peaks corresponds with the elution position of testosterone. (Modified from Goymann *et al*.[10])

stonechats, a different picture emerges. Females have peak levels of plasma testosterone approximately 12.5 times lower than those of males during March (0.08 ng/mL; E. Gwinner, unpublished data). Surprisingly, and in striking contrast to these low plasma testosterone levels, the excreted "testosterone metabolite" levels of females in March were identical to those of males (FIG. 6). This finding suggests that our testosterone RIA does not reliably measure testosterone metabolite patterns in female stonechats. I suspect that during the reproductive season, females produce other steroids that are metabolized in a way that they cross-react with the testosterone assay. Other androgens, such as dihydrotestosterone and androstenedione levels,

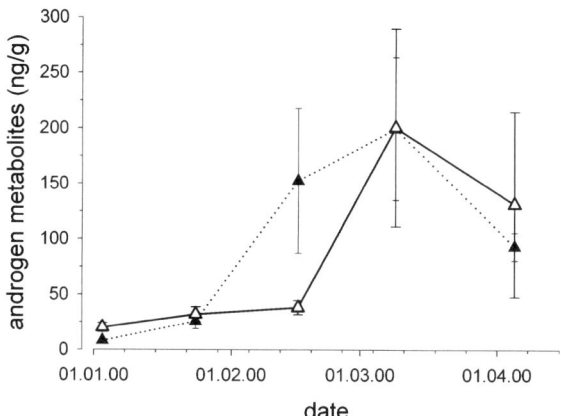

FIGURE 6. Seasonal testosterone metabolite pattern in 9 male (*closed triangles*) and 7 female (*open triangles*) European stonechats. In males, testosterone metabolite levels correspond very well both to validations using a GnRH challenge and blood plasma levels with a peak of 1.0 ng/mL testosterone in March. Female plasma levels of testosterone are much lower (0.08 ng/mL) at the same time, despite similar levels of testosterone metabolites in females and males. Most likely, "testosterone metabolite" levels in female stonechats reflect seasonal gonadal activitybut not true testosterone metabolism.

are also usually very low in female stonechats (E. Gwinner, unpublished data), so that it is unlikely that they contribute much to the "testosterone metabolite" peak in March. But the seasonal pattern with high "testosterone metabolite" levels in spring (FIG. 6) suggests that reproductive steroids like progestins or possibly other (so far unmeasured) androgens such as dehydroepiandrosterone (DHEA) are potential candidates,[22] although plasma DHEA levels are likely to be low in stonechats. This effect—that hormone metabolites, which are not of conventional androgenic origin, are measured with an assay for testosterone metabolites—may be specific for stonechats. So far, however, stonechats are the only passerine bird species for which a physiological validation for testosterone metabolites exists,[10] and hence the phenomenon may occur in other bird species as well. For this reason, great caution should be used when presenting and interpreting "testosterone metabolite" data from droppings of female birds, especially when they have not been physiologically validated, such as in Schwabl[47] and Langmore *et al.*[48]

DIET AND HORMONE METABOLITE CONCENTRATIONS

So far, only a few studies have investigated the potential impact of diet on hormone metabolites in excreta. For free-ranging mammals and birds, the composition and availability of food may change seasonally and may affect hormone metabolite measurements. In baboons, an increase in dietary fiber decreased the amount of excreted progesterone metabolites per unit mass.[49] In dairy cattle, changing the amount of dry matter in the diet did not affect fecal progesterone metabolite concen-

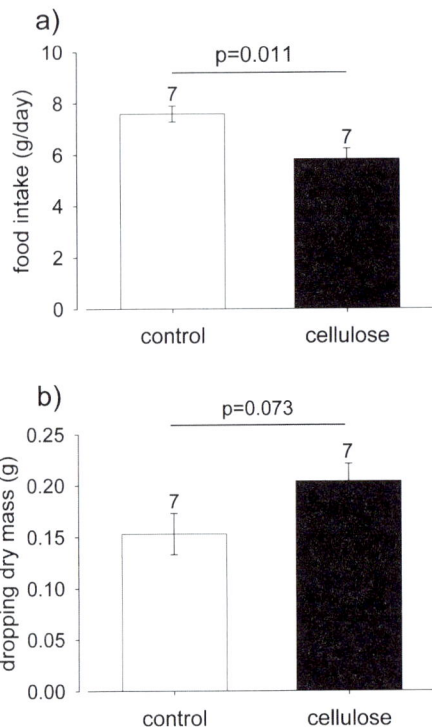

FIGURE 7. Food intake (**a**) and dry mass (**b**) of droppings in European stonechats that were fed with (*black bars*) and without (*white bars*) 20% cellulose-enriched diet. Food intake was significantly lower in the cellulose-fed group [Mann-Whitney U-test, $U_{7,7} = 6$, $P = .011$, effect size: $\delta = 1.746$ (CI_δ: -0.511; $+2.936$)], whereas the dry mass of droppings tended to be lower in the cellulose-fed group [$U_{7,7} = 10.5$, $P = .073$, effect size: $\delta = -1.1014$ (CI_δ: -2.117; $+0.126$)].

trations.[50] In birds, dietary fibers may increase gut passage time and at the same time increase fecal bulk.[51] I tested whether an increase in dietary fibers affects excreted hormone metabolite levels in male European stonechats. Adding 20% cellulose for 1 week to our standard stonechat food[52] significantly decreased food intake (FIG. 7a). At the same time, the dry weight of droppings tended to increase (FIG. 7b), which may be due to the nondigestible fibers. Assuming a constant output of excreted hormones, one would expect a lower concentration of hormone metabolite levels in the cellulose-fed group, because the same amount of hormone metabolites is excreted into droppings of larger mass. Indeed, both corticosteroid and androgen metabolite concentrations were lower in the cellulose-fed group (TABLE 1). In the case of corticosteroid metabolites, the difference was significant; for androgen metabolites, it represented a strong trend. Because the experiment was carried out in winter, production of androgens might not have been high enough to detect differences of larger magnitude. If the data were expressed as the amount of steroid excreted per

TABLE 1. Mean concentrations (ng/g) of testosterone and corticosteroid metabolite levels in male European stonechats fed with 20% cellulose-enriched diet ($N = 7$ for corticosteroids and $N = 6$ for androgens) and without cellulose ($N = 7$)

Metabolites	Cellulose group	Control group	Test statistics
Androgen	3.63 (−0.56; +0.66)	5.75 (−1.29; +1.66)	$t_{10.7} = 1.584$, $P = .071$, $\delta = 0.847$ [−0.269; +1.931]
Corticosteroid	13.18 (−2.00; +3.04)	24.55 (−4.87; +6.07)	$t_{11.9} = 2.061$, $P = .031$, $\delta = 1.102$ [−0.052; +2.217]

NOTE: Tests were performed using log-transformed hormone concentrations, and data are presented as backtransformed means, with negative and positive standard errors in brackets. The statistical results refer to one-tailed tests, as the cellulose-fed group was expected to have lower concentrations of hormone metabolites than the control group. δ indicates standardized effect size, followed in brackets by the confidence interval for standardized effect size.

TABLE 2. Mean amount (pg/h) of testosterone and corticosteroid metabolites in male European stonechats fed with 20% cellulose-enriched diet ($N = 7$) and without cellulose ($N = 7$)

Metabolites	Cellulose group	Control group	Test statistics
Androgen	0.42 (−0.11; +0.14)	0.35 (−0.50; +0.58)	$t_{7.0} = 0.796$, $P = .452$, $\delta = 0.227$ [−0.829; +1.274]
Corticosteroid	1.31 (−0.21; +0.24)	1.75 (−0.14; +0.15)	$t_{11.9} = 1.517$, $P = .164$, $\delta = 0.829$ [−0.300; +1.891]

NOTE: Tests were performed using log-transformed hormone concentrations, and data are presented as backtransformed means, with negative and positive standard errors in brackets. The statistical results refer to two-tailed tests, as no *a priori* expectation existed for hourly hormone metabolite concentrations. δ indicates standardized effect size, followed in brackets by the confidence interval for standardized effect size.

unit of time instead of the concentration per gram, there was no difference in the amount of corticosterone or androgen metabolites between cellulose-fed and control-fed stonechats (TABLE 2). This indicates that the dietary effect was really caused by a change in the amount of dropping mass and not by a change in overall hormone levels. I should point out that TABLE 1 and TABLE 2 refer to the results of the same experiment and only use different measures, the concentration (ng/g) and the amount (pg/h) of hormone metabolites, respectively.

In a second experiment, I added elderberries to our standard diet[52] for stonechats. In this experiment, neither food intake [elderberry: 8.73 ± 0.54 g, control: 8.20 ± 0.54 g; $U_{7,8} = 22$, $P = .486$, effect size: $\delta = 0.360$ (CI$_\delta$: −0.671; +1.376)] nor the dry weight of droppings [elderberry: 0.16 ± 0.03 g, control: 0.19 ± 0.02 g; $U_{7,8} = 22$, $P = .487$, effect size: $\delta = 514$ (CI$_\delta$: −0.528; +1.538)] differed between groups. Also, there were no differences in androgen or corticosterone metabolite levels (TABLE 3) between the elderberry-fed group and the control group. Similar results were obtained when using the amount of excreted steroid metabolites (pg/h) instead of the concentration (ng/g; data not shown).

TABLE 3. Mean concentrations (ng/g) of androgen and corticosteroid metabolite levels in male European stonechats fed with ($N = 8$) and without elderberries ($N = 7$)

Metabolites	Elderberry group	Control group	Test statistics
Androgen	5.89 (−1.52; 2.05)	4.07 (−0.98; 1.30)	$t_{13} = -0.843$, $P = .415$, $\delta = -0.430$ [−1.450; +0.605]
Corticosteroid	12.73 (−3.70; 5.21)	12.59 (−2.00; 2.37)	$t_{7.5} = -0.041$, $P = .967$, $\delta = 0.025$ [−1.115; +1.066]

NOTE: Tests were performed using log-transformed hormone concentrations, and data are presented as backtransformed means, with negative and positive standard errors in brackets. The statistical results refer to two-tailed tests, as no *a priori* expectation regarding hormone levels existed. δ indicates standardized effect size, followed in brackets by the confidence interval for standardized effect size.

In summary, there is now some evidence that changes in dietary fiber content may affect hormone metabolite measurements in stonechats. Most likely, this is true for other birds and mammals as well. We urgently need more information about how dietary changes affect hormone metabolite levels. This is especially important for studies of free-ranging animals, for which large differences in dietary composition can be expected on a seasonal (and individual) basis.

Changes in energy metabolism may affect hormone metabolite concentrations in excreta, too. For example, the energy requirements for a small temperate zone songbird to maintain resting metabolism are probably much higher in winter than in summer. To maintain basic body functions, the bird needs to metabolize a larger amount of food in winter than in summer. How does this affect hormone metabolite levels? I am currently conducting experiments in which European stonechats are kept at low- and high-temperature regimes. Food intake and dropping output of birds kept at 5°C is approximately twice as high as that of birds kept at 22°C. The next step will be to examine how these differences in food intake and dropping output affect hormone metabolite levels. To my knowledge, similar studies have been done in only dairy cattle.[53]

CONCLUSIONS

The first major goal of this article was to describe and emphasize the importance of physiological validations for hormone metabolite measurements. Physiological validations are absolutely essential. Without such validations, measurements of excreted hormone patterns can mean anything and, in the worst case, may have nothing to do with the hormone in question. The noninvasiveness and superficial ease of the method are certainly appealing to many investigators. This is reflected in the cumulating amount of publications that are based on noninvasive hormone measurements (for an overview, see also Ref. 9). However, with an increasing general acceptance of the method, investigators, referees, and journal editors may become less critical about the necessity of physiological validations. Even now, there are already a number of papers in highly reputed journals that lack such validations. This is a dangerous practice and should be discouraged because such a practice may lead to erroneous results.

The second major goal of this article was to describe the need for further, more detailed investigations that should be conducted in at least a number of model species. These studies should include potential sex, individual, and population differences in hormone metabolism. For example, using a testosterone antibody that reliably measures androgen metabolites in male stonechats, I have demonstrated a striking discrepancy between plasma concentrations of testosterone and excreted levels of androgen metabolites in female stonechats. I presented this example for two reasons: first, to remind us that there may be large sex differences in hormone metabolism and second to warn us regarding any premature interpretation of "antibody immunoreactivity" in excreta (which could mean anything) as "testosterone metabolite concentration" without any physiological validation.

Furthermore, there is still little knowledge regarding the impact of diet on excreted hormone metabolite levels, but the articles of Klasing,[2] and Wasser et al.[49] on baboons and this study on stonechats suggest that dietary influences should be seriously taken into consideration. Also, seasonal changes in energy requirements that go along with changes in metabolic throughput of food may represent a confounding factor. I am currently investigating this in further detail in the stonechat. We definitely need more research regarding potential sex, individual, and population differences, and the impact of diet and energy metabolism on hormone metabolites.

ACKNOWLEDGMENTS

I thank Ingrid Schwabl and Christina Wolf for assistance with the RIAs and feeding experiments reported in this article. Many thanks also to Barbara Helm, Susanne Jenni-Eiermann, Erich Möstl, and Rupert Palme, who made important contributions to improve previous versions of the manuscript, and to Nicole Hoiss for editorial work. Furthermore, I thank all participants of the European Science Foundation (ESF) technical meeting on "Measurement of hormones in bird droppings and egg yolk" for the countless stimulating discussions during the meeting.

REFERENCES

1. WINGFIELD, J.C., C.M. VLECK & M.C. MOORE. 1992. Seasonal changes of the adrenocortical response to stress in birds of the Sonoran desert. J. Exp. Zool. **264:** 419–428.
2. KLASING, K.C. 2005. Potential impact of nutritional strategy on noninvasive measurements of hormones in birds. Ann. N.Y. Acad. Sci. **1046:** 5–16
3. PALME, R., et al. 2005. Stress hormones in mammals and birds: comparative aspects regarding metabolism, excretion, and noninvasive measurement in fecal samples. Ann. N.Y. Acad. Sci. **1040:** 162–171.
4. PALME, R., F. WETSCHER & C. WINCKLER. 2003. Measuring faecal cortisol metabolites: a noninvasive tool to assess animal welfare in cattle? Proc. of the IVth Central European Buiatric Congress in Lovran, Croatia, April 23–27, 2003: 145–150.
5. BUCHANAN, K.L. & A.R. GOLDSMITH. 2004. Noninvasive endocrine data for behavioural studies: the importance of validation. Anim. Behav. **67:** 183–185.
6. MILLSPAUGH, J.J. & B.E. WASHBURN. 2004. Use of fecal glucocorticoid metabolite measures in conservation biology research: considerations for application and interpretation. Gen. Comp. Endocrinol. **138:** 189–199.
7. PALME, R. 2005. Measuring fecal steroids: guidelines for practical application. Ann. N.Y. Acad. Sci. **1046:** 75–80.

8. MÖSTL, E., S. RETTENBACHER & R. PALME. 2005. Measurement of corticosterone metabolites in birds' droppings: an analytical approach. Ann. N.Y.Acad. Sci. **1046:** 17–34.
9. TOUMA, C. & R. PALME. 2005. Measuring fecal glucocorticoid metabolites in mammals and birds: the importance of alidation. Ann. N.Y. Acad. Sci. **1046:** 54–74.
10. GOYMANN, W., E. MÖSTL & E. GWINNER. 2002. Non-invasive methods to measure androgen metabolites in excrements of European stonechats, *Saxicola torquata rubicola*. Gen. Comp. Endocrinol. **129:** 80–87.
11. GOYMANN, W., E. MÖSTL & E. GWINNER. 2002. Corticosterone metabolites can be measured noninvasively in excreta of European stonechats (*Saxicola torquata rubicola*). Auk **119:** 1167–1173.
12. BALTIC, M. *et al.* 2005. A noninvasive technique to evaluate human-generated stress in the Black Grouse. Ann. N.Y. Acad. Sci. **1046:** 81–95.
13. THIEL, D., S. JENNI-EIERMANN & R. PALME. 2005. Measuring corticosterone metabolites in droppings of Capercaillies (*Tetrao urogallus*). Ann. N.Y. Acad. Sci. **1046:** 96–108.
14. DLONIAK, S.M. *et al.* 2004. Non-invasive monitoring of fecal androgens in spotted hyenas (*Crocuta crocuta*). Gen. Comp. Endocrinol. **135:** 51–61.
15. CHARD, T. 1995. An introduction to radioimmunoassay and related techniques. Elsevier. Amsterdam, The Netherlands.
16. TOUMA, C. *et al.* 2003. Effects of sex and time of day on metabolism and excretion of corticosterone in urine and feces of mice. Gen. Comp. Endocrinol. **130:** 267–278.
17. PALME, R. *et al.* 1996. Excretion of infused 14C steroid hormones via feces and urine in domestic livestock. Anim. Reprod. Sci. **43:** 43–63.
18. GRAHAM, L.H. & J.L. BROWN. 1996. Cortisol metabolism in the domestic cat and implications for non-invasive monitoring of adrenocortical function in endangered felids. Zoo Biol. **15:** 71–82.
19. WASSER, S.K. *et al.* 1996. Excretory fate of estradiol and progesterone in the African elephant (*Loxodonta africana*) and patterns of fecal steroid concentrations throughout the estrous cycle. Gen. Comp. Endocrinol. **102:** 255–262.
20. MONFORT, S.L. *et al.* 1997. Steroid-metabolism and validation of noninvasive endocrine monitoring in the African wild dog (*Lycaon pictus*). Zoo Biol. **16:** 533–548.
21. TELL, L.A. 1997. Excretion and metabolic fate of radiolabeled estradiol and testosterone in the cockatiel (*Nymphicus hollandicus*). Zoo Biol. **16:** 505–518.
22. MÖHLE, U. *et al.* 2002. Characterization of urinary and fecal metabolites of testosterone and their measurement for assessing gonadal endocrine function in male nonhuman primates. Gen. Comp. Endocrinol. **129:** 135–145.
23. HIRSCHENHAUSER, K., K. KOTRSCHAL & E. MÖSTL. 2005. Synthesis of measuring steroid metabolites in goose feces. Ann. N.Y. Acad. Sci. **1046:** 138–153.
24. BROWN, J.L. *et al.* 1994. Comparative aspects of steroid hormone metabolism and ovarian activity in felids, measured noninvasively in feces. Biol. Reprod. **51:** 776–786.
25. COCKREM, J. F. & J. R. ROUNCE. 1994. Faecal measurements of oestradiol and testosterone allow the non-invasive estimation of plasma steroid concentrations in the domestic fowl. Br. Poult. Sci. **35:** 433–443.
26. CROFOOT, M. *et al.* 2003. Reproductive assessment of the great hornbill (*Buceros bicornis*) by fecal hormone analysis. Zoo Biol. **22:** 135–145.
27. ROEDL, T. *et al.* 2004. Excremental androgen metabolite concentrations and gonad sizes in temperate zone vs. tropical Stonechats (*Saxicola torquata ssp.*). Gen. Comp. Endocrinol. **139:** 124–130.
28. HIEBERT, S.M. *et al.* 2000. Noninvasive methods for measuring and manipulating corticosterone in hummingbirds. Gen. Comp. Endocrinol. **120:** 235–247.
29. KOTRSCHAL, K., K. HIRSCHENHAUSER & E. MOESTL. 1998. The relationship between social stress and dominance is seasonal in Graylag geese. Anim. Behav. **55:** 171–176.
30. PALME, R. & E. MOESTL. 1997. Measurement of cortisol metabolites in feces of sheep as a parameter of cortisol concentration in blood. Z. Saeugetierk. **62**(Suppl. 2): 192–197.
31. GOYMANN, W. *et al.* 1999. Noninvasive fecal monitoring of glucocorticoids in spotted hyenas, *Crocuta crocuta*. Gen. Comp. Endocrinol. **114:** 340–348.

32. WASSER, S. K. et al. 2000. A generalized fecal glucocorticoid assay for use in a diverse array of nondomestic mammalian and avian species. Gen. Comp. Endocrinol. **120:** 260–275.
33. FRIGERIO, D. et al. 2004. Excreted corticosterone metabolites co-vary with ambient temperature and air pressure in male Graylag geese (*Anser anser*). Gen. Comp. Endocrinol. **137:** 29–36.
34. SOUSA, M.B.C. & T.E. ZIEGLER. 1998. Diurnal-variation on the excretion patterns of fecal steroids in common marmoset (*Callithrix jacchus*) females. Am. J. Primatol. **46:** 105–117.
35. MILLSPAUGH, J.J. & B.E. WASHBURN. 2003. Within-sample variation of fecal glucocorticoid measurements. Gen. Comp. Endocrinol. **132:** 21–26.
36. RETTENBACHER, S. et al. 2004. Measurement of corticosterone metabolites in chicken droppings. Br. Poult. Sci. **45:** 704–711.
37. WASSER, S. K. et al. 1997. Noninvasive physiological measures of disturbance in the northern spotted owl. Conserv. Biol. **11:** 1019–1022.
38. DEHNHARD, M. et al. 2003. Measurement of plasma corticosterone and fecal glucocorticoid metabolites in the chicken (*Gallus domesticus*), the great cormorant (*Phalacrocorax carbo*), and the goshawk (*Accipiter gentilis*). Gen. Comp. Endocrinol. **131:** 345–352.
39. WASHBURN, B.E. et al. 2004. Factors related to fecal estrogens and fecal testosterone in California spotted owls. Condor **106:** 567–579.
40. TEMPEL, D.J. & R.J. GUTTIERREZ, 2004. Factors related to fecal corticosterone levels in California spotted owls: implications for assessing chronic stress. Conserv. Biol. **18:** 538–547.
41. BRETT, M.T. 2004. When is a correlation between non-independent variables "spurious"? Oikos **105:** 647–656.
42. GOYMANN, W. & J.C. WINGFIELD. 2004. Competing females and caring males. Sex steroids in African black coucals, *Centropus grillii*. Anim. Behav. **64:** 733–740.
43. KHAN, M.Z. et al. 2002. A matter of time: evaluating the storage of fecal samples for steroid analysis. Gen. Comp. Endocrinol. **128:** 57–64.
44. WASHBURN, B.E. & J.J. MILLSPAUGH. 2002. Effects of simulated environmental conditions on glucocorticoid metabolite measurements in white-tailed deer feces. Gen. Comp. Endocrinol. **127:** 217–222.
45. HUNT, K.E. & S.K. WASSER. 2003. Effect of long-term preservation methods on fecal glucocorticoid concentrations of grizzly bear and African elephant. Physiol. Biochem. Zool. **76:** 918–928.
46. LYNCH, J.W. et al. 2003. Concentrations of four fecal steroids in wild baboons: short-term storage conditions and consequences for data interpretation. Gen. Comp. Endocrinol. **132:** 264–271.
47. SCHWABL, H. 1996. Environment modifies the testosterone levels of a female bird and its eggs. J. Exp. Zool. **276:** 157–163.
48. LANGMORE, N.E., J.F. COCKREM & E.J. CANDY. 2002. Competition for male reproductive investment elevates testosterone levels in female dunnocks, *Prunella modularis*. Proc. R. Soc. B. Biol. Sci. **269:** 2473–2478.
49. WASSER, S.K. et al. 1993. Effects of dietary fibre on faecal steroid measurements in baboons (*Papio cynocephalus cynocephalus*). J. Reprod. Fertil. **97:** 569–574.
50. RABIEE, A.R. et al. 2002. Progesterone clearance rate in lactating dairy cows with two levels of dry matter and metabolisable energy intakes. Anim. Reprod. Sci. **72:** 11–25.
51. KLASING, K.C. 1998. Comparative Avian Nutrition. CAB International Publishing. New York.
52. GWINNER, E., S. KOENIG & C. HALEY. 1995. Genetic and environmental factors influencing clutch size in equatorial and temperate zone stonechats (*Saxicola torquata axillaris* and *S. t. rubicola*): an experimental study. Auk **112:** 748–755.
53. RABIEE, A.R., K.L. MACMILLAN & F. SCHWARZENBERGER. 2001. The effect of level of feed intake on progesterone clearance rate by measuring faecal progesterone metabolites in grazing dairy cows. Anim. Reprod. Sci. **67:** 205–214.

Measuring Fecal Glucocorticoid Metabolites in Mammals and Birds: The Importance of Validation

CHADI TOUMA[a] AND RUPERT PALME[b]

[a]*Department of Behavioral Neuroendocrinology, Max Planck Institute of Psychiatry, D-80804 Munich, Germany*

[b]*Institute of Biochemistry, Department of Natural Sciences, University of Veterinary Medicine, Vienna, Austria*

ABSTRACT: In recent years, the noninvasive monitoring of steroid hormone metabolites in feces of mammals and droppings of birds has become an increasingly popular technique. It offers several advantages and has been applied to a variety of species under various settings. However, using this technique to reliably assess an animal's adrenocortical activity is not that simple and straightforward to apply. Because clear differences regarding the metabolism and excretion of glucocorticoid metabolites (GCMs) exist, a careful validation for each species and sex investigated is obligatory. In this review, general analytical issues regarding sample storage, extraction procedures, and immunoassays are briefly discussed, but the main focus lies on experiments and recommendations addressing the validation of fecal GCM measurements in mammals and birds. The crucial importance of scrutinizing the physiological and biological validity of fecal GCM analyses in a given species is stressed. In particular, the relevance of the technique to detect biologically meaningful alterations in adrenocortical activity must be shown. Furthermore, significant effects of the animals' sex, the time of day, season, and different life history stages are discussed, bringing about the necessity to seriously consider possible sex differences as well as diurnal and seasonal variations. Thus, comprehensive information on the animals' biology and stress physiology should be carefully taken into account. Together with an extensive physiological and biological validation, this will ensure that the measurement of fecal GCMs can be used as a powerful tool to assess adrenocortical activity in diverse investigations on laboratory, companion, farm, zoo, and wild animals.

KEYWORDS: stress hormones; glucocorticoids; cortisol; corticosterone; HPA axis; noninvasive monitoring; feces/faeces; validation; ACTH challenge test; dexamethasone suppression test; sex differences; diurnal variation; seasonal variation; life history stages; review

Address for correspondence: Chadi Touma, Ph.D., Max Planck Institute of Psychiatry, Department of Behavioral Neuroendocrinology, Kraepelinstrasse 2-10, D-80804 Munich, Germany. Voice: +49-(0)89-30622-228; fax: +49-(0)89-30622-569.
touma@mpipsykl.mpg.de

INTRODUCTION

Hormones, Stress, and the Hypothalamic–Pituitary–Adrenal Axis

Hormones are of great interest to scientists from various fields because they are largely involved in virtually all bodily functions in health and disease, including the regulation of reproduction, development, and the expression of behavior. A wide variety of endocrine factors has been linked to genetic, environmental, and social variation, including gonadal and adrenal steroids, pituitary peptides, growth factors, and biogenic amines. In particular, the so-called stress hormones and the concepts of stress have a very long history of research (going back to the ancient Greeks), as they deal with the daily social and nonsocial stimuli that are challenging or threatening to the survival, health, and reproductive success of animals (for reviews, see Refs. 1–9).

Stress in its broadest sense is well known to have a substantial impact on a variety of bodily functions. Its disruptive effects, for example, on the immune system, reproduction, cognition, and behavior of vertebrates have been broadly demonstrated (for reviews, see Refs. 2, 4, 6, and 8–12). Furthermore, stress hormones have been implicated in a wide range of human disorders, including depression, anxiety, cancer, asthma, cardiovascular diseases, diabetes, and dementia.[2–7,9,11,13,14] Assessing physiological parameters related to stress is therefore essential for the understanding and improvement of animal welfare, health, and reproduction.

When confronted with a stressor (environmental, physiological, or psychological), an individual typically displays a stress response consisting of a suite of physiological and behavioral alterations to cope with the challenge. One of the main mediators of this response is the hypothalamic–pituitary–adrenal (HPA) axis, which is responsive not only to stressors but also to other types of activity that are associated with emotional arousal (e.g., courtship or sexual behaviors).[4,5,8,9] Within minutes of the onset of a perceived stressor, the adrenal cortex begins to secrete glucocorticoids (GCs), mainly cortisol or corticosterone. Which GC is predominantly produced depends largely on the species and should be considered when choosing an appropriate assay system.[15–17] The major GC in most primates, carnivores, and ungulates, for example, is cortisol, whereas in most rodents, birds, and reptiles it is corticosterone. These GCs orchestrate the organism's response to challenges, acting on many organ systems, including the brain, to modulate physiology and behavior.[2,4–9,12,18] The secretion of GCs from the adrenal cortex is regulated by the release of adrenocorticotropic hormone (ACTH) from the anterior pituitary gland, which in turn is stimulated by corticotropin-releasing hormone (CRH) and vasopressin (AVP) derived from neurons of the paraventricular nucleus (PVN) of the hypothalamus.[1,2,5–7,9,18] Plasma GC concentrations are therefore widely used to assess stress responses in various species.[2,4,5,8,9,12,19,20,26]

However, constraints of the blood sampling procedure pose some limitations to this approach, particularly for small animals, such as most rodents and birds, or for free-ranging animals. A further limitation of invasive sampling techniques is that circulating hormone levels are affected rapidly in response to the stress of handling, physical restraint, and the blood sampling procedure itself, which can substantially alter physiological and behavioral parameters investigated in the experiment.[4,8,21] An additional drawback of measurements in the plasma is that blood samples represent concentrations at only a single point. Because steroid hormones and especially GCs may exhibit regular as well as episodic changes over time (i.e., circadian variations and pulsatile secretion patterns),[1,4,5,8] hormone levels representing a very narrow time frame might be biased. However, alternative techniques of measuring steroid hormone metabolites in excreta like feces of mammals or droppings of birds offer a possible solution to overcome some of these problems.

NONINVASIVE MONITORING OF HORMONES

In general, circulating steroid hormones are metabolized by the liver and excreted as conjugates via the kidneys into the urine or via the bile into the gut.[22–26] Although steroids in the gut are subjected to some extent to an enterohepatic circulation (i.e., reabsorption into the blood stream) and are intensively metabolized by the microbial flora, the sterane skeletal structure is not degraded.[22,27,28] Therefore, specific steroid metabolites can be detected in the feces of mammals and in droppings of birds.[16,25,26,29,30] When a lag time between hormonal events in the plasma and the appearance of the respective signal in the feces is considered, a similar pattern to that found in the plasma is reflected in the feces.[16,22,24,26,30] This lag time, which depends mainly on the intestinal transit time from the duodenum to the rectum, is largely species-specific and must be taken into account when comparing endocrine patterns found in plasma and feces.[16,23,26,31] Considering these points, it is possible to use fecal hormone metabolite analyses as a noninvasive tool to assess various endocrine functions in mammals and birds.

In recent years, this completely noninvasive technique has been established in an increasing number of species, ranging from laboratory animals, companion and farm animals, to wild animals (in zoos and in the field). It is now widely used to investigate hormone–behavior relationships as well as various questions in the realms of stress and animal welfare, reproductive physiology, behavioral ecology, conservation biology, and biomedical research (see TABLE 1 and studies discussed in the text).

Using fecal samples offers several advantages. Feces can be collected very easily, and the sampling is feedback free, because there is no need to capture and handle the animal. Therefore, repeated sampling of the same individual is possible without affecting the animal's behavior or its endocrine status. This method allows the monitoring of short-term hormonal changes in reaction to specific situations, social encounters, or treatments, as well as assessing day-to-day changes or even long-term endocrine profiles. In addition, circulating hormone levels in the feces are integrated over a certain period. Hence, rather than the actual steroid concentration, fecal hormone metabolite levels reflect the production rate, that is, the cumulative secretion and elimination of hormones, over several hours.[16,23,26,30] Therefore, unlike blood samples, fecal samples are less affected by episodic fluctuations or the pulsatility of hormone secretion. Consequently, depending on the research question, steroid metabolite concentrations measured in feces might represent the hormonal status of an animal more accurately than a single plasma sample. However, dampening of short peaks of hormone secretion can also be a disadvantage if fecal samples are used to monitor these short-term alterations.

TABLE 1. Selected studies describing experiments to physiologically and/or biologically validate techniques to measure fecal glucocorticoid metabolites as a noninvasive tool to assess adrenocortical activity in different species of mammals and birds

Species		Sample size		Physiological validation (pharmacological treatment)	Biological validation (challenging procedure/situation/condition)	Ref.
Scientific name	Common name	Males (M)	Females (F)			
Mammalia						
Primates						
Macaca fascicularis	long-tailed macaque		10	ACTH (data for 6 shown)		32
Papio cynocephalus	yellow baboon		2	ACTH		32
Pan troglodytes	chimpanzee	1	3		anesthesia	33
Lemur catta	ring-tailed lemur		12		capture+blood sampling, correlation plasma–feces	34
Rodentia						
Mus musculus f. domesticus	laboratory mouse	36	36	ACTH, Dex (2 doses each, 12M/12F)	saline injection, diurnal variation (6M/6F, each)	35
Rattus norvegicus f. domesticus	laboratory rat	24		ACTH, Dex (6M, each)	saline injection, handling (6M, each, no effects)	36
		10	9		effects of the estrous cycle, diurnal variation	37
Clethrionomys gapperi	red-backed vole	10 (5)			novelty, cold stress	38
		5			confinement in a trap (4/12 h)	39
Peromyscus maniculatus	deer mouse	18/17/16			confinement in a trap (<4/4–8/>8 h)	39
Lagomorpha						
Lepus europaeus	European hare	10	10		effects of rousing	40
Carnivora						
Crocuta crocuta	spotted hyena	4	2	ACTH	translocation (1M) agonistic interactions (1M/1F)	41
Ursus arctos horribilis	grizzly bear	1	1	ACTH		42

TABLE 1. (*continued*) Selected studies describing experiments to physiologically and/or biologically validate techniques to measure fecal glucocorticoid metabolites as a noninvasive tool to assess adrenocortical activity in different species of mammals and birds

Species		Sample size		Physiological validation	Biological validation	Ref.
Scientific name	Common name	Males (M)	Females (F)	(pharmacological treatment)	(challenging procedure/situation/condition)	
Carnivora (*continued*)						
Helarctos malayanus	Malayan sun bear		1	ACTH		32
Ursus thibetanus	Himalayan black bear	1		ACTH	aggression (1F), no consistent effects	43
Canis lupus f. familiaris	domestic dog	5	5	ACTH, Dex		44
Canis lupus	wolf	>40 (3 packs)		ACTH (data not shown)		45
			2		effects of rank, aggression	45
Canis rufus	red wolf	1		ACTH	various stressors (1M/1F), inconsistent effects	43
Lycaon pictus	African wild dog	2	3	ACTH		46
		34	22		effects of social status	47
Acinonyx jubatus	cheetah	2	2	ACTH		48
			7		immobilization, translocation, mate introduction	48
		2	2	ACTH (data for 1 shown)		32
			1	ACTH	anesthesia	43
Neofelis nebulosa	clouded leopard	2	2	ACTH		49
		36	36		effects of housing condition	49
		2	1	ACTH	anesthesia (1F)	43
		3	2	ACTH (data for 1 shown)		32
Felis silvestris f. catus	domestic cat	5	5	ACTH, Dex		44
		1	1	ACTH	anesthesia (1F), no clear effect	43
Suricata suricatta	Slender-tailed meerkat	2		ACTH		43
Enhydra lutris kenyoni	Alaskan sea otter	1		ACTH		32
Mustela nigripes	black-footed ferret	4	6	ACTH (partly inconsistent)	restraint, saline injection (inconsistent effects)	50
Eumetopias jubatus	Steller sea lion	1	2	ACTH (feces + plasma)		51

TABLE 1. (*continued*) Selected studies describing experiments to physiologically and/or biologically validate techniques to measure fecal glucocorticoid metabolites as a noninvasive tool to assess adrenocortical activity in different species of mammals and birds

Species		Sample size		Physiological validation	Biological validation	Ref.
Scientific name	Common name	Males (M)	Females (F)	(pharmacological treatment)	(challenging procedure/situation/condition)	
Proboscidea						
Loxodonta africana	African elephant	1		ACTH		52
			4 (juv)	ACTH (feces + plasma)		53
			2	ACTH (data for 1 shown)	effects of enclosure size (in total, 23F)	32
Perissodactyla						
Equus caballus	domestic horse	3	3	ACTH, Dex (feces + plasma)		54
		10			anesthesia + surgery (castration)	55
Diceros bicornis	black rhinoceros	4		ACTH		56
		1	1	ACTH (data for 1 shown)		32
Artiodactyla						
Bos taurus	cattle	3	3	ACTH (3 doses), Dex (feces + plasma)		57
			8 + 4		loading into cattle carrier, transportation	58
			5	ACTH (feces + plasma)		59
			10 + 8		novel environment, transportation	59
Ovis ammon f. aries	domestic sheep	3	3	ACTH, Dex (feces + plasma)		57
Cervus elaphus	elk/red deer		6	ACTH	saline injection, disturbance	60
Cervus elaphus roosevelti	Roosevelt elk	1	1	ACTH (data for 1 shown)		32
Capreolus capreolus	roe deer	5		ACTH, Dex	capture, injection, transportation (4M)	61
Litocranius walleri	gerenuk	2		ACTH (data for 1 shown)		32
Oryx dammah	Scimitar-horned oryx	2		ACTH (data for 1 shown)		32
Sus scrofa f. domestica	domestic pig	3	3	ACTH, Dex (partly inconsistent)		54

TABLE 1. (*continued*) **Selected studies describing experiments to physiologically and/or biologically validate techniques to measure fecal glucocorticoid metabolites as a noninvasive tool to assess adrenocortical activity in different species of mammals and birds**

Species		Sample size		Physiological validation (pharmacological treatment)	Biological validation (challenging procedure/situation/condition)	Ref.
Scientific name	Common name	Males (M)	Females (F)			
Aves						
Anseriformes						
Anser domesticus	domestic goose	24	10	ACTH (8M)	social stimulation/confrontation (8M per group)	62
Anser anser	greylag goose	43			effects of higher competition (mating season)	63
		37			correlation with agonistic/courtship behavior	64
Galliformes						
Gallus domesticus	domestic chicken	10	10	ACTH, Dex (no Dex response)		65
			15	ACTH, Dex (feces + plasma, 5F, each)	saline injection (5F, Dex + saline, no clear effects)	66
Tetrao urogallus	capercaillie	3	2	ACTH		67
Tetrao tetrix	black grouse	2	2	ACTH		68
Stringiformes						
Strix occidentalis caurina	northern spotted owl		1	ACTH	translocation (feces + plasma)	69, 32
Passeriformes						
Parus major	great tit	10 + 6			social challenge, clear effects in one line (6M)	70
Saxicola torquata rubicola	European stonechat	7	1	ACTH		71
Spiza americana	dickcissel	10			capture, radio-tagging (partly inconsistent)	72
Columbiformes						
Zenaida macroura	mourning dove	9	7	ACTH (2 doses, 2M/2F each)	saline injection (saline + low dose ACTH, no effect)	73
Sphenisciformes						
Pygoscelis adeliae	Adelie penguin	1	1	ACTH		74

Measuring Fecal Glucocorticoid Metabolites

Prior to the analysis, fecal glucocorticoid metabolites (GCMs) must be extracted from the fecal matrix. Because fecal steroid metabolites are a mixture of several different metabolites with a wide range of polarities, the selection of an appropriate extraction procedure is a serious issue.[15,16,25,30] We recommend extracting fecal steroid metabolites simply by suspending (and shaking) a certain amount of homogenized feces (e.g., a portion of 0.5 g of fresh or dry weight) in a fixed volume (e.g., 5 mL) of methanol (80% in mammals and 60% in birds proved to work best).[15,17,26] This very simple extraction technique is highly practical (no complex apparatuses or evaporation steps are needed) and yielded good recovery levels for virtually all species tested so far.[15,16,26]

Usually, the quantification of fecal steroid metabolites is then performed by using an aliquot of the extract in a radioimmunoassay (RIA) or an enzyme immunoassay (EIA).[15,17,25,26,29] Often, commercially available cortisol or corticosterone kits are applied (for examples, see Refs. 32 and 43). However, the antibodies used in these assays might have some shortcomings, because they are produced primarily to measure the respective unmetabolized steroid in the plasma. As alternatives to these assay kits, different EIAs have been developed that are especially designed for measuring groups of steroid metabolites usually present in the feces (e.g., 11,17-dioxoandrostanes[15,26,75]). These EIAs use so-called group-specific antibodies, which have several advantages for the analysis of fecal GCMs.[15,16,26]

However, there are considerable species- and sex-specific differences in the types of GCMs formed, resulting in a characteristic pattern of GCMs present in the feces of a given species.[15,16,30] Accordingly, it is important to select an appropriate assay system that includes an antibody capable of detecting most, or at least a considerable proportion, of the respective GCMs present in the feces of the species investigated.[15–17,25,26,29,30]

In addition, after defecation, several factors, such as temperature, humidity, and other environmental conditions, may influence concentrations of immunoreactive GCMs in the sample.[15,17,29,54,76–78] Moreover, bacterial enzymes are reported to increase or decrease levels of immunoreactive fecal GCMs if samples are not frozen shortly after voidance.[15,29,54,59,76,77,79] Heat, alcohol, or other preservatives, such as acids, are therefore frequently used, especially in the field, where direct freezing of the samples to avoid further metabolism of the steroids is difficult.[25,29,42,76–81] However, because adding alcohol already starts the extraction, and because fecal GCM concentrations of samples preserved in this manner were also reported to change over time,[29,42,76,81] a careful evaluation of each sampling and storage protocol is necessary and strongly recommended.[17]

Taken together, a careful validation of all protocols, including sample storage, extraction procedures, and the immunoassays used to analyze fecal steroid metabolites, is crucial. In particular, the assay systems (including the respective antibody) should be characterized in analytical terms regarding sensitivity, accuracy, precision, and cross-reactivity with the reduced steroid metabolites present in the feces.

It is also necessary to use the correct nomenclature. In the literature, the term "fecal cortisol/corticosterone" or "fecal glucocorticoid" is often used for the substances measured by immunoassays. However, this implies that the measured substances are biologically active, which is neither known nor proved. Hence, the term is misleading and should be avoided, because the native, unmetabolized GC (cortisol or corticosterone) circulating in the blood is not present in the feces, but rather their 5α/5β reduced metabolites.[15,16] Instead, these metabolites should be referred to as "fecal glucocorticoid metabolites," or they should be labeled according to the group of metabolites detected by the respective antibody used in the assay.[15,17,26]

IMPORTANCE OF VALIDATION

For a reliable monitoring of adrenocortical activity in mammals and birds using fecal GCM analyses, it is of crucial importance to carefully validate the techniques used.

As pointed out previously, technical and analytical issues, such as sample preservation and stability, extraction procedures, and antibodies used in the assays (RIA or EIA), must be considered for each species (including both sexes). Furthermore, the importance of proving the biological relevance of the technique—that is, if the assay system can detect biologically meaningful alterations in the endocrine status of the animals—cannot be overestimated. As is pointed out in the following paragraphs, experiments dealing with the physiological and biological validity of fecal GCM analyses are essential and must be performed before applying the technique in a given species.

Physiological and Biological Validation

Physiological validation of the technique means to pharmacologically induce physiological changes in circulating GC levels and to evaluate whether these changes are reflected in measured concentrations of fecal GCMs afterward. In this respect, the most widely used experiment to stimulate adrenocortical activity (i.e., increase circulating GC levels) is the so-called ACTH challenge test (see TABLE 1). Ideally, fecal samples are collected frequently a certain time before and after the injection of ACTH, which should result in a significant increase of plasma GC concentrations. This pattern of sharply increasing (and decreasing again later) GC levels should be clearly reflected in the concentration of fecal GCMs after a certain lag time. Examples of studies describing ACTH challenge experiments involving various species of mammals and birds are compiled in TABLE 1. Although there are more than 140 articles published in peer-reviewed journals dealing with fecal GCMs in more than 70 species of mammals and birds, by the time of the writing of this review, convincing validation experiments have been performed on only a few species (see TABLE 1). This situation is especially dramatic in primates, for which only a single article describes ACTH challenge tests in long-tailed macaques and yellow baboons.[32] On the other hand, many such studies have been described in carnivores (including some of the earliest[82]) and ruminants (mainly domestic livestock). In birds, ACTH challenge tests have been performed very recently on only a few species (see TABLE 1).

Similar to findings in the plasma, authors who investigated a substantial number of animals reported considerable individual variation, both in basal and ACTH-induced levels of fecal GCMs (e.g., long-tailed macaque,[32] laboratory mouse,[35] domestic dog,[44] cattle and domestic sheep,[57] domestic goose,[62] domestic chicken[65] see also TABLE 1). In fact, there are a few examples in the literature stating that in some species, certain individuals showed the expected pattern of fecal GCM concentrations after stimulation, whereas inconsistent results were obtained in others (e.g., long-tailed macaque,[32] black-footed ferret,[50] domestic pig[54]). Thus, for a proper physiological validation of the technique, it is strongly recommended to use enough individuals (of both sexes) and not to rely upon results obtained from only one or two individuals of a given species. Furthermore, each animal can be used as its own control, thereby minimizing the problems of individual differences in basal and peak levels of fecal GCMs (i.e., absolute differences or percent increases can be calculated; see also recommendations in Ref. 8).

A second experiment to physiologically validate the measurement of fecal GCMs is to perform the so-called dexamethasone (Dex) suppression test (see examples compiled in TABLE 1). Dex is an artificial steroid that mimics endogenous GCs and reduces circulating corticosteroid levels via the negative-feedback mechanism of the HPA axis.[1,4,5] Therefore, after injection of Dex, a suppression of adrenocortical activity (i.e., decreased concentrations of circulating GCs) is expected and should be reflected in reduced fecal GCM concentrations for a certain period (largely depending on the dosage of Dex).

The Dex suppression test is also very important to analytically discriminate between true GCM measurements and blank values. Because synthetic GCs and their fecal metabolites usually do not cross-react with the antibodies used in the respective immunoassays, the concentrations of naturally occurring GCMs in the feces should be very low after Dex treatment. Therefore, measured concentrations in those samples can be referred to as blank values, or they can reflect contributions of other cross-reacting steroid metabolites, probably of gonadal origin. Although only successfully included in a small fraction of studies in mammals (laboratory mouse,[35] domestic dog,[44] domestic cat,[44] domestic horse,[54] cattle,[57] domestic sheep,[57] roe deer[61]), a physiological validation of the technique using the Dex suppression test is strongly recommended. In birds, a Dex suppression test has been performed in the domestic chicken only. However, for elusive reasons, the expected effects could be detected neither in plasma nor in feces.[65,66] Whether this was caused by a different biological activity of Dex in birds or whether the injected dose was too low still needs to be investigated.

Besides these two pharmacological treatments, it is often argued that a strong positive correlation between concentrations of plasma GCs and GCMs measured in the feces indicates that the applied assay system is valid. However, although this may be true for the relatively slowly changing plasma levels of progesterone, the situation is quite different for GCs and androgens. Diurnal rhythms, as well as episodic fluctuations, result in considerable changes of circulating GC concentrations, even within short periods.[4,5,8] In addition, as will be discussed below, a potentially varying time delay of fecal excretion renders it difficult to select pairs of samples for a meaningful correlation.

Another aspect that can be covered by physiological validation experiments is to investigate the biological sensitivity of an assay used for fecal GCM measurements in a given species; that is, to evaluate which (small) alterations in adrenocortical activity, and thus plasma GC levels, can be reliably detected in the feces. This can be achieved by administration of different dosages of ACTH or Dex, respectively (dose-response effects). However, up to now, such experiments have been described only for laboratory mice (low and high doses of ACTH and Dex, respectively[35]), cattle (a range of different ACTH doses, 0.06–3 mg[57,83]), and mourning doves (two different doses of ACTH[73]). In cows, Palme et al.[57,83] also reported that the percent increase of fecal GCMs above basal levels (but not the absolute values of fecal GCM concentrations, and neither absolute nor increase of plasma cortisol) was correlated with the administered dose of ACTH. This finding suggests that fecal GCM concentrations reflected the amount of secreted GCs better than plasma levels. This is especially important because very high plasma GC levels (induced by extremely high dosages of ACTH, as have been used in several studies in the literature) may very well be reflected in the concentration of fecal GCMs, but this might not be the case for smaller or moderate stimulations ("iceberg effect").

In addition to a careful physiological validation, experiments proving the biological validity of the technique are also important. That is, serial samples before and after a known stressful event like capture, immobilization, or transportation can be used to evaluate the biological relevance of an established technique. Such experiments have been described for a number of species from various taxa (see TABLE 1). Besides the previously mentioned procedures, others, such as anesthesia,[33,43] confinement/restraint,[39,61,72] disturbances caused by the presence of humans,[40,60,84] novelty,[38] agonistic encounters/social challenges,[41,48,62,63,85] different housing conditions,[53,83,86] and translocation[41,48,59,87] were reported to influence fecal GCM levels (see also TABLE 1). As an experiment to investigate the biological validity of the technique, it can also be useful to assess effects of injection procedures (e.g., injecting saline solution) or blood sampling.[35,60] Furthermore, measuring whether the naturally occurring diurnal variation of GCs is also reflected in the feces of a given species can indicate biological relevance.[35,37]

In endangered or intractable species, however, a rigorous physiological validation might not be possible. Nevertheless, even under these constraints, at least experiments to biologically validate the assay technique must be performed to produce reliable results (see also recommendations in Refs. 17 and 30).

Taken together, immunoassays for the assessment of fecal GCMs must be extensively validated. Besides analytical issues, such as demonstrating that the antibodies used cross-react to a considerable extent with GCMs present in the feces, experiments scrutinizing the physiological and biological validity of the technique must be performed. For a physiological validation, it is recommended to use pharmacological stimulation and suppression of adrenocortical activity (by injecting different doses of ACTH and Dex, respectively), inducing specific changes in circulating GC concentrations that should be reflected in fecal GCM concentrations afterward. Furthermore, a biological validation should be performed using different stressors relevant to the animal (e.g., restraint, injections, blood sampling, transportation, agonistic interactions), demonstrating that the technique can detect biologically meaningful changes in circulating GC levels.

As described in the Introduction, various factors can affect the levels of GCs in plasma and concentrations of GCMs in feces of animals. Besides differences between species, effects of age, social status, or early life experiences (prenatal or postnatal), gender differences as well as diurnal and seasonal variations (including life history stages) on GC levels are of special importance for the noninvasive monitoring, and are therefore addressed further in the following sections.

Gender Differences

Males and females differ with respect to various physiological and behavioral aspects. Several studies have shown pronounced gender differences regarding baseline levels of GCs as well as the reactivity of the HPA axis to stressors (e.g., laboratory mouse,[88] laboratory rat,[89] rabbit,[4] arctic ground squirrel,[90] guinea pig,[91] domestic sheep,[92] European starling,[93] Inca dove[19]).

Consequently, these differences in plasma concentrations also affect concentrations of GCMs assessed in fecal samples. Similar to findings in the plasma, several studies investigating fecal GCM levels in both sexes report higher concentrations in females (common marmoset,[94] northern muriqui,[95] laboratory mouse,[35] European hare,[40] domestic dog,[44] African wild dog,[47] domestic cat,[44] cheetah[86]), males (laboratory rat,[37] Steller sea lion,[96] domestic chicken[65]), or no difference between the sexes (wolf,[45] black rhinoceros,[56] white rhinoceros,[56] elk/red deer,[97] mourning dove[73]).

Different factors may be responsible for these gender-specific differences. First, higher plasma values (observed mostly in females) should also result in higher fecal GCM concentrations. This effect, with females having higher plasma GC levels than males, is thought to be brought about by a higher capacity of steroid-binding globulins expressing certain affinities to GCs;[98] that is, in females, circulating GCs are also bound to a considerable extent to gonadal steroid binding globulins, and therefore the total GC concentration can be higher. This might also be why GC concentrations vary significantly across the female's estrous cycle, as changing concentrations of estrogens and progesterone are known to influence the expression and the occupancy of plasma steroid-binding globulins.[3,98] In laboratory rats, for example, significant alterations of plasma GCs as well as fecal GCMs were reported across the estrous cycle, with highest levels occurring on the day of proestrus.[37]

Second, the level of metabolites excreted via the urine or via the feces might differ significantly between males and females. Touma et al.,[99] for example, showed in laboratory mice that males excreted about 73% of radioactive corticosterone via the feces, whereas females excreted only about 53% via the feces. Hence, females eliminated a larger fraction via the urine. Similar findings were reported for horses and cats.[23,44]

Third, the GCMs formed might differ significantly between males and females. High-performance liquid chromatography immunograms performed with fecal samples from different species revealed considerable gender-specific differences regarding the structure as well as the quantity of fecal GCMs (laboratory mouse,[99] laboratory rat,[37] domestic chicken,[65] European stonechat[71]). Because the cross-reactivity of the antibodies used in a given assay strongly depends on the biochemical structure of the steroid, these differences are likely to bring about different concentrations of immunoreactive GCMs in males and females (see also Ref. 16).

Moreover, especially when males with high plasma levels of androgens are investigated (e.g., dogs[44] or elephants in musth[52,100]), it should be carefully monitored that the antibody used in the respective assay does not cross-react with androgen metabolites present in the feces. In such cases it might be advisable to use different assays for males and females (e.g., domestic cat,[43] African elephant[52,53]).

Taken together, possible gender-specific effects should be carefully evaluated and gender differences should be seriously considered when measuring fecal GCMs in males and females of a given species. This is especially important in field studies, when samples from unknown individuals are collected, and therefore fecal GCM concentrations might be biased by the gender of the animals.

Diurnal Variations

Well-defined circadian rhythms of plasma GCs (with peak levels 5–10 times higher than trough levels) have been described in most vertebrate species (e.g., laboratory mouse,[101] laboratory rat,[102] tree shrew,[4] squirrel monkeys,[103] domestic chicken,[104] white-crowned sparrow,[105] house sparrow,[106] European starling[93]). Usually, the peak of hormone secretion occurs toward the end of the dark period in primates and other diurnal animals, whereas in primarily nocturnal animals like most rodents and cats, there is a peak toward the end of the light period. Therefore, it is obviously important to sample GCs at the same time of day if repeated measurements are to be made on different days or if comparing different groups/populations of animals.

This diurnal variation of GCs should also be taken into account for the monitoring of hormone metabolites from fecal samples. So far, only a few studies have addressed this point, but diurnal variations of GC metabolites have been observed in fecal samples of some mammalian and bird species (common marmoset,[94] long-tailed macaque,[108] laboratory mouse,[35] laboratory rat,[36,37,107] domestic goose,[62] great tit[70]). Particularly in small animals, which usually defecate more frequently (i.e., providing a higher temporal resolution in the feces), a distinct circadian rhythm of fecal GCM excretion is expected and has been documented in detail for mice and rats.[35,37] However, in species with a relatively long gut passage time (e.g., hind-gut fermenters) or animals that defecate rather infrequently (e.g., most carnivores and reptiles), it might be impossible to detect diurnal changes of circulating GC levels in the feces.

Thus, information on the animals' activity rhythm, gut passage time, and defecation rate should be considered when planning the fecal sampling regimen for a given species. Complex interactions between these parameters may also exist. For example, in mice, the amount of feces produced varied during the course of the day (in accordance with the animals' activity pattern) and thereby influenced the lag time of fecal GCM excretion.[99] Similar effects are likely to exist in other species as well.

To consider diurnal changes and to avoid possible effects of the time of sampling, an option is to collect and combine all samples defecated by an individual over 24 h. These pooled samples are likely to represent the hormonal status of an animal more accurately, because individual differences or shifts in activity patterns and/or excretion profiles are compensated for (cf. Ref. 109).

In other cases, however, the diurnal variation of GCs might be an important parameter to monitor. In humans, for example, several pathologic states have been associated with alterations in the circadian rhythm of different endocrine parameters including GCs (e.g., depression, anxiety disorders, Parkinson's disease, or Alzheimer's disease[3,110,111]). Perturbations of the GC rhythm have also been reported well before other symptoms of the disease appeared.[112] Thus, the noninvasive technique to assess similar changes by means of fecal hormone metabolite analyses in, for example, laboratory mice and rats, which are the most commonly used animal models for human diseases, can be a unique opportunity and might open new perspectives in biomedical research (cf. Ref. 109).

Taken together, circadian rhythms of GC secretion should be seriously taken into account. Fecal samples should be collected at the same time each day, or all samples voided over a 24-h period should be pooled to avoid fluctuations caused by diurnal variations in GCM concentrations.

Seasonal Variations and Life History Stages

Under natural conditions, GC concentrations vary significantly in most vertebrate species studied so far.[20] The basal activity as well as the reactivity of the HPA axis to stressors is modulated; that is, baseline GC concentrations and the magnitudes of stress responses might vary depending upon the time of year (e.g., squirrel monkey,[103] rabbit,[4] yellow-pine chipmunk,[113] arctic ground squirrel,[90] mountain chickadee,[114] snow bunting,[115] house sparrow[106]). Although the underlying mechanisms and the functional significance of the annual GC rhythm are still poorly understood, in reptiles, amphibians, birds, and at least some mammals, the annual cycle of GCs tends to peak during the breeding season, indicating biological relevance and effects of life history stages.[20,116]

Besides effects of variations in the level of plasma corticosteroid binding globulin (CBG) and a changing sensitivity of target tissues to GCs, three potential explanations for the seasonal modulation of GCs are discussed in detail by Romero (the energy mobilization hypothesis, the behavior hypothesis, and the preparative hypothesis[20]).

Therefore, alterations of the general activity and reactivity of the HPA axis in response to seasonal changes or different life history stages should be carefully taken into account when assessing GC concentrations in plasma as well as in fecal samples of wild animals.

Significant effects of season or weather conditions such as temperature, humidity, and availability of food and water on fecal GCM concentrations have been shown for several species of mammals and birds, with most studies reporting higher levels during harsher conditions in winter or during the dry season (chacma baboon,[117] northern muriqui,[118] ring-tailed lemur,[34] red-backed vole,[39] deer mouse,[39] grizzly bear,[119] African elephant,[120] elk/red deer,[84,97,121] greylag goose,[122] mourning dove[73]).

Similarly, times with higher intraspecific competition for food or mating partners, associated with higher levels of aggression, also correlated with elevated fecal GCM concentrations (northern muriqui,[95] capuchin monkey,[123] ring-tailed lemur,[34] wolf,[45] greylag goose[63]).

The reproductive status of a female is another seasonal/life history event that significantly influences the concentration of fecal GCMs. Similar to findings in plasma samples, the phase of the estrous cycle or the stage during pregnancy was found to be associated with alterations in fecal GCM levels (chacma baboon,[117] ring-tailed lemur,[34] laboratory rat,[37] spotted hyena[85]). Increased metabolic demands—for example, during late pregnancy and lactation in mammals or during egg production, laying, and incubation periods in birds—are likely to influence GCM concentrations as well (cf. Refs. 30 and 116). GCs are also known to rise significantly near term in most mammalian species because they actually trigger the cascade resulting in parturition.[26,87] Furthermore, the placenta can produce large amounts of androgens or their derivatives, which can influence the levels of immunoreactive metabolites in the feces (due to their cross-reactivities with the antibody used in the assay[87]).

Taken together, fecal GCM concentrations in mammals and birds often vary seasonally and can be largely influenced by life history stages. Therefore, knowledge about these seasonal variations should be carefully incorporated in the study design and considered for the interpretation of the results.

CONCLUSION

The monitoring of adrenocortical activity by means of fecal GCM analysis offers several advantages and has been successfully applied to various species of mammals and birds. Because the sampling is completely noninvasive, the animal's behavior and endocrine state as well as physiological functions, like the circadian GC rhythm, are not affected by stress responses associated with capture, restraint, or blood sampling. Therefore, frequent sampling of the same individual is possible (even over extended periods), allowing the monitoring of short-term as well as long-term endocrine changes. In addition, due to pooling effects in the gut, concentrations of fecal GCMs represent a more integrated measure of adrenocortical activity, dampening episodic fluctuations, or pulsatile secretion patterns of GCs. Because only the unbound fraction of circulating GCs is readily metabolized by the liver and excreted via the bile into the gut, levels of GCMs measured in feces might also reflect the biologically active fraction more accurately. Thus, this noninvasive technique has tremendous potential for diverse investigations in laboratory, companion, farm, zoo, and wild animals.

However, because clear differences regarding the metabolism and excretion of GCMs exist, the technique needs to be extensively validated for each species and gender investigated. In analytical terms, the protocols of sample storage, extraction procedure, and immunoassay performance should be carefully evaluated (including sensitivity, accuracy, and precision of the assay used, as well as the cross-reactivity of the respective antibody with fecal steroid metabolites).

Furthermore, it is crucial to scrutinize the physiological as well as the biological validity of fecal GCM measurements in a given species, that is, the relevance of the technique to detect biologically meaningful alterations in adrenocortical activity. Because significant effects of the animals' gender, the time of day, season, and different life history stages (involving various behavioral and physiological alterations) have been shown in plasma and feces, possible differences between the sexes, diurnal, and seasonal variations should be seriously considered when measuring fecal GCMs.

Thus, besides analytical and technical issues, comprehensive information on the animals' biology and stress physiology should be carefully taken into account and thoroughly included in the study design. Together with an extensive physiological and biological validation of the technique, this will ensure that the measurement of fecal GCMs can be used as a powerful tool to assess adrenocortical activity in mammals and birds.

ACKNOWLEDGMENTS

We thank Susi Jenni-Eiermann and Wolfgang Goymann for organizing an excellent European Science Foundation Technical Meeting in Seewiesen (facilitating very stimulating discussions and networking among the participants) and for initiating the publication of the workshop proceedings. Furthermore, we are indebted to Sophie Rettenbacher, Matthias Asher, and André Ganswindt for helpful comments on drafts of the manuscript.

REFERENCES

1. AXELROD, J. & T.D. REISINE. 1984. Stress hormones: their interaction and regulation. Science **224**: 452–459.
2. MUNCK, A., P.M. GUYRE & N.J. HOLBROOK. 1984. Physiological functions of glucocorticoids in stress and their relation to pharmacological actions. Endocr. Rev. **5**: 25–44.
3. DE KLOET, E.R., E. VREUGDENHIL, M.S. OITZL & M. JOELS. 1998. Brain corticosteroid receptor balance in health and disease. Endocr. Rev. **19**: 269–301.
4. VON HOLST, D. 1998. The concept of stress and its relevance for animal behavior. Adv. Study Behav. **27**: 1–131.
5. SAPOLSKY, R.M., L.M. ROMERO & A.U. MUNCK. 2000. How do glucocorticoids influence stress responses? Integrating permissive, suppressive, stimulatory, and preparative actions. Endocrine Rev. **21**: 55–89.
6. ENGELMANN, M., R. LANDGRAF & C.T. WOTJAK. 2004. The hypothalamic-neurohypophysial system regulates the hypothalamic-pituitary-adrenal axis under stress: an old concept revisited. Front. Neuroendocrinol. **25**: 132–149.
7. LANDGRAF, R. & I. NEUMANN. 2004. Vasopressin and oxytocin release within the brain: a dynamic concept of multiple and variable modes of neuropeptide communication. Front. Neuroendocrinol. **25**: 150–176.
8. ROMERO, L.M. 2004. Physiological stress in ecology: lessons from biomedical research. Trends Ecol. Evol. **19**: 249–255.
9. KORTE, S.M. *et al.* 2005. The Darwinian concept of stress: benefits of allostasis and costs of allostatic load and the trade-off in health and disease. Neurosci. Behav. Physiol. **29**: 3–38.
10. RILEY, V. 1981. Psychoneuroendocrine influences on immunocompetence and neoplasia. Science **212**: 1100–1109.
11. MCEWEN, B.S. & R.M. SAPOLSKY 1995. Stress and cognitive function. Curr. Opin. Neurobiol. **5**: 205–216.
12. WINGFIELD, J.C. & R.M. SAPOLSKY. 2003. Reproduction and resistance to stress: when and how. J. Neuroendocrinol. **15**: 711–724.
13. HOLSBOER, F. 2000. The corticosteroid receptor hypothesis of depression. Neuropsychopharmacology **23**: 477–501.
14. KORTE, S.M. 2001. Corticosteroids in relation to fear, anxiety and psychopathology. Neurosci. Biobehav. Rev. **25**: 117–142.
15. MÖSTL, E., S. RETTENBACHER & R. PALME. 2005. Measurement of corticosterone metabolites in birds' droppings: an analytical approach. Ann. N.Y. Acad. Sci. **1046**: 17–34.
16. PALME, R. *et al.* 2005. Stress hormones in mammals and birds: comparative aspects regarding metabolism, excretion and noninvasive measurement in fecal samples. Ann. N.Y. Acad. Sci. **1040**:.
17. PALME, R. 2005. Measuring fecal steroids: guidelines for practical application. Ann. N.Y. Acad. Sci. **1046**: 75–80.
18. HERMAN, J. P. & W.E. CULLINAN. 1997. Neurocircuitry of stress: central control of the hypothalamo-pituitary-adrenocortical axis. Trends Neurosci. **20**: 78–84.
19. WINGFIELD, J.C., C.M. VLECK & M.C. MOORE. 1992. Seasonal changes of the adrenocortical response to stress in birds of the Sonora Desert. J. Exp. Zool. **264**: 419–428
20. ROMERO, L.M. 2002. Seasonal changes in plasma glucocorticoid concentrations in free-living vertebrates. Gen. Comp. Endocrinol. **128**: 1–24.

21. HENNESSY, M.B. & S. LEVINE. 1978. Sensitive pituitary-adrenal responsiveness to varying intensities of psychological stimulation. Physiol. Behav. **21:** 295–297.
22. TAYLOR, W. 1971. The excretion of steroid hormone metabolites in bile and feces. Vitam. Horm. **29:** 201–285.
23. PALME, R. et al. 1996. Excretion of infused ^{14}C-steroid hormones via faeces and urine in domestic livestock. Anim. Reprod. Sci. **43:** 43–63.
24. SCHWARZENBERGER, F. et al. 1996. Faecal steroid analysis for non-invasive monitoring of reproductive status in farm, wild and zoo animals. Anim. Reprod. Sci. **42:** 515–526.
25. WHITTEN, P.L., D.K. BROCKMAN & R. STAVISKY. 1998. Recent advances in noninvasive techniques to monitor hormone-behavior interactions. Yrbk. Phys. Anthropol. **41:** 1–23.
26. MÖSTL, E. & R. PALME. 2002. Hormones as indicators of stress. Dom. Anim. Endocrinol. **23:** 67–74.
27. LINDNER, H.R. 1972. Enterohepatic circulation and patterns of urinary excretion of cortisol metabolites in the ewe. J. Endocrinol. **52:** xix–xx.
28. MACDONALD, I.A., et al. 1983. Degradation of steroids in the human gut. J. Lipid Res. **24:** 675–700.
29. MILLSPAUGH, J.J. & B.E. WASHBURN. 2004. Use of fecal glucocorticoid metabolite measures in conservation biology research: considerations for application and interpretation. Gen. Comp. Endocrinol. **138:** 189–199.
30. GOYMANN, W. 2005. Noninvasive monitoring of hormones in bird droppings: physiological validations, sampling, extraction, sex differences, and the influence of diet on hormone metabolite levels. Ann. N.Y. Acad. Sci. **1046:** 35–53.
31. KLASING, K.C. 2005. Potential impact of nutritional strategy on noninvasive measurements of hormones in birds. Ann. N.Y. Acad. Sci. **1046:** 5–16.
32. WASSER, S.K. et al. 2000. A generalized fecal glucocorticoid assay for use in a diverse array of nondomestic mammalian and avian species. Gen. Comp. Endocrinol. **120:** 260–275.
33. WHITTEN, P.L., R. STAVISKY, F. AURELI & E. RUSSELL. 1998. Response of fecal cortisol to stress in captive chimpanzees (*Pan troglodytes*). Am. J. Primatol. **44:** 57–69.
34. CAVIGELLI, S.A. 1999. Behavioural patterns associated with faecal cortisol levels in free-ranging female ring-tailed femurs, *Lemur catta*. Anim. Behav. **57:** 935–944.
35. TOUMA, C., R. PALME & N. SACHSER. 2004. Analyzing corticosterone metabolites in fecal samples of mice: a noninvasive technique to monitor stress hormones. Horm. Behav. **45:** 10–22.
36. BAMBERG, E., R. PALME & J.G. MEINGASSNER. 2001. Excretion of corticosteroid metabolites in urine and faeces of rats. Lab. Anim. **35:** 307–314.
37. CAVIGELLI, S.A. et al. 2005. Frequent serial fecal corticoid measures from rats reflect circadian and ovarian corticosterone rhythms. J. Endocrinol. **184:** 153–163.
38. HARPER, J.M. & S.N. AUSTAD. 2000. Fecal glucocorticoids. A noninvasive method of measuring adrenal activity in wild and captive rodents. Physiol. Biochem. Zool. **73:** 12–22.
39. HARPER, J.M. & S.N. AUSTAD. 2001. Effect of capture and season on fecal glucocorticoid levels in deer mice (*Peromyscus maniculatus*) and red-backed voles (*Clethrionomys gapperi*). Gen. Comp. Endocrinol. **123:** 337–344.
40. TESKEY-GERSTL, A., E. BAMBERG, T. STEINECK & R. PALME. 2000. Excretion of corticosteroids in urine and faeces of hares (*Lepus europaeus*). J. Comp. Physiol. [B] **170:** 163–168.
41. GOYMANN, W. et al. 1999. Noninvasive fecal monitoring of glucocorticoids in spotted hyenas, *Crocuta crocuta*. Gen. Comp. Endocrinol. **114:** 340–348.
42. HUNT, K. & S.K. WASSER. 2003. Effect of long-term preservation methods on fecal glucocorticoid concentrations of grizzly bear and African elephant. Physiol. Biochem. Zool. **76:** 918–928.
43. YOUNG, K.M. et al. 2004. Noninvasive monitoring of adrenocortical activity in carnivores by fecal glucocorticoid analyses. Gen. Comp. Endocrinol. **137:** 148–165.
44. SCHATZ, S. & R. PALME. 2001. Measurement of faecal cortisol metabolites in cats and dogs: a noninvasive method for evaluating adrenocortical function. Vet. Res. Commun. **25:** 271–287.

45. SANDS, J. & S. CREEL. 2004. Social dominance, aggression and faecal glucocorticoid levels in a wild population of wolves, *Canis lupus.* Anim. Behav. **67:** 387–396.
46. MONFORT, S.L., K.L. MASHBURN, B.A. BREWER & S.R. CREEL. 1998. Evaluating adrenal activity in African wild dogs (*Lycaon pictus*) by fecal corticosteroid analysis. J. Zoo Wildl. Med. **29:** 129–133.
47. CREEL, S., N.M. CREEL & S.L. MONFORT. 1997. Radiocollaring and stress hormones in African wild dogs. Conserv. Biol. **11:** 544–548.
48. TERIO, KA., S.B. CITIN & J.L. BROWN. 1999. Fecal cortisol metabolite analysis for noninvasive monitoring of adrenocortical function in the cheetah (*Acinonyx jubatus*). J. Zoo Wildl. Med. **30:** 484–491.
49. WIELEBNOWSKI, N.C. *et al.* 2002. Noninvasive assessment of adrenal activity associated with husbandry and behavioral factors in the North American clouded leopard population. Zoo Biol. **21:** 77–98.
50. YOUNG, K.M., J.L. BROWN & K.L. GOODROWE. 2001. Characterization of reproductive cycles and adrenal activity in the black-footed ferret (*Mustela nigripes*) by fecal hormone analysis. Zoo Biol. **20:** 517–536.
51. MASHBURN, K.L. & S. ATKINSON. 2004. Evaluation of adrenal function in serum and feces of Steller sea lions (*Eumetopias jubatus*): influences of molt, gender, sample storage, and age on glucocorticoid metabolism. Gen. Comp. Endocrinol. **136:** 371–381.
52. GANSWINDT, A. *et al.* 2003. Noninvasive assessment of adrenocortical function in the male African elephant (*Loxodonta africana*) and its relation to musth. Gen. Comp. Endocrinol. **134:** 156–166.
53. STEAD, S.K., D.G. MELTZER & R. PALME. 2000. The measurement of glucocorticoid concentrations in the serum and faeces of captive African elephants (*Loxodonta africana*) after ACTH stimulation. J. South Afr. Vet. Assoc. **71:** 192–196.
54. MÖSTL, E. *et al.* 1999. Measurement of glucocorticoid metabolite concentrations in faeces of domestic livestock. Zentralbl. Veterinarmed. A **46:** 621–632.
55. MERL, S. *et al.* 2000. Pain causes increased concentrations of glucocorticoid metabolites in equine faeces. J. Equine Vet. Sci. **20:** 586–590.
56. BROWN, J.L. *et al.* 2001. Comparative analysis of gonadal and adrenal activity in the black and white rhinoceros in North America by noninvasive endocrine monitoring. Zoo Biol. **20:** 463–486.
57. PALME, R. *et al.* 1999. Measurement of faecal cortisol metabolites in ruminants: a noninvasive parameter of adrenocortical function. Wien. Tierärztl. Mschr. **86:** 237–241.
58. PALME, R., C. ROBIA, W. BAUMGARTNER & E. MÖSTL. 2000. Transport stress in cattle as reflected by an increase in faecal cortisol metabolites. Vet. Rec. **146:** 108–109.
59. MORROW, C.J., E.S. KOLVER, G.A. VERKERK & L.R. MATTHEWS. 2002. Fecal glucocorticoid metabolites as a measure of adrenal activity in dairy cattle. Gen. Comp. Endocrinol. **126:** 229–241.
60. HUBER, S. *et al.* 2003. Non-invasive monitoring of adrenocortical response in red deer. J. Wildl. Manag. **67:** 258–266.
61. DEHNHARD, M. *et al.* 2001. Non-invasive monitoring of adrenocortical activity in the roe deer (*Capreolus capreolus*) by measuring faecal cortisol metabolites. Gen. Comp. Endocrinol. **123:** 111–120.
62. KOTRSCHAL, K. *et al.* 2000. Effects of physiological and social challenges in different seasons on fecal testosterone and corticosterone in male domestic geese (*Anser domesticus*). Acta Ethol. **2:** 115–122.
63. KOTRSCHAL, K., K. HIRSCHENHAUSER & E. MÖSTL. 1998. The relationship between social stress and dominance is seasonal in greylag geese. Anim. Behav. **55:** 171–176.
64. HIRSCHENHAUSER, K. *et al.* 2000. Endocrine and behavioural responses of male greylag geese (*Anser anser*) to pairbond challenges during the reproductive season. Ethology **106:** 63–77.
65. RETTENBACHER, S. *et al.* 2004. Measurement of corticosterone metabolites in chicken droppings. Br. Poult. Sci. **45:** 704–711.
66. DEHNHARD, M. *et al.* 2003. Measurement of plasma corticosterone and fecal glucocorticoid metabolites in the chicken (*Gallus domesticus*), the great cormorant (*Phalacrocorax carbo*), and the goshawk (*Accipiter gentilis*). Gen. Comp. Endocrinol. **131:** 345–352.

67. THIEL, D., S. JENNI-EIERMANN & R. PALME. 2005. Measuring corticosterone metabolites in droppings of capercaillies (*Tetrao urogallus*). Ann N.Y. Acad. Sci. **1046:** 96–108.
68. BALTIC, M. *et al.* 2005. A noninvasive technique to evaluate human-generated stress in the black grouse. Ann. N.Y. Acad. Sci. **1046:** 81–95.
69. WASSER, S.K. *et al.* 1997. Noninvasive physiological measures in the northern spotted owl. Conserv. Biol. **11:** 1019–1022.
70. CARERE, C. *et al.* 2003. Fecal corticosteroids in a territorial bird selected for different personalities: daily rhythm and the response to social stress. Horm. Behav. **43:** 540–548.
71. GOYMANN, W., E. MÖSTL & E. GWINNER. 2002. Corticosterone metabolites can be measured non-invasively in excreta of European stonechats, *Saxicola torquata rubicola*. Auk **119:** 1167–1173.
72. WELLS, K.M. *et al.* 2003. Effects of radio-transmitters on fecal glucocorticoid levels in captive Dickcissels. Condor **105:** 805–810.
73. WASHBURN, B.E. *et al.* 2003. Using fecal glucocorticoids for stress assessment in mourning doves. Condor **105:** 696–706.
74. NAKAGAWA, S., E. MÖSTL & J.R. WAAS. 2003. Validation of an enzyme immunoassay to measure faecal glucocorticoid metabolites from Adélie penguins (*Pygoscelis adeliae*): a noninvasive tool for the measurement of stress? Polar Biol. **26:** 491–493.
75. PALME, R. & E. MÖSTL. 1997. Measurement of cortisol metabolites in faeces of sheep as a parameter of cortisol concentration in blood. Int. J. Mammal. Biol. **62**(Suppl. II): 192–197.
76. KHAN, M.Z., J. ALTMANN, S.S. ISANI & J. YU. 2002. A matter of time: evaluating the storage of fecal samples for steroid analysis. Gen. Comp. Endocrinol. **128:** 57–64.
77. TERIO, K.A. *et al.* 2002. Comparison of different drying and storage methods on quantifiable concentrations of fecal steroids in the cheetah. Zoo Biol. **21:** 215–222.
78. WASHBURN, B.E. & J.J. MILLSPAUGH. 2002. Effects of simulated environmental conditions on glucocorticoid metabolite measurements in white-tailed deer feces. Gen. Comp. Endocrinol. **127:** 217–222.
79. BEEHNER, J.C. & P.L. WHITTEN. 2004. Modifications of a field method for fecal steroid analysis in baboons. Physiol. Behav. **82:** 269–277.
80. LYNCH, J.W. *et al.* 2003. Concentrations of four fecal steroids in wild baboons: short-term storage conditions and consequences for data interpretation. Gen. Comp. Endocrinol. **132:** 264–271.
81. MILLSPAUGH, J.J. *et al.* 2003. Effects of heat and chemical treatments on fecal glucocorticoid measurements: implications for sample transport. Wildl. Soc. Bull. **31:** 399–406.
82. GRAHAM, L.H. & J.L. BROWN. 1996. Cortisol metabolism in the domestic cat and implications for noninvasive monitoring of adrenocortical function in endangered felids. Zoo Biol. **15:** 71–82.
83. PALME, R., F. WETSCHER & C. WINCKLER. 2003. Measuring faecal cortisol metabolites: a noninvasive tool to assess animal welfare in cattle? *In* Proc. of the IVth Central European Buiatric Congress in Lovran, April 23–27, 2003. Josip Kos, Ed.: 145–150. Faculty of Veterinary Medicine, Zagreb and Croation Veterinary Chamber.
84. CREEL, S. *et al.* 2002. Snowmobile activity and glucocorticoid stress responses in wolves and elk. Conserv. Biol. **16:** 809–814.
85. GOYMANN, W. *et al.* 2001. Social, state-dependent and environmental modulation of faecal corticosteroid levels in free-ranging female spotted hyenas. Proc. R. Soc. London B: Biol. Sci. **268:** 2453–2459.
86. WIELEBNOWSKI, N.C. *et al.* 2002. Impact of social management on reproductive, adrenal and behavioural activity in the cheetah (*Acinonyx jubatus*). Anim. Conserv. **5:** 291–301.
87. MÖSTL, E. *et al.* 2002. Measurement of cortisol metabolites in faeces of ruminants. Vet. Res. Commun. **26:** 127–139.
88. JONES, B.C. *et al.* 1998. Contribution of sex and genetics to neuroendocrine adaptation to stress in mice. Psychoneuroendocrinology **23:** 505–517.
89. HANDA, R.J., L.H. BURGESS, J.E. KERR & J.A. O'KEEFE. 1994. Gonadal steroid hormone receptors and sex differences in the hypothalamo-pituitary-adrenal axis. Horm. Behav. **28:** 464–476.

90. BOONSTRA, R. *et al.* 2001. Seasonal changes in glucocorticoid and testosterone concentrations in free-living arctic ground squirrels from the boreal forest of the Yukon. Can. J. Zool. **79:** 49–58.
91. KAISER, S., M. KIRTZECK, G. HORNSCHUH & N. SACHSER. 2003. Sex-specific difference in social support—a study in female Guinea pigs. Physiol. Behav. **79:** 297–303.
92. TURNER, A.I. *et al.* 2002. Influence of sex and gonadal status of sheep on cortisol secretion in response to ACTH and on cortisol and LH secretion in response to stress: importance of different stressors. J. Endocrinol. **173:** 113–122.
93. ROMERO, L.M. & L. REMAGE-HEALEY. 2000. Daily and seasonal variation in response to stress in captive starlings (*Sturnus vulgaris*): corticosterone. Gen. Comp. Endocrinol. **119:** 52–59.
94. RAMINELLI, J.L. *et al.* 2001. Morning and afternoon patterns of fecal cortisol excretion among reproductive and non-reproductive male and female common marmosets, *Callitrix jacchus*. Biol. Rhythm Res. **32:** 159–167.
95. STRIER, K.B., J.W. LYNCH & T.E. ZIEGLER. 2003. Hormonal changes during the mating and conception seasons of wild northern muriquis (*Brachyteles arachnoides hypoxanthus*). Am. J. Primatol. **61:** 85–99.
96. HUNT, K.E., A.W. TRITES & S.K. WASSER. 2004. Validation of fecal glucocorticoid assay for Steller sea lions (*Eumetopias jubatus*). Physiol. Behav. **80:** 595–601.
97. HUBER, S., R. PALME & W. ARNOLD. 2003. Effects of season, sex, and sample collection on concentrations of fecal cortisol metabolites in red deer (*Cervus elaphus*). Gen. Comp. Endocrinol. **130:** 48–54.
98. BREUNER, C.W. & M. ORCHINIK. 2002. Plasma binding proteins as mediators of corticosteroid action in vertebrates. J. Endocrinol. **175:** 99–112.
99. TOUMA, C., N. SACHSER, E. MÖSTL & R. PALME. 2003. Effects of sex and time of day on metabolism and excretion of corticosterone in urine and feces of mice. Gen. Comp. Endocrinol. **130:** 267–278.
100. 100. GANSWINDT, A., H.B. RASMUSSEN, M. HEISTERMANN & J.K. HODGES. 2005. The sexually active states of free-ranging male African elephants (*Loxodonta africana*): defining musth and non-musth using endocrinology, physical signals, and behavior. Horm. Behav. **47:** 83–91.
101. HALBERG, F., P.G. ALBRECHT & J.J. BITTNER. 1959. Corticosterone rhythm of mouse adrenal in relation to serum corticosterone and sampling. Am. J. Physiol. **197:** 1083–1085.
102. ATKINSON, H.C. & B.J. WADDELL. 1997. Circadian variation in basal plasma corticosterone and adrenocorticotropin in the rat: sexual dimorphism and changes across the estrous cycle. Endocrinology **138:** 3842–3848.
103. COE, C.L. & S. LEVINE. 1995. Diurnal and annual variation of adrenocortical activity in the squirrel monkey. Am. J. Primatol. **35:** 283–292.
104. DE JONG, I.C. *et al.* 2001. Determination of the circadian rhythm in plasma corticosterone and catecholamine concentrations in growing broiler breeders using intravenous cannulation. Physiol. Behav. **74:** 299–304.
105. BREUNER, C.W., J.C. WINGFIELD & L.M. ROMERO. 1999. Diel rhythms of basal and stress-induced corticosterone in a wild, seasonal vertebrate, Gambel's white-crowned sparrow. J. Exp. Zool. **284:** 334–342.
106. RICH, E.L. & L.M. ROMERO. 2001. Daily and photoperiod variation of basal and stress-induced corticosterone concentrations in house sparrows (*Passer domesticus*). J. Comp. Physiol. B **171:** 543–547.
107. PIHL, L. & J. HAU. 2003. Faecal corticosterone and immunoglobulin A in young adult rats. Lab. Anim. **37:** 166–171.
108. STAVISKY, R.C., P.L. WHITTEN, D.H. HAMMETT & J.R. KAPLAN. 2001. Lake pigments facilitate analysis of fecal cortisol and behavior in group-housed macaques. Am. J. Phys. Anthropol. **116:** 51–58.
109. TOUMA, C. *et al.* 2004. Age- and sex-dependent development of adrenocortical hyperactivity in a transgenic mouse model of Alzheimer's disease. Neurobiol. Aging **25:** 893–904.
110. MAGRI, F. *et al.* 1997. Changes in endocrine circadian rhythms as markers of physiological and pathological brain aging. Chronobiol. Int. **14:** 385–396.

111. SWAAB, D.F. 1999. Biological rhythms in health and disease: the suprachiasmatic nucleus and the autonomic nervous system. *In* Handbook of Clinical Neurology, The Autonomic Nervous System—Part I. Normal Functions. P.J. Vinken & G.W. Bruyn, Eds.: Vol. 74: 467–521. Elsevier. Amsterdam.
112. HARTMANN, A. *et al.* 1997. Twenty-four hour cortisol release profiles in patients with Alzheimer's and Parkinson's disease compared to normal controls: ultradian secretory pulsatility and diurnal variation. Neurobiol. Aging **18:** 285–289.
113. KENAGY, G.J. & N.J. PLACE. 2000. Seasonal changes in plasma glucocorticosteroids of free-living female yellow-pine chipmunks: effects of reproduction and capture and handling. Gen. Comp. Endocrinol. **117:** 189–199.
114. PRAVOSUDOV, V.V. *et al.* 2002. The effect of photoperiod on adrenocortical stress response in mountain chickadees (*Poecile gambeli*). Gen. Comp. Endocrinol. **126:** 242–248.
115. ROMERO, L.M., K.K. SOMA & J.C. WINGFIELD. 1998. Changes in pituitary and adrenal sensitivities allow the snow bunting (*Plectrophenax nivalis*), an arctic-breeding song bird, to modulate corticosterone release seasonally. J. Comp. Physiol. [B] **168:** 353–358.
116. ROMERO, L.M., J.M. REED & J.C. WINGFIELD. 2000. Effects of weather on corticosterone responses in wild free-living passerine birds. Gen. Comp. Endocrinol. **118:** 113–122.
117. WEINGRILL, T., D.A. GRAY, L. BARRETT & S.P. HENZI. 2004. Fecal cortisol levels in free-ranging female Chacma baboons: relationship to dominance, reproductive state and environmental factors. Horm. Behav. **45:** 259–269.
118. STRIER, K.B., T.E. ZIEGLER & D.J. WITTWER. 1999. Seasonal and social correlates of fecal testosterone and cortisol levels in wild male muriquis (*Brachyteles arachnoides*). Horm. Behav. **35:** 125–134.
119. VON DER OHE, C.G., S.K. WASSER, K.E. HUNT & C. SERVHEEN. 2004. Factors associated with fecal glucocorticoids in Alaskan brown bears (*Ursus arctos horribilis*). Physiol. Biochem. Zool. **77:** 313–320.
120. FOLEY, C.A., S. PAPAGEORGE & S.K. WASSER. 2001. Noninvasive stress and reproductive measures of social and ecological pressures in free ranging African elephants. Conserv. Biol. **15:** 1134–1142.
121. MILLSPAUGH, J.J. *et al.* 2001. Fecal glucocorticoid assays and the physiological stress response in elk. Wildl. Soc. Bull. **29:** 899–907.
122. FRIGERIO, D., J. DITTAMI, E. MÖSTL & K. KOTRSCHAL. 2004. Excreted corticosterone metabolites co-vary with air temperature and air pressure in male greylag geese (*Anser anser*). Gen. Comp. Endocrinol. **137:** 29–36.
123. LYNCH, J.W., T.E. ZIEGLER & K.B. STRIER. 2002. Individual and seasonal variation in fecal testosterone and cortisol levels of wild male tufted capuchin monkeys, *Cebus apella nigritus*. Horm. Behav. **41:** 275–287.

Measuring Fecal Steroids

Guidelines for Practical Application

RUPERT PALME

Institute of Biochemistry, Department of Natural Sciences,
University of Veterinary Medicine, A-1210 Vienna, Austria

ABSTRACT: During the past 20 years, measuring steroid hormone metabolites in fecal samples has become a widely appreciated technique, because it has proved to be a powerful, noninvasive tool that provides important information about an animal's endocrine status (adrenocortical activity and reproductive status). However, although sampling is relatively easy to perform and free of feedback, a careful consideration of various factors is necessary to achieve proper results that lead to sound conclusions. This article aims to provide guidelines for an adequate application of these techniques. It is meant as a checklist that addresses the main topics of concern, such as sample collection and storage, time delay extraction procedures, assay selection and validation, biological relevance, and some confounding factors. These issues are discussed briefly here and in more detail in other recent articles.

KEYWORDS: steroid hormones; estrogens; gestagens; androgens; glucocorticoids; noninvasive monitoring; feces/faeces; validation; sex differences; stress

INTRODUCTION

Noninvasive methods of measuring fecal steroid metabolites to assess an animal's endocrine status were pioneered in the late 1970s (birds[1]) and early 1980s (mammals[2,3]) and have been established during the past two decades in an increasing number of species. These methods are now widely used to investigate hormone–behavior relationships, as well as questions in the fields of reproduction, animal welfare, ecology, conservation biology, and biomedicine (for a review, see Refs. 4–9). Because metabolism and excretion of steroids differ significantly between species, and sometimes even between sexes, these noninvasive methods must be rigorously validated for each species before application. Researchers who are not familiar with these endocrine techniques and who want to use them as a noninvasive tool in their field of research need to be especially aware of this caveat. Therefore, the following guidelines highlight the main points of concern and serve as a kind of checklist that briefly addresses these topics.

Address for correspondence: Associate Professor Dr. Rupert Palme, Institute of Biochemistry, Department of Natural Sciences, University of Veterinary Medicine, Veterinaerplatz 1, A-1210 Vienna, Austria. Voice: +43-1-25077-4103; fax: +43-1-25077-4190.
  Rupert.Palme@vu-wien.ac.at

MAIN POINTS OF CONCERN

Sample Collection and Storage

Undoubtedly, the best option is to collect a sample shortly after defecation and to freeze it immediately.[7-9] However, this is not always possible, especially in field studies. Because several authors have highlighted that fecal steroids are not stable and undergo metabolism,[6-8,10] different preservatives were evaluated to find one that does not compromise results. Bacterial enzymes are the main source for the observed metabolism; therefore, water has to be removed (by drying or lyophilization) or heat or chemical substances, such as alcohol or acids (the latter is necessary when import regulations demand disinfection7), have to be applied.[9,10] However, it is important to notice that, for example, the addition of alcohol to the sample will initiate steroid extraction, and its loss will bias the results. It is therefore advisable to standardize the amounts of both the feces and alcohol, and to use tightly sealed vials to avoid any loss by leakage. Because increased levels of fecal steroid metabolites have been reported in mammals,[7,11] caution is advised when samples are stored in ethanol for long periods of time. However, storage experiments also should be performed on bird feces.

Because steroids are often not evenly distributed within fecal samples,[7,12] homogenization of samples is recommended. In birds, feces and urine are excreted together in a species-dependent way[8-9,13-15] in the form of droppings; the relative proportion of both may vary between samples. This may increase sample-to-sample variation; therefore, a standardized protocol (e.g., homogenization or fecal portion alone) for the analysis of droppings is important.

Wet or dry feces are used for analysis, and both measures usually correlate quite well.[5,16] Wet samples are easier to handle and are therefore favored by many researchers.[6,8,10] However, when fresh samples are not available or when undigested materials need to be sorted out, dry feces are used.[7] Furthermore, it is advisable to test the stability of fecal steroids under the same environmental conditions expected during the experiment.[13,14] Fecal bacteria may differ from animal to animal, both qualitatively and quantitatively, which will result in individual differences concerning stability of fecal steroids.[10] Therefore, it is necessary to perform storage experiments with samples from several animals (and both sexes) of a species under investigation, if samples cannot be frozen right away. In addition, attention must be paid to the fact that contamination with water (rain[7]) or urine in mammals may affect concentrations of measured steroid metabolites.

Time Delay of Fecal Excretion

Steroids are metabolized in the liver and excreted via the bile into the gut. Therefore, measured amounts of fecal steroids reflect an event a certain time ago, which allows separation of the experimental and sampling phase. As postulated earlier, on the basis of radiometabolism studies,[12] the gut passage time (confluence of the bile fluid to the rectum) reflects that time delay quite well and therefore provides a rough estimate of the expected delay. This lag time may range from less than 30 min to more than one day, depending on species, sometimes even within species, depending on the activity rhythms of animals.[8,9,13,14,16,17] Knowledge of those delay times of fecal excretion is crucial for the experimental setup, because these times will determine sampling intervals, depending on the events that should be monitored (e.g., basal values or acute vs. chronic stress). In addition, the number of samples necessary will differ from only a few (e.g., in the case of a pregnancy diagnosis) to many samples per individual (e.g., in behavioral studies to evaluate stress).[6,18]

Extraction Procedures

Before analysis, steroid metabolites must be extracted from the feces. Selection of extraction procedures is a significant concern, because fecal steroid metabolites are a mixture of several metabolites with different polarities.[8–10,15] Nevertheless, extraction should be kept as simple as possible. Additional steps increase the variation of determined concentrations; however, low amounts of fecal metabolites demand more sophisticated extraction procedures. To measure fecal glucocorticoid (GC) metabolites in mammals (and progesterone and androgen metabolites), the recommended procedure is to shake a portion of the wet feces (e.g., 0.5 g) suspended in 5 mL of 80% methanol. This percentage of methanol yielded the highest recovery of naturally occurring metabolites in all species tested so far (for a review, see Refs. 6, 8, 10, and 19). This method is practicable, because no evaporation step is needed; at the same time, it yields high recoveries. So far, similar experiments in birds are lacking, but because of the high portion of polar metabolites, 60% methanol is used for extraction of GC metabolites by some authors.[10,13,14] In most studies reported, no additional hydrolysis step is performed in the course of the extraction (for details, see Ref. 10). Although some authors have described extraction procedures with boiling ethanol, most by now have shifted to protocols using a high percentage of methanol for extraction of the fecal samples.[5,7–10] In some studies, radiolabeled steroids (e.g., cortisol or progesterone) were added to estimate the efficiency of extraction procedures, but their results do not reflect the actual recoveries, because these steroids are normally not present in the feces (for a detailed discussion, see Refs. 9, 10, 15, and 20).

Sample volumes of less than 0.05 g were reported to bias the results.[7] However, this may be caused mainly by spurious correlations, and thus it is unlikely to be meaningful.[15] Nevertheless, measurement error (depending on the sensitivity of the balance used to weigh the samples) may increase with small samples and can be avoided by pooling samples over a longer period of time.[7,9,16]

Selection of an Appropriate Immunoassay

Most steroids, particularly GCs, are heavily metabolized by the liver and in the gut.[6,10] Therefore, cortisol or corticosterone itself is virtually absent in the feces. The same is true for other steroids such as progesterone and testosterone. This is demonstrated by almost all radiometabolism studies conducted so far (for a review, see Refs. 8 and 20), which report only very small amounts, if any at all, of radioactive substances with chromatographic properties similar to the steroids present in the blood. Although the terms "fecal cortisol/corticosterone" and "fecal GCs" (and in analogy, the same for gonadal steroids) are often used in the literature, they are incorrect and should therefore be avoided. Instead, the group of metabolites recognized by the respective immunoassay should be mentioned (e.g., 11,17-

dioxoandrostanes[6,10,19]). Alternatively, the measured metabolites should be called "cortisol/corticosterone (glucocorticoid) metabolites" or simply "immunoreactive cortisol or corticosterone.

On the basis of the diverse array of metabolites present in the feces, it is advisable to apply group-specific immunoassays for their measurement.[4,8,10,19] Radioimmunoassays or enzyme immunoassays are most commonly used. Because of several advantages, the latter are becoming more popular.[10] However, all immunoassays must be validated analytically with regard to sensitivity, accuracy, precision, and cross-reactions with the $5\alpha5\beta$-reduced metabolites present in the feces, as described in detail by Möstl et al.[10] To characterize fecal metabolites, high-performance liquid chromatography immunograms should be performed. After chromatographic separation, the presence of immunoreactive metabolites in collected fractions is determined with different assays and, in the case of radiometabolism studies, the radiolabeled metabolites present are measured. This procedure helps to clarify whether and which metabolites derived from the plasma hormones of interest are measured by the applied immunoassays.[8,10,15]

Biological Relevance

After the analytical validation of an immunoassay, a physiological validation must be conducted to demonstrate that an assay technique is capable of detecting changes in the levels of fecal steroid metabolites compared to respective changes of steroid concentrations in the blood. In the case of GCs, a widely used method is an adrenocorticotrophic hormone (ACTH) challenge test.[5,9,15] Many authors who have performed such a stimulation of the adrenocortex in a larger number of animals described a considerable variation between individuals (for a review, see Ref. 9). Therefore, a sufficient number of animals of both sexes should be used. A suppression of the adrenocortical activity (e.g., by dexamethasone) also might be performed for a more profound validation.[9] Different assays are available and should be tested to select the one that shows the most expressed differences between basal and peak concentrations (e.g., after an ACTH test), and thus yields the highest signal-to-noise ratio.[10,15]

However, rigorous physiological validations (and radiometabolism studies) sometimes are not possible (e.g., in endangered species). Even under these constraints, a biological validation should be performed.[9,15] In the case of GC metabolites, serial samples before and after some known stressful events such as immobilization or transportation and for gonadal steroids samples from different reproductive stages can be used to evaluate the biological relevance of such an established noninvasive method.

Confounding Factors

Many factors influencing blood GC levels (and those of other steroids' levels) are expected to be reflected in the concentrations of the metabolites in the feces. These include individual and species differences; daily rhythms; seasonal variations; effects of weather, sex, and age; life-history stages (such as molt or reproductive status); sensitization; and habituation.[7,9,15,16,21] These factors should be kept in mind when designing an experiment to produce meaningful, biologically relevant results.

In general, episodic fluctuations reported in the blood plasma (e.g., androgens, GCs) are smoothed in the feces.[9]

In addition, sex was reported to play a role in the metabolism and excretion of steroids. This is probably the reason why some assays for measuring fecal steroid metabolites yielded good results in one sex, but not in the other.[9,15] The influence of changing diets or metabolic rates (e.g., because of season), especially in birds, has been little evaluated so far,[15] but must be taken into consideration.

CLOSING REMARKS

Even though the highest standards need to remain, compromises cannot always be avoided (especially in field studies). In this case, these compromises should be addressed frankly. Last but not least, it is advisable to search and read the literature carefully. Currently, approximately 140 articles that deal with fecal GC metabolites (for a review, see Refs. 6–9) and an even larger number dealing with fecal metabolites of gonadal steroids[1,14,16] have been published in peer-reviewed journals. The quality of analytical procedures differs significantly among those articles. Analyzing hormone metabolites in the feces may appear to be a quick and easy solution for many problems. In fact, it is not, and there is a strong need to carefully validate assays analytically, physiologically, and biologically in each new species under investigation. Hence, to save time and money and to avoid some of the many pitfalls, it is advisable to contact an experienced laboratory before the start of the experiments.

Applying the recommendations addressed by these guidelines and considering the conclusions of the respective articles in this volume will help to keep the standards of the methods high to get the best out of these noninvasive methods. This will help to establish, confirm, and spread these helpful tools to answer questions related to the endocrine status of an animal (or even a population). In addition, more research is needed to clarify unresolved problems (e.g., standardization among laboratories[10]) or to address open questions (e.g., the influence of diet or plasma binding proteins[9,10,15]).

ACKNOWLEDGMENTS

The stimulating discussions among the participants of the European Science Foundation (ESF) technical workshop and the helpful comments on this manuscript by many of the participants (especially Kurt Kotrschal, Chadi Touma, Erich Möstl, Sophie Rettenbacher, Wolfgang Goymann, Katherine Buchanan and Kirk Klasing) and Elmar Bamberg are gratefully acknowledged.

REFERENCES

1. CZEKALA, N.M. & B.L. LASLEY. 1977. A technical note on sex determination in monomorphic birds using fecal steroid analysis. Int. Zoo Yearb. **17:** 209–211.
2. MÖSTL, E. *et al.* 1983. Trächtigkeitsdiagnose bei der Stute mittels Östrogenbestimmung im Kot. Prakt. Tierarzt **64:** 491–492.
3. MÖSTL, E. *et al.* 1984. Pregnancy diagnosis in cows and heifers by determination of oestradiol-17α in faeces. Br. Vet. J. **140:** 287–291.

4. SCHWARZENBERGER, F. *et al.* 1996. Faecal steroid analysis for non-invasive monitoring of reproductive status in farm, wild and zoo animals. Anim. Reprod. Sci. **42:** 515–526.
5. WASSER, S.K. *et al.* 2000. A generalized fecal glucocorticoid assay for use in a diverse array of nondomestic mammalian and avian species. Gen. Comp. Endocrinol. **120:** 260–275.
6. MÖSTL, E. & R. PALME. 2002. Hormones as indicators of stress. Dom. Anim. Endocrinol. **23:** 67–74.
7. MILLSPAUGH, J.J. & B.E. WASHBURN. 2004. Use of fecal glucocorticoid metabolite measures in conservation biology research: considerations for application and interpretation. Gen. Comp. Endocrinol. **138:** 189–199.
8. PALME, R. *et al.* 2005. Stress hormones in mammals and birds: comparative aspects regarding metabolism, excretion, and noninvasive measurement in fecal samples. Ann. N.Y. Acad. Sci. **1040:** 162–171.
9. TOUMA, C. & R. PALME. 2005. Measuring fecal glucocorticoid metabolites in mammals and birds: the importance of a biological validation. Ann. N.Y. Acad. Sci. **1046:** 54–74.
10. MÖSTL, E., S. RETTENBACHER & R. PALME. 2005. Measurement of corticosterone metabolites in birds' droppings: an analytical approach. Ann. N.Y. Acad. Sci. **1046:** 17–34.
11. HUNT, K. & S.K. WASSER. 2003. Effect of long-term preservation methods on fecal glucocorticoid concentrations of Grizzly bear and African elephant. Physiol. Biochem. Zool. **76:** 918–928.
12. PALME, R. *et al.* 1996. Excretion of infused ^{14}C-steroid hormones via faeces and urine in domestic livestock. Anim. Reprod. Sci. **43:** 43–63.
13. BALTIC, M. *et al.* 2005. A noninvasive technique to evaluate human-generated stress in the black grouse. Ann. N.Y. Acad. Sci. **1046:** 81–95.
14. THIEL, D., S. JENNI-EIERMANN & R. PALME. 2005. Measuring corticosterone metabolites in droppings of Capercaillies (*Tetrao urogallus*). Ann. N.Y. Acad. Sci. **1046:** 96–108.
15. GOYMANN, W. 2005. Noninvasive monitoring of hormones in bird droppings: physiological validation, sampling, extraction, sex differences, and the influence of diet on hormone metabolite levels. Ann. N.Y. Acad. Sci. **1046:** 35–53.
16. WASSER, S.K. & K.E. HUNT. 2005. Noninvasive measures of reproductive function and disturbance in the barred owl, great horned owl, and northern spotted owl. Ann. N.Y. Acad. Sci. **1046:** 109–137.
17. KLASING, K.C. 2005. Potential impact of nutritional strategy on noninvasive measurements of hormones in birds. Ann. N Y Acad. Sci. **1046:** 5–16.
18. SCHEIBER, I.B.R., S. KRALJ & K. KOTRSCHAL. 2005. Sampling effort/frequency necessary to infer individual acute stress responses from fecal analysis in greylag geese (*Anser anser*). Ann. N.Y. Acad. Sci. **1046:** 154–167.
19. PALME, R. & E. MÖSTL. 1997. Measurement of cortisol metabolites in faeces of sheep as a parameter of cortisol concentration in blood. Int. J. Mammal. Biol. **62**(Suppl. II): 192–197.
20. PALME, R. *et al.* 1997. Faecal metabolites of infused ^{14}C-progesterone in domestic livestock. Reprod. Domest. Anim. **32:** 199–206.
21. FRIGERIO, D. *et al.* 2004. Excreted corticosterone metabolites co-vary with ambient temperature and air pressure in male Greylag geese (*Anser anser*). Gen. Comp. Endocrinol. **137:** 29–36.

A Noninvasive Technique to Evaluate Human-Generated Stress in the Black Grouse

MARJANA BALTIC,[a] SUSANNE JENNI-EIERMANN,[b] RAPHAËL ARLETTAZ,[a,c] AND RUPERT PALME[d]

[a]*Zoological Institute, Division of Conservation Biology, University of Bern, Baltzerstrasse 6, CH–3012 Bern, Switzerland*

[b]*Swiss Ornithological Institute, CH–6204 Sempach, Switzerland*

[c]*Swiss Ornithological Institute, Valais Field Station, Nature Centre, CH–3970 Salgesch, Switzerland*

[d]*Institute of Biochemistry, Department of Natural Sciences, University of Veterinary Medicine, Veterinärplatz 1, A–1210 Vienna, Austria*

ABSTRACT: The continuous development of tourism and related leisure activities is exerting an increasingly intense pressure on wildlife. In this study, a novel noninvasive method for measuring stress in the black grouse, an endangered, emblematic species of European ecosystems that is currently declining in several parts of its European range, is tested and physiologically validated. A radiometabolism study and an ACTH challenge test were performed on four captive black grouse (two of each sex) in order to get basic information about the metabolism and excretion of corticosterone and to find an appropriate enzyme-immunoassay (EIA) to measure its metabolites in the feces. Peak radioactivity in the droppings was detected within 1 to 2 hours. Injected ^{3}H-corticosterone was excreted as polar metabolites and by itself was almost absent. A cortisone-EIA was chosen from among seven tested EIAs for different groups of glucocorticoid metabolites, because it cross-reacted with some of the formed metabolites and best reflected the increase of excreted corticosterone metabolites, after the ACTH challenge test. Concentrations of the metabolites from fecal samples collected from snow burrows of free-ranging black grouse were within the same range as in captive birds. The noninvasive method described may be appropriate for evaluating the stress faced by free-living black grouse populations in the wild, particularly in mountain ecosystems where human disturbance, especially by winter sports, is of increasing conservation concern.

KEYWORDS: corticosterone metabolism; noninvasive endocrine monitoring; conservation biology; *Tetrao tetrix*; wildlife management; ecology

Address for correspondence: Prof. Dr Raphaël Arlettaz, Zoological Institute, Division of Conservation Biology, Baltzerstrasse 6, CH–3012 Bern, Switzerland. Voice: +41-31-631-3161; fax: +41-31-631-4535.

Raphael.Arlettaz@nat.unibe.ch

INTRODUCTION

In addition to habitat degradation, the intensification of human leisure activities exerts a negative pressure on wildlife.[1–5] This is of particular concern as regards species that are otherwise (e.g., through habitat degradation and fragmentation) already threatened and vulnerable. Yet, until the recent development of appropriate analytical tools, it has remained difficult to quantify properly how human-generated disturbances affect animals' constitution, physiological condition and, ultimately, reproductive fitness. Not surprisingly, until recently, most studies have therefore instead focused on variations in time budgets and juvenile survival.[1,3,6,7] Rarely, however, could changes in time allocation to various behaviors and activities be associated with the actual additional physiological costs they entail. The emergence of noninvasive techniques for estimating stress in free-ranging animals opens new avenues for a proper quantification of the impact of human-generated stress onto wildlife.[8–15]

The black grouse (*Tetrao tetrix*), an emblematic game bird with great economic and cultural value, is endangered and declining in several parts of its Palearctic distribution range.[16] Several authors have identified human activities and infrastructures as the probable main cause. This includes habitat loss and fragmentation,[16] hunting,[17,18] collisions with aerial cables and fences,[19,20] as well as increasing disturbance through popular leisure activities such as winter sports and activities.[1,21,22] Despite the knowledge, gathered from research on different species, to the effect that prolonged stress generated either by human or natural factors can have deleterious consequences on an individual,[23–25] physiological constraints imposed by human activities are only poorly explored in the black grouse.[26,27] In order to propose appropriate mitigation measures for this endangered species in those areas where leisure activities are a potential threat, a proper quantification of the impact of human-elicited stress appears a first necessary step. Additionally, however, since black grouse are endangered, a noninvasive method, which does not alter or constrain bird behavior, is another prerequisite in any such study.

The first physiological response of an organism to different stressful stimuli is a cascade of hormone secretions, starting with the release of catecholamines (epinephrine) from the adrenal medulla within seconds after the stimulus, triggering the hypothalamic–adenohypophysal–adrenocortical axis within a few minutes, which is followed by synthesis and secretion of glucocorticosteroids (corticosterone in the case of birds) from the adrenal cortex, as well as cytokines from cells of the immune system.[23] The blood sampling techniques usually applied in stress research are invasive. They are thus not convenient for the study of threatened free-ranging animals such as the black grouse.[28] Instead, noninvasive techniques for monitoring stress on free-living populations have therefore been developed extensively in recent years. In birds, feces are often used for this purpose, as they are easy to collect. However, it must be taken into account that metabolites are excreted with a species-specific time delay of about a few hours in birds.[9–11,29,30]

Until now, noninvasive methods for measuring adrenal activity by measuring fecal glucocorticoid metabolites by group-specific enzyme-immunoassays (EIAs) have been mainly developed on domestic and captive animals for purposes of research on animal welfare[28,29,31] and behavioral ecology.[10,12,32] However, such methods also have a big application potential in conservation biology.[9,13,15,22] Metabolism and excretion of glucocorticoids differ between species, and sometimes even between sexes and individuals within a given species.[30,31] Therefore, it is not possible to draw analogous conclusions from other bird species. Thus, the aims of this study were to get basic information about the metabolism and excretion of corticosterone, to characterize fecal metabolites of black grouse, and to select an EIA for measuring the corticosterone metabolites (CM) in feces of free-ranging individuals. In addition, the stability of the CM was tested within natural settings.

For this purpose, we performed first a radiometabolism study of corticosterone on birds in captivity. We also used the same birds for physiological validation of several EIAs by testing immunoreactivity of the CM excreted after adrenal stimulation with adrenocorticotropic hormone (ACTH). Yet, it should be noted that the stability of glucocorticoid metabolites in feces could be significantly affected by environmental conditions, due, for instance, to bacterial metabolism. This can give misleading results.[33] In order to both define a suitable sampling protocol of black grouse feces in nature and to achieve accuracy in the hormone assay, we first had to find out if the concentration of measured CM significantly changes in feces exposed to a variety of winter environmental conditions, and, second, at which time intervals between defecation and sampling of fecal pellets these changes take place. This was accomplished by incubating feces at different time intervals and at temperatures above 0°C (below this threshold, metabolic activity was assumed to be insignificant), which might occur in winter on sun-exposed snow surfaces. Also, we must stress that our radiometabolism study and the hormonal stimulation experiment for selecting the most appropriate EIA were carried out on captive birds that might have different stress reactions than free-ranging individuals, as well as distinct digestive tract size (ceca) and transit durations due to a different diet.[34–36] Since corticosterone metabolism can be further affected by these factors, we analyzed samples from free-living birds with the selected EIA to finally confirm the suitability of the method for assaying stress in wild black grouse during the winter season.

MATERIALS AND METHODS

Birds and Experimental Setup

Permission for the animal experiment was given by the Department of Veterinary Medicine, canton Lucerne. Two male and two female black grouse, all 7 months old, were used in the experiments. The females came from private raisers, the males from the Bern Zoological Garden. Between experiments, the birds were kept in a large outdoor aviary at the Hasli Ethological Station of the Zoological Institute, University of Bern. Because we intended to apply this method to free-living birds during the cold season, we conducted laboratory experiments by the end of November 2002. At that time of year, the birds had already reached adult plumage and body mass, whereas reproductive mechanisms, which could induce a stress state in displaying males,[10] had not yet matured. During experiments, the birds were placed in individual cages with water and food provided *ad libitum*. Special cages (80 × 80 × 80 cm in size), with a double bottom, were constructed. The top of the cage as well as the back and one sidewall were made out of green textile mesh (0.8 mm in diameter) to protect the birds from injuries due to the reduced space available. The front wall with the entrance, as well as one sidewall, consisted of wooden plates that isolated the birds from the researcher during sampling. The bottom of the cage was composed of a wire-mesh floor (1-cm mesh diameter). This enabled droppings to fall through into a removable, exchangeable drawer. Cages and birds were exposed to natural light and temperature conditions, but protected from precipitation.

Habituation of the birds to this setup lasted 4 days. By that time, they were feeding regularly and defecating normal droppings. As they would under natural conditions, birds had at least two main feeding periods during the day[37,38] and defecated almost hourly 1–3 solid feces. To reduce the risk of a possible influence of different dietary compounds on digestion,[34] only homogenous grouse food (article 872.4, Protector SA, CH-1522 Lucens) was provided.

Hormone Administration and Sampling Pattern

Each bird was injected in the ulnar vein with 1.85 MBq (=50 µCi) of radiolabeled corticosterone ([1,2,6,7-^{3}H(N)]-corticosterone, specific activity: 76.5 Ci/mmol, NET-399, Perkin-Elmer Life Sciences, Boston, MA), which was dissolved in 0.5 mL of a physiological solution (0.9% NaCl) containing 10% ethanol. Birds were injected between 8:00 and 8:15 A.M. Manipulation with each individual lasted less than 5 minutes.

Fecal samples were collected one hour before injection, to determine the background radioactivity, and after injection, once per hour during the following 24 hours. The feces were collected from the exchangeable drawers. They were frozen immediately at −22°C until further analyses, and the drawers were cleaned with 70% ethanol and water after each sampling event.

In order to physiologically validate different EIAs, the same birds were injected intravenously with 0.5 mg of adrenocorticotropic hormone (ACTH; Synacthen; Novartis Pharma AG, Basel, Switzerland). On the day before injection, the fecal samples were collected hourly during a total of 24 hours in order to get a control group (pretreatment group). On the following day birds were injected between 8:00 and 8:15 A.M., and the fecal samples were again collected hourly during the next 24 hours (treatment group). During the third and fourth day, feces were collected early in the morning, at midday, and in late afternoon in order to control for posttreatment levels (posttreatment groups I and II, respectively). All samples were immediately stored at −22°C until analyses.

Extraction and Characterization of the Excreted ^{3}H-Corticosterone Metabolites

A total of 0.5 g of each well-homogenized fecal sample was mixed with 3 mL methanol and 2 mL water and vortexed for 30 min. After centrifugation (2500 g; 10 min), aliquots (0.5 mL) of the supernatant (in duplicate) were transferred into scintillation vials (Article 6008117, Packard Instruments, Meriden, CT), each containing 6 mL scintillation fluid (Quicksafe, A, 100800, Zinsser Analytic, Maidenhead, UK). The radioactivity of each sample was measured (5 min) by a liquid scintillation counter (Tri-Carb 2100 TR, Packard Instruments) with a quench compensation program. Radioactivity is expressed as kBq per g of feces.

In order to characterize the CM, samples containing peak radioactivity were extracted, the radioactive substances purified by a Sep-pak C_{18} cartridge and subjected to reverse-phase high-performance liquid chromatography (RP-HPLC), as described by Rettenbacher et al.[30] Briefly, steroids were separated on a Novapak C_{18} column (3.9 × 150 mm, Millipore Corporation, Milford, MA) with a methanol/water solvent. A linear gradient from 20% to 100% methanol with a flow rate of 1 mL/min was applied. A total of 96 fractions were collected (three per min). Radioactivity in an aliquot (50 μL) of each fraction was determined (Top Count; Packard Instruments, Meriden, CT).

Immunoreactivity of ^{3}H-Corticosterone Metabolites and the Physiological Validation of Assays

An array of different, previously established EIAs was tested to select the one best suited for black grouse. Among the seven EIAs were a corticosterone-EIA,[39] a tetrahydrocorticosterone-EIA,[40] a 5α-pregnane-3β,11β,21-triol-20-one-EIA,[31] an 11β-hydroxyetiocholanolone-EIA,[41] an 11-oxoetiocholanolone-EIA,[42] a cortisone-EIA,[30] and a so far unpublished 20β-dihydrocorticosterone-EIA. Aliquots of each HPLC fraction of males were measured using the different EIAs to check if radiolabeled metabolites are recognized. The EIA procedure was described in detail by Palme and Möstl[39] and Touma et al.[31] The antibody of the 20β-dihydrocorticosterone-EIA (working dilution 1:80000) was raised against 20β-dihydrocorticosterone-3-CMO:BSA in a rabbit. The label (20β-dihydrocorticosterone-3-CMO-biotinyl-LC; 1:5000000) was produced as described by Möstl et al.[42] The standard (20β-dihydrocorticosterone) curve ranged from 0.33 pg/well to 80 pg/well. Only the cortisone-EIA was applied on HPLC fractions of females, but the later three EIAs, which were able to detect significant amounts of immunoreactive substances in the HPLC fractions of males, were used for the analyses of the samples from the stimulation experiment (ACTH) of both sexes. Therefore an aliquot of the supernatant (after diluting 1:10 with assay buffer) was subjected to the EIAs after extraction (5mL of 60% methanol) of the droppings (0.5 g).

Stability of ^{3}H-Corticosterone Metabolites in the Feces

In order to optimize the collection of fecal samples from free-living black grouse, it was important to know whether and how the concentration of metabolites changes with time, especially when ambient temperature increases above zero and allows activation of fecal bacteria. We did this with captive birds, two males and one female, from which we collected feces on the third day (posttreatment control group II, see earlier) after the ACTH injection. All feces excreted by a single bird (ca. 20 g) were pooled, homogenized, and divided into four equal subsamples. A subsample was frozen immediately at −22°C (control), while the three other samples were incubated in a refrigerator at 6–7°C for 24 h, 48 h, and 72 h, respectively. Temperature in the refrigerator was set up as the highest measured temperature in the snow burrow, 6.5°C (unpublished personal data). During the night, as well as during the day, black grouse dig burrows into the snow to protect themselves from the cold during resting periods. These snow burrows or "igloos" are rebuilt anew each time. From each subsample, 10 aliquots of 0.5 g each were extracted and the concentration of the metabolites measured with the cortisone-EIA, as described earlier.

Concentrations of ^{3}H-Corticosterone Metabolites in Feces from Free-Ranging Birds

In a preliminary field experiment, amounts of CM present in free-living birds were measured in order to get information on interindividual and intraindividual variances. We collected samples from the snow burrows of four free-living black grouse males, in February 2003, at Verbier and Les Diablerets (southwestern Swiss Alps). The birds were flushed from their diurnal burrows early in the afternoon and fecal material that had accumulated within burrows (9–15 separate droppings each) was collected. In order to gather information about variance of CM concentrations of an individual bird within the burrow, each dropping was analyzed separately. The cortisone-EIA, which gave the best results in the ACTH challenge test, was selected to further analyze of the feces from the free-living birds. All feces were extracted as described previously (5 mL of 60% methanol), and an aliquot (diluted 1:10) of the supernatant analyzed in the EIA.

Statistical Analyses

Results of the ACTH test (physiological validation) of the three EIAs, which cross-reacted significantly with radioactive metabolites, were analyzed by an ANOVA standard least-square fit model. We tested for the following effects and interaction term: individual bird ($n = 4$), treatment (vs. control day, i.e., data from the day prior to ACTH administration), bird*treatment.

The within-individual (i.e., within-burrow fecal sample) and among-individual (between burrows) variations in CM concentrations of the feces from free-ranging black grouse males were analyzed by one-way ANOVA after controlling for variance homoscedasticity (Levene's test). All statistical analyses were performed with JMP 4.04 (SAS Institute Inc., 1989–2001). Test rejection probability levels were set at 5% throughout.

RESULTS

Excretion and Characterization of the ^{3}H-Corticosterone Metabolites

The main portion of radioactivity was quickly excreted. Peak concentrations (75 to 139 kBq/g feces) were reached after one (one male and one female) or two (the other two animals) hours (FIG. 1). Radioactivity decreased almost continuously afterwards (only one male had a second, somewhat smaller peak after 5 h). When sampling was stopped 24 h after the injection, radioactivity was low, but background levels were not yet reached.

Injected ^{3}H-corticosterone was heavily metabolized, as demonstrated by HPLC separation of CM in the peak radioactivity samples of the four individual birds (FIG. 2). Three to four main metabolites were present, all eluting between fractions 20 and 45, thus resembling conjugated, or polar unconjugated steroids. Males had more polar metabolites if compared with females. In all samples only small amounts,

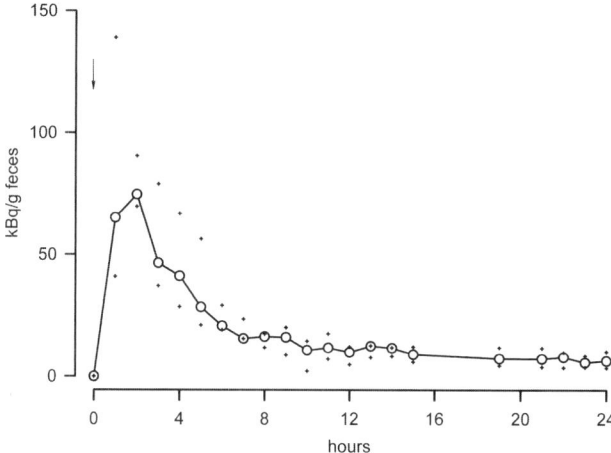

FIGURE 1. Excreted radioactivity (kBq/g feces; min., max., median) in the droppings of four black grouse birds after ^{3}H-corticosterone administration. Animals were injected intravenously between 8:00 and 8:15 A.M. (0 h, *arrow*).

if at all, of unmetabolized corticosterone could be detected. In the HPLC fractions, significant amounts of CM could be measured with three of the seven EIAs tested (FIG. 2), that is, concentrations were higher than the detection limit of the respective EIAs. The cortisone-EIA, measuring metabolites with a common 3,11-dione structure, yielded the highest amounts of immunoreactivity. The most prominent metabolite peaked at fractions 31/32. The 11-oxoetiocholanolone- and the 11β-dihydrocorticosterone-EIA measured only smaller amounts of immunoreactivity.

Physiological Validation of the EIAs–ACTH Challenge

The concentrations measured with the cortisone-EIA (3,11-dioxo-CM) differed significantly between the experimental groups (ANOVA, $df = 1$, $F = 16.41$, $P < .0001$), but no difference among individuals (ANOVA, $df = 3$, $F = 1.933$, $P = .128$) was found, although males tended to have higher basal values. A *post hoc* test showed that this difference was due to differences between the treatment group (i.e., data obtained during the first 24 h after adrenal stimulation) and the pretreatment group, and the posttreatment I and II control groups (Dunnett's test). Note that 24 h after injecting ACTH, the concentration of the CM had returned approximately to

FIGURE 2. Immunoreactivity (nmol per fraction) of ^{3}H-corticosterone metabolites (Bq per fraction) evaluated by reverse-phase high-performance liquid chromatography (RP-HPLC) with three different EIAs (cortisone-, 11-oxoetiocholanolone-, and 11β-dihydrocorticosterone) in two male and two female black grouse. Elution positions of corticosterone, cortisol, 17β-estradiol-disulfate ($E_2\beta$-diSO$_4$), estrone-glucuronide (E_1G), and estrone-sulfate (E_1S) are marked. A gradient solvent system with a water/methanol ratio changing from 20% to 100% was applied.

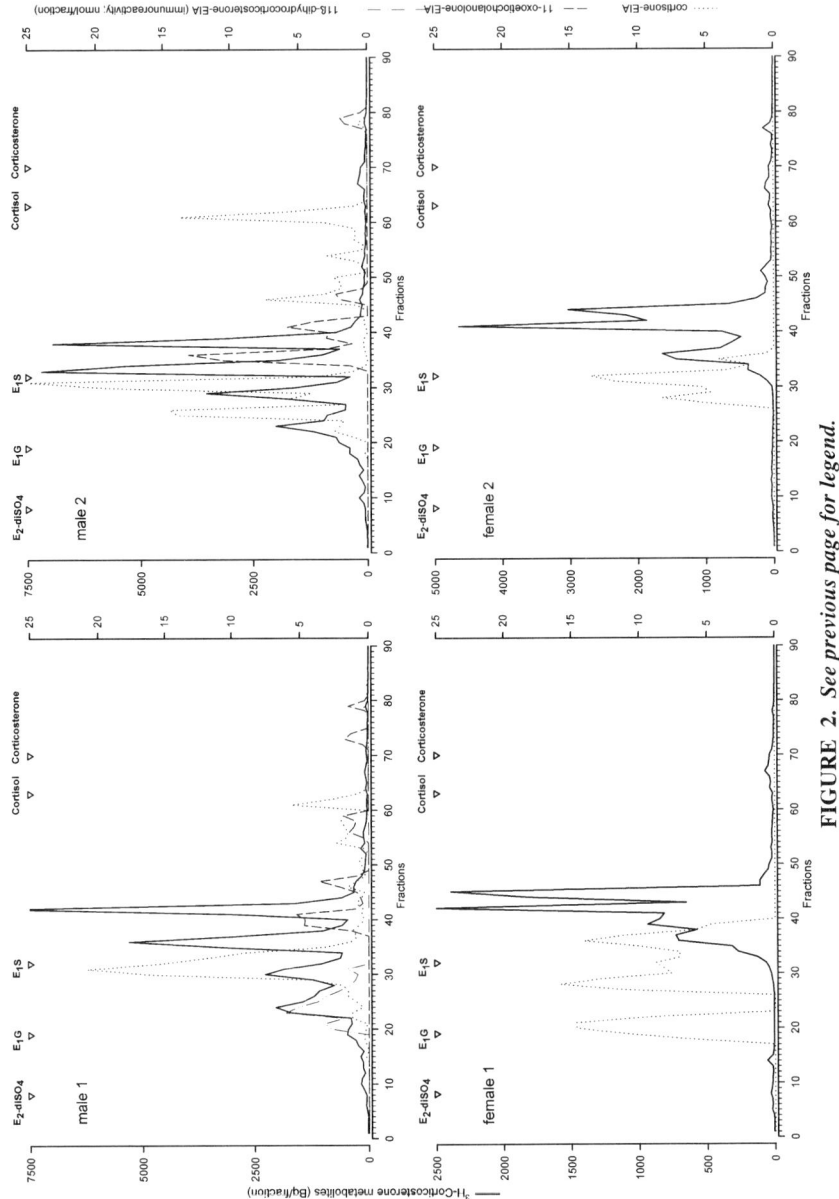

FIGURE 2. See previous page for legend.

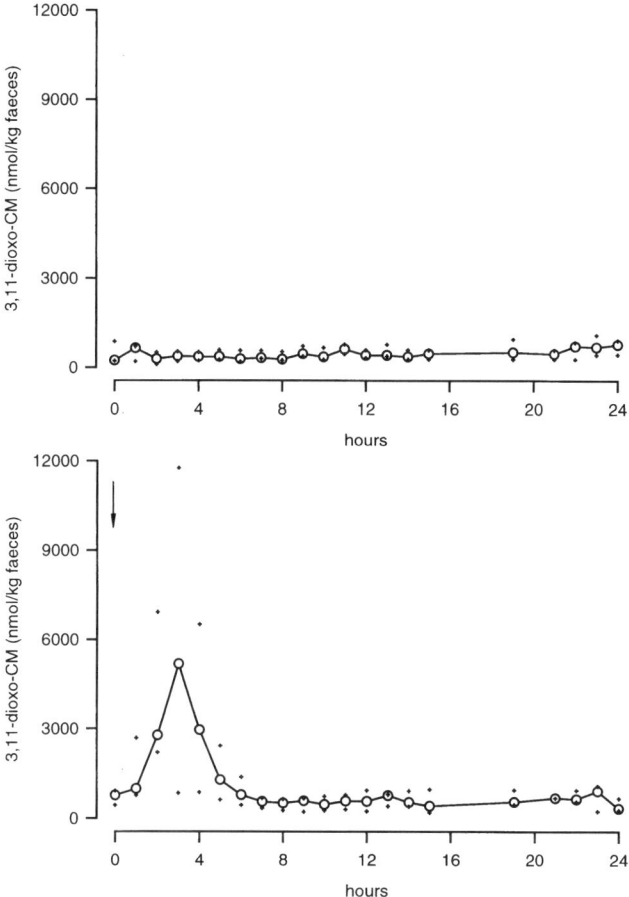

FIGURE 3. Concentrations (min., max., median) of corticosterone metabolites (3,11-dioxo-CM; nmol/kg feces) before (**Upper Panel**: pretreatment group) and after stimulation with ACTH (**Lower Panel**: treatment group), measured by a cortisone-EIA in two male and two female black grouse. Injections took place at 8:00–8:15 A.M.

the levels recorded before the experimental treatment (*post hoc* Dunnett's test, not significant).

During the pretreatment day (pretreatment control group), the mean (±SE) concentration of CM in droppings of all four birds was 454 ± 31 nmol/kg feces (FIG. 3). On the second day, after the ACTH injection at 8:00–8:15 A.M., the CM concentration reached its maximum (3 to 12 μmol/kg) during the first three hours, which represents a 13-fold increase in comparison with the control, pretreatment baseline values. Within the following four hours, the concentration decreased again to 594 ± 25 nmol/kg, a level that remained more or less constant over the next 16 hours. On the third day of the experiment (posttreatment group I), concentrations returned to levels similar to pretreatment (453 ± 41 nmol/kg feces) and remained similar on the fourth day (posttreatment group II; 438 ± 56 nmol/kg).

The four birds reacted differently to the ACTH injection. The strongest response (24-fold magnitude in comparison with control values) was observed in female 2, but it should be mentioned that this bird showed the lowest mean baseline concentration of CM. The highest peak of metabolites, amounting to 12 µmol/kg feces, was found in the male, which already exhibited the highest average concentration of metabolites in the pretreatment control group.

As with the cortisone-EIA, there was a significant difference in the concentrations of the CM measured with the 11-oxoaetiocholanolone- and the 11β-dihydrocorticosterone-EIA between the pretreatment control group and the treatment group during the 48 hours following the ACTH injection (ANOVA, 11-oxoaetiocholanolone-EIA: $df = 1$, $F = 24.28$, $P < .0001$; 20β-dihydrocorticosterone-EIA: $df = 1$, $F = 57.49$, $P < .0001$), although no prominent peak was recognizable. In addition, concentrations of these CM were 5 to 20 times lower than the 3,11-dioxo-CM measured with the cortisone-EIA and showed statistical differences among individuals (ANOVA, 11-oxoaetiocholanolone-EIA: $df = 3$, $F = 41.13$, $P < .0001$; 20β-dihydrocorticosterone-EIA: $df = 3$, $F = 26.61$, $P < .0001$).

Stability of ^3H-Corticosterone Metabolites in the Feces

After incubation of feces at 6–7°C in a refrigerator, a slight decrease of CM (by 16%) was noticed after 24 hours, which, however, was not statistically significant (360 nmol/kg vs. 430 nmol/kg for the control sample; FIG. 4). Concentrations re-

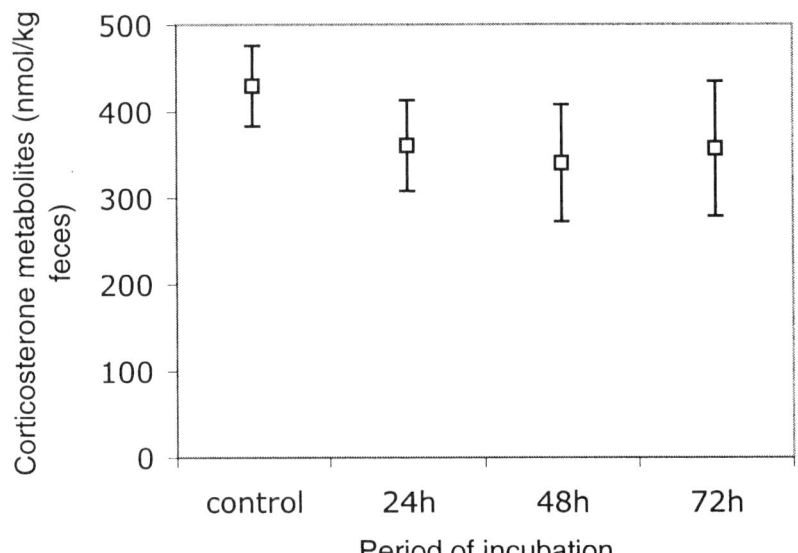

FIGURE 4. Concentrations (mean ± SE; $n = 3$) of corticosterone metabolites (3,11-dioxo-CM; nmol/kg feces) after incubation at 6–7°C for 24, 48, and 72 h, respectively. Controls stem from samples that were immediately frozen at −22°C.

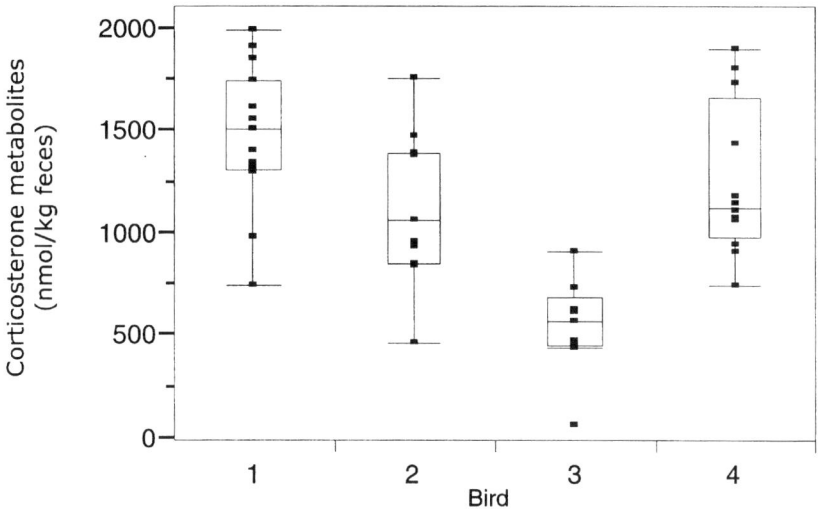

FIGURE 5. Boxplots of 3,11-dioxo-CM concentrations (nmol/kg feces) from samples from the snow burrows of four free-ranging black grouse males in February 2003.

mained close to that level after 48 h and 72 h (average decrease by 21% and 17%, respectively, from the control sample).

Concentrations of ^3H-Corticosterone Metabolites from Feces of Free-Ranging Birds

Concentrations of CM in free-ranging black grouse assayed by the cortisone-EIA ranged from 62 nmol/kg to 1993 nmol/kg feces (FIG. 5). There was a significant interindividual variation in metabolite concentrations (ANOVA, $F_{[3,43]} = 14.852$, $I < .0001$), but within the burrow variances did not differ significantly (Levene's test, $F_{[3,43]} = 1.048$, $P = .381$). This range of concentrations was similar to, or somewhat higher than that of captive birds on the control day prior to the ACTH challenge experiment, but much lower than concentrations induced by ACTH (FIG. 3) assayed with the same EIA.

DISCUSSION

In this study, we successfully tested and physiologically validated a noninvasive method for evaluating adrenocortical activity in the black grouse. This method offers the novel possibility of quantifying the level of stress, for instance, induced by human disturbance, in free-ranging populations of this endangered species, since feces can easily be collected from snow burrows in winter. In line with previous investigations, we could demonstrate that assaying CM by group-specific EIAs gives an accurate picture of the adrenocortical activity.[28] This is the first time that a noninvasive technique for evaluating disturbances has been described via corticosterone metabolites measured by EIA in fecal material of a species of Phasianidae.

Birds' excreta consist of an inhomogeneous cloacal mixture of urine and feces. Steroid metabolites are excreted in urine and feces at different time intervals, first in urine, and in feces some hours later.[30,31,42] In the black grouse, as in chicken,[30] the first peak of radioactivity appeared within the first two hours after injecting 3H-corticosterone (FIG. 1). This corresponded to an initial period when most excreta in all four birds were rather liquid, containing a large number of urine components. Yet, contrary to what was found in other studies,[11,30,44] a second peak, reflecting excreted fecal metabolites, roughly corresponding to the timing of gut passage, was not detected in black grouse (except in one male), where the concentration of metabolites decreased almost continuously. The reason for this may be a more expressed urinary excretion, thus concealing the smaller amounts of fecal metabolites or some mixing of excreta in the cloaca.

Our HPLC analysis of CM from the feces of two captive males and females demonstrated that corticosterone is heavily metabolized mainly to polar metabolites, whereas corticosterone itself was almost absent. This corresponds to the findings of Goymann et al.[11] in European stonechats, Carere et al.[32] in great titmice, and Rettenbacher et al.[30] in chicken. Altogether, three to four prominent peaks, probably representing conjugated CM, were present in all four black grouse individuals. There were also some apparent sex differences in the pattern of metabolites formed, similar to those shown for other vertebrate species.[30,31]

Only three out of seven group-specific EIAs tested cross-reacted significantly with the ^{3}H-CM present in the HPLC fractions (FIG. 2). In order to choose the one best suited to assess adrenocortical activity, all three of those EIAs were physiologically validated. This was achieved by the ACTH challenge test. Blood samples were not taken, as corticosterone concentrations in plasma are known to correlate well with the concentration of metabolites in the feces.[45,46] Because we were interested in the pattern of CM in the droppings, it was important to avoid confounding effects of the stress experienced by the blood sampling procedure itself. In addition, we did not want to apply an invasive, frequent blood-sampling regime to the endangered birds.

Injection of ACTH resulted in a distinct increase in measured 3,11-dioxo-CM concentrations in all four birds. After the initial peak, which took place during the first 2 to 3 h after ACTH administration, there was a rapid decrease in CM values within 4 h. Peak concentrations were approximately 13 times higher than baseline values (FIG. 3). This increase was more pronounced than in chicken,[30] and makes it more probable that some less stressful events also can be monitored by fecal analysis.

Although sex differences were observed in the pattern of formed ^{3}H-CM, there were no statistically significant differences in measured levels of 3,11-dioxo-CM. This is another advantage of the cortisone-EIA, because the gender of an animal can be neglected in comparative analyses of stress levels faced by birds under various environmental conditions. The other two assays (11-oxoetiocholanolone and 11β-dihydrocorticosterone-EIA) were found to be unsuitable for evaluating stress properly; as the measured concentrations were much lower, no distinctive peaks could be recognized after ACTH injection, and pronounced individual differences were observed.

As Washburn and Millspaugh[33] and Morrow et al.[47] showed for "even-toed" ungulate feces, environmental conditions, particularly moisture in combination with higher temperatures, can significantly affect CM degradation in the feces, as they would favor the activity of bacteria. This could be a serious source of bias in the quantification and interpretation of stress levels. One advantageous trait of black grouse is that in winter they roost in snow burrows, in which they defecate. This enables the collection of fecal samples, which are, so to say, naturally stored at optimal temperature conditions (<0°C), which is lower than in our incubation experiment(6–7°C, where degradation is only slight).

Another question is whether there is variation in CM between the droppings of the same individual within a short time. This variation could, for example, be due to a diurnal rhythm. In our preliminary field experiment, the concentrations of CM taken from samples of free-ranging birds varied considerably within the individual. However, the mean of the individuals still differed significantly. It is therefore advisable to take all droppings from a snow burrow and homogenize them before analysis. We think that this is the best way to characterize the level of CM over the time the feces were excreted.

We conclude that measuring corticosterone metabolites from feces of black grouse with the cortisone-EIA provides a suitable, novel tool for quantifying noninvasively adrenocortical activity, and thus stress, in free-ranging black grouse populations. This enables one to investigate properly the levels of stress, acclimation, and facilitation actually faced by this endangered bird species, especially in mountain habitats, where the increasing intrusion of human leisure activities might potentially represent a serious additional source of threat to the fauna in general. In the future, we can envision modeling tolerance thresholds toward human disturbance within black grouse populations. This might be an essential step for proposing sound, targeted conservation measures to mitigate the impact of man on that emblematic species.

ACKNOWLEDGMENTS

This research was funded by a grant from the Swiss National Science Foundation to R. Arlettaz (31-67186.01). We thank B. Schmutz for providing a black grouse female for the experiments and O. Roth for helping with the birds' maintenance in captivity. We are also grateful to A. Kuchar for performing HPLC analyses and A. Zechner for patient laboratory assistance.

REFERENCES

1. INGOLD, P. et al. 1992. Recreational activities—a serious problem for animals? Ornithol. Beob. **89:** 205–216.
2. FOWLER, G.S. 1999. Behavioral and hormonal responses of Magellanic penguins, *Spheniscus magellanicus* to tourism and nest site visitation. Biol. Conserv. **90:** 143–149.
3. MCCLUNG, M.R. et al. 2004. Nature-based impact on yellow-eyed penguins *Megadyptes antipodes*: does unregulated visitors access affect fledging weight and juvenile survival? Biol. Conserv. **119:** 279–285.
4. MÜLLNER, A., E. LINSENMAIR & M. WIKELSKI. 2003. Exposure to ecotourism reduces survival and affects stress response in hoatzin chicks (*Opisthocomus hoazin*). Biol. Conserv. **118:** 549–558.

5. WATSON, A. & R. MOSS. 2004. Impact of ski-development on ptarmigan (*Lagopus mutus*) at Cairn Gorm, Scotland. Biol. Conserv. **116:** 267–275.
6. BÉLANGER, L. & J. BÉDARD. 1990. Energetic cost of man-induced disturbance to staging snow geese. J. Wildl. Manage. **54:** 36–41.
7. FRID, A. & L.M. DILL. 2002. Human-caused disturbance stimuli as a form of predation risk. Conserv. Ecol. **6:** 11.
8. CREEL, S., N.M. CREEL & S.L. MONFORT. 1997. Radiocollaring and stress hormones in African wild dogs. Conserv. Biol. **11:** 544–548.
9. WASSER, S.K. *et al.* 1997. Noninvasive physiological measures of disturbance in the northern spotted owl. Conserv. Biol. **11:** 1019–1022.
10. KOTRSCHAL, K., K. HIRSCHENHAUSER & E. MÖSTL. 1998. The relationship between social stress and dominance is seasonal in greylag geese. Anim. Behav. **55:** 171–176.
11. GOYMANN, W., E. MÖSTL & E. GWINNER. 2002. Corticosterone metabolites can be measured noninvasively in excreta of European stonechats, *Saxicola torquata rubicola*. Auk **119:** 1167–1173.
12. HUBER, S. *et al.* 2003. Non-invasive monitoring of adrenocortical response in red deer. J. Wildl. Manage. **67:** 258–266.
13. NAKAGAWA, S., E. MÖSTL & J.R. WAAS. 2003. Validation of an enzyme immunoassay to measure fecal glucocorticoid metabolites from Adélie penguins (*Pygoscelis adeliae*): a noninvasive tool for the measurement of stress? Polar Biol. **26:** 491–493.
14. ROMERO, L.M. 2004. Physiological stress in ecology: lessons from biomedical research. Trends Ecol. Evol. **19:** 249–255.
15. TEMPEL, D.J. & R.J. GUTIÉRREZ. 2004. Factors related to fecal corticosterone levels in California spotted owls: implications for assessing chronic stress. Conserv. Biol. **18:** 538–547.
16. STORCH, I. 2000. Grouse status survey and conservation action plan 2000-2004. WPA/BirdLife/SSC Grouse Specialist Group. IUCN, Gland, Switzerland and Cambridge, UK and the World Pheasant Association. Reading, UK.
17. ELLISON, L.N., P. LEONARD & E. MENONI. 1988. Effect of shooting on a black grouse population in France. Suppl. Ric. Biol. Selvaggina **14:** 117–128.
18. CAIZERGUES, A. & L.N. ELLISON. 1998. Impact of radio-tracking on black grouse *Tetrao tetrix* reproductive success in the French Alps. Wildl. Biol. **4:** 205–212.
19. BEVANGER, K. 1995. Estimates and population consequences of tetraonid mortality caused by collisions with high tension power lines in Norway. J. Appl. Ecol. **32:** 745–753.
20. BAINES, D. & R.W. SUMMERS. 1997. Assessment of bird collisions with deer fences in Scottish forests. J. Appl. Ecol. **34:** 941–948.
21. MEILE, P. 1981. Skiing facilities in Alpine habitat of black grouse and capercaillie. Proc. 2nd Int. Symp Grouse Essex Printers Ltd. Suffolk, UK.
22. CREEL, S. *et al.* 2002. Snowmobile activity and glucocorticoid stress responses in wolves and elk. Conserv. Biol. **16:** 809–814.
23. WINGFIELD, J.C., C. BREUNER & J. JACOBS. 1997. Corticosterone and behavioural responses to unpredictable events. *In* Perspectives in Avian Endocrinology. S. Harvey & R.J. Etches, Eds.: 267–278. Journal of Endocrinology Ltd. Bristol, UK.
24. WINGFIELD, J.C. *et al.* 1997. Environmental stress, field endocrinology, and conservation biology. *In* Behavioral Approaches to Conservation in the Wild. J.R. Clemmons & R. Buchholz, Eds.: 95–131. Cambridge Univ. Press. Cambridge.
25. HOFER, H. & M.L. EAST. 1998. Biological conservation and stress. *In* Advances in the Study of Behavior. P.J.B. Slater, J.S. Rosenblatt, C.T. Snowdon & M. Milinski, Eds.: 405–525. Academic Press. San Diego.
26. RINTAMÄKI, H. *et al.* 1983. Summer and winter temperature regulation in the black grouse *Lyrurus tetrix*. Physiol. Zool. **56:** 152–159.
27. ANGELSTAM, P. 1984. Sexual and seasonal differences in mortality of the Black Grouse *Tetrao tetrix* in boreal Sweden. Ornis Scan. **15:** 123–134.
28. MÖSTL, E. & R. PALME. 2002. Hormones as indicators of stress. Domest. Anim. Endocrinol. **23:** 67–74.

29. WASSER, S.K. et al. 2000. A generalized fecal glucocorticoid assay for use in a diverse array of nondomestic mammalian and avian species. Gen. Comp. Endocrinol. **120:** 260–275.
30. RETTENBACHER, S. et al. Measurement of corticosterone metabolites in chicken droppings. Br. Poult. Sci. **45:** 704–711.
31. TOUMA, C. et al. 2003. Effects of sex and time of day on metabolism and excretion of corticosterone in urine and feces in mice. Gen. Comp. Endocrinol. **130:** 267–278.
32. CARERE, C. et al. 2003. Fecal corticosteroids in a territorial bird selected for different personalities: daily rhythm and response to social stress. Horm. Behav. **43:** 540–548.
33. WASHBURN, B.E. & J.J. MILLSPAUGH. 2002. Effects of simulated environmental conditions on glucocorticoid metabolite measurements in white-tailed deer feces. Gen. Comp. Endocrinol. **127:** 217–222.
34. ZBINDEN, N. 1980. Zur Verdaulichkeit und umsetzbaren Energie von Tetraoniden-Winternahrung und zum Erhaltungsbedarf des Birkhuhns, *Tetrao tetrix* in Gefangenschaft mit Hinweisen auf Verdauungsversuche. Vogelwelt. **101:** 1–18.
35. DEGOLIER, T.F., S. MAHONEY & G.E. DUKE. 1999. Relationships of avian cecal lengths to food habits, taxonomic position, and intestinal lengths. Condor **101:** 622–634.
36. LIUKKONEN-ANTTILA, T. 2002. Kanalinnut tarhassa— ongelmina ravinto ja jopa syntyperä. (Adaptation of galliform birds to life in captivity: nutritional and genetic approach). Suom. Riista **48:** 46-54.
37. PAULI, H.R. 1974. Zur Winterökologie des Birkhuhns *Tetrao tetrix* in den Schweizer Alpen. Ornithol. Beob. **71:** 247–278.
38. PAULI, H.R. 1980. Nahrungsökologische Untersuchungen am Birkhuhn in der Schweizer Alpen. Beih. Veröff. Naturschutz Landschaftspflege Bad. Würt. **16:** 23–35.
39. PALME, R. & E. MÖSTL. 1997. Measurement of cortisol metabolites in feces of sheep as a parameter of cortisol concentration in blood. Int. J. Mamm. Biol. **62:** 192–197.
40. QUILLFELDT, P. & E. MÖSTL. 2003. Resource allocation in Wilson's storm-petrels *Oceanites oceanicus* determined by measurement of glucocorticoid excretion. Acta Ethol. **5:** 115–122.
41. FRIGERIO, D., E. MÖSTL & K. KOTRSCHAL. 2001. Excreted metabolites of gonadal steroid hormones and corticosterone in greylag geese, *Anser anser,* from hatching to fledging. Gen. Comp. Endocrinol. **124:** 246–255.
42. MÖSTL, E. et al. 2002. Measurement of cortisol metabolites in feces of ruminants. Vet. Res. Commun. **26:** 127–139.
43. PALME, R. et al. 1996. Extraction of infused ^{14}C-steroid hormones via feces and urine in domestic livestock. Anim. Reprod. Sci. **43:** 43–63.
44. SCHATZ, S. & R. PALME. 2001. Measurement of fecal cortisol metabolites in cats and dogs: a non-invasive method for evaluating adrenocortical function. Vet. Res. Commun. **25:** 271–287.
45. PALME, R. et al. 1999. Measurement of fecal cortisol metabolites in ruminants: a noninvasive parameter of adrenocortical function. Wien. Tierärztl. Monschr. **86:** 237–241.
46. DEHNHARD, M. et al. 2003. Measurement of plasma corticosterone and fecal glucocorticoid metabolites in the chicken,*Gallus domesticus,* the great cormorant, *Phalacrocorax carbo,* and the goshawk, *Accipiter gentilis.* Gen. Comp. Endocrinol. **131:** 345–352.
47. MORROW, C.J. et al. 2002. Fecal glucocorticoid metabolites as a measure of adrenal activity in dairy cattle. Gen. Comp. Endocrinol. **126:** 229–241.

Measuring Corticosterone Metabolites in Droppings of Capercaillies *(Tetrao urogallus)*

DOMINIK THIEL,[a] SUSANNE JENNI-EIERMANN,[a] AND RUPERT PALME[b]

[a]*Swiss Ornithological Institute, 6204 Sempach, Switzerland*

[b]*Institute of Biochemistry, Department of Natural Sciences, University of Veterinary Medicine, Vienna, Austria*

ABSTRACT: The capercaillie *(Tetrao urogallus)*, the largest grouse species in the world, is decreasing in numbers in major parts of its distribution range. Disturbances by human outdoor activities are discussed as a possible reason for this population decline. An indicator for disturbances is the increase of the glucocorticoid corticosterone, a stress hormone, which helps to cope with life-threatening situations. However, repeated disturbances might result in a long-term increase of the basal corticosterone concentration, which can result in detrimental effects like reduced fitness and survival of an animal. To measure corticosterone metabolites (CMs) noninvasively in the droppings of free-living capercaillies, first an enzyme immunoassay (EIA) in captive birds had to be selected and validated. Therefore, the excretion pattern of intravenously injected radiolabeled corticosterone was determined and ^{3}H metabolites were characterized. High-performance liquid chromatography (HPLC) separations of the samples containing peak concentrations revealed that corticosterone was extensively metabolized. The HPLC fractions were tested in several EIAs for glucocorticoid metabolites. The physiological relevance of this method was proved after pharmacological stimulation of the adrenocortical activity. Only the recently established cortisone assay, measuring CMs with a 3,11-dione structure, detected an expressed increase of concentrations following ACTH stimulation. To set up a sampling protocol suited for the field, we examined the influence of various storage conditions and time of day on concentrations of CMs.

KEYWORDS: capercaillie; noninvasive measurement; disturbance; conservation biology; ecology; field endocrinology

INTRODUCTION

The capercaillie *(Tetrao urogallus)*, the largest Galliform bird species in the Palearctic, is decreasing in numbers in most parts of its distribution range.[1] This decrease is particularly strong in central Europe, where many populations are already extinct or threatened with extinction.[2] Consequently, the capercaillie is classified as an endangered species on the Red List of Switzerland,[3] and as critically endangered in

Address for correspondence: Dominik Thiel, Swiss Ornithological Institute, 6204 Sempach, Switzerland. Voice: 0041 41 462 97 00; fax: 0041 41 462 97 10.

dominik.thiel@vogelwarte.ch

Germany[4] and Austria.[5] In Switzerland, numbers of capercaillies were reduced by more than half during the last 30 years, and the decline is still going on[6] The decrease is caused mainly by habitat loss and habitat degradation.[2,6,7] The decline of the capercaillie in areas with increasing human outdoor activities has led to the assumption that capercaillies are negatively affected by human disturbance.[2,8] First, human presence may lead to the avoidance of the disturbed area and to a change into a habitat of lower quality, a reaction that may negatively affect survival,[9] as was suggested for the capercaillies in Germany.[10] Second, human-induced disturbance can have substantial energetic consequences, particularly during winter, and may disturb the normal activity pattern and cause an energy deficit, as was shown for snow geese (*Chen caerulescens*[11]). If prolonged or repeated, human disturbance causes repeated physiological stress reactions, which may result in long-term negative effects, such as reduced reproduction or reduced immuncompetence.[12,13] However, in the Black Forest (Germany), some capercaillies live near paths, ski trails, or ski runs.[14] Whether these birds are stressed or adapted to human encounters is not known.

One means to determine whether an individual shows physiological stress reactions, even in the absence of an obvious behavioral response, is to analyze the levels of stress hormones released. The organism reacts to stress with the activation of the hypothalamo–pituitary–adrenal (HPA) axis, resulting in a release of glucocorticoids (in birds: corticosterone) into the blood, which triggers adjustments in physiology and behavior to help the organism survive.[13,15] Plasma glucocorticoid concentrations are therefore widely used to diagnose a physiological stress response.[16–19]

In free-living capercaillies, it is impossible to sample blood without causing severe stress by capture and handling. Therefore, the noninvasive method to quantify hormone production by the measurement of the hormone metabolites excreted in droppings collected in the field allows tracking the metabolic response to disturbances.[20] Because this method is feedback free, repeated measurements in the same individual are possible.[21]

Corticosterone metabolites (CMs) in droppings have been quantified in a few avian species.[22–28] Glucocorticoids are extensively metabolized before excretion, and native corticosterone was not found in the droppings.[28] The metabolism of corticosterone varies between species (and even gender), and therefore the best-suited immunoassay has to be chosen and the method validated for each species anew.[29,30] When using droppings from a free-living species, additional studies are necessary to assess whether the particular sampling conditions in the field, often suboptimal, affect the concentration of CMs in the droppings. Because of their low population densities and cryptic behavior, capercaillies cannot be followed and observed easily.[8] Droppings cannot be collected shortly after defecation, and therefore the exact time of defecation in droppings found in the field is unknown. Therefore, it needs to be evaluated whether and under which ambient conditions CMs degrade after voidance[31,32] and whether the excretion of CMs varies during the day.[33]

The aim of this study was thus to select and validate an enzyme immunoassay (EIA) for the quantification of CMs in capercaillie droppings collected in the field and to test the influence of various storage conditions. This was done by the following steps: In a radiometabolism study, the excreted CMs were characterized with reversed-phase high-performance liquid chromatography (RP-HPLC). Then, several antibodies were tested to select the best-suited EIA for the quantification of the CMs. Finally, the method was physiologically validated by inducing a corticosterone release through an adrenocorticotropic hormone (ACTH) injection. Experimentally, we tested the influence of time of day, temperature, and duration of storage on the concentration of CMs in droppings. This procedure is in accordance with the guidelines recommended previously.[34]

MATERIAL AND METHODS

Animals

All experiments with capercaillies were conducted with captive birds in the Max Planck Institute for Ornithology, Radolfzell, Germany. The birds were housed in outdoor aviaries, exposed to natural light and temperature conditions, but protected from precipitation. Each aviary had one male, whereas the much smaller females could freely move between aviaries through small openings. During the cold season, the period when we did our experiments, the birds were supplied with water, conifer needles, and maize ad libitum. The age of the birds was at least 1 year. Body mass ranged between 3 and 4 kg in males and 1.5 and 2 kg in females.

For the experiments, birds were transferred singly to aviaries with dimensions of $3 \times 3 \times 2.5$ m. The floor was covered with a plastic sheet so that droppings could be easily collected and the floor cleaned. Capercaillies feeding on needles and maize void brown, nearly nitrogen-free, dry, and compact droppings.[35,36] One female reacted very nervously when workers entered the aviary for sampling droppings during the radiometabolism study, causing her to void more liquid droppings. The droppings of the ceca, which are of a pasty consistency, a different shape, and a penetrating odor, were not sampled. They are voided only once per day, are hard to find in the field, and decay within a few days under frozen conditions.

Radiometabolism Study

The experiment was carried out during April 13–16, 2003. The day before the experiment, two males and two females were placed in separate aviaries to ensure that the droppings were not mixed up. The next day at 8 A.M., each bird was injected with 1.85 MBq (50 µCi) of radiolabeled corticosterone ($[1,2,6,7,-^{3}H(N)]$-corticosterone; specific activity 76.5 Ci/mmol, Perkin-Elmer Life Sciences, Boston, MA) dissolved in 0.5 mL of 0.9% NaCl solution containing 10% ethanol into the *vena ulnaris*.

One hour before injection, droppings of each bird were collected to determine background levels of radioactivity. During the first day after injection, droppings were collected every hour until 11 P.M. and frozen immediately at –23°C. Thereafter they were collected every second hour until 7 A.M. During the second day after injection, droppings were collected in 3-h intervals until 5 P.M. and during the third and fourth day at 8 A.M., 1 P.M., and 6 P.M.

Radioactivity in the droppings was measured as described previously.[28] In brief, 0.5 g of the homogenized sample was extracted with 5 mL of 60% methanol by shaking for 30 min. After centrifugation, aliquots (0.5 mL in duplicates) of the supernatant were mixed with 6 mL of a scintillation fluid (Quicksafe A, No. 100800; Zinsser Analytic, Maidenhead, UK) and measured in a liquid scintillation counter (Packard

Tri-carb 2100TR; Meriden, CT). Radioactivity was expressed as kilobecquerels (kBq) per kilogram of sample.

To characterize the excreted metabolites, we performed RP-HPLC separations with the samples containing the highest concentrations of radioactive metabolites. Cleanup and separations were performed as described previously.[28,33,37]

Analysis of Metabolites

The immunoreactivity of the 90 fractions eluted from the RP-HPLC (diluted 1:5 with assay buffer) was measured in different EIAs. Three assays were tested: an 11-oxoetiocholanolone,[38] a cortisone,[28] and a 20-dihydrocorticosterone EIA.[39] EIAs were performed as described previously[40] on microtiter plates coated with anti-rabbit immunoglobulin G (IgG) using a double-antibody technique and biotinylated steroids as labels.

Administration of ACTH

The physiological relevance of the method was evaluated by stimulating adrenocortical activity with ACTH. Three males and two females were injected with 1 mL (0.25 mg) of ACTH (Synacthen; Novartis Pharma Schweiz SA, Bern, Switzerland) in the *vena ulnaris*.

Experiments were conducted April 13–16, 2003 (one male, one female), and February 8–12, 2004 (two males, one female). To obtain basal values, we collected droppings 1 h before injection. After injection, droppings were collected for 4 days, once per hour for the first 48 h, and thereafter sampling frequency was gradually reduced to three times per day on the last day. In 2004, droppings were also collected over the 28 h before injection to assess the diurnal rhythm of CMs in droppings.

All samples were immediately stored at –23°C until analysis. The samples were extracted as previously described, except the 2004 samples, which were lyophilized before extraction. Accounting for the water loss of lyophilization, we used a reduced weight of sample for the extraction. Results were expressed in nanomoles per kilogram of fresh weight. The aliquots of the supernatant (diluted 1:10 with assay buffer) were measured with the cortisone EIA and the 11-oxoetiocholanolone EIA[28,38] to determine the levels of CMs in the droppings.

Effect of Storage Duration, Temperature, and Gender on CMs

Droppings were collected in three different aviaries within 24 h after cleaning the floor, each holding one male but several females, the latter freely moving between aviaries. Therefore, only male droppings were specific to the individual. One hundred grams of the droppings of each gender and aviary was pooled, homogenized, and divided into subsamples of 0.5 g each. Fifteen samples of each gender and aviary were frozen immediately. Five samples of each gender and aviary were stored at 8°C for 1, 7, and 21 days, respectively, and another five samples of each gender and aviary at 21°C for 1, 7, and 21 days, respectively. CM levels were analyzed following the procedure previously described.

We tested for effects of these factors on $\log_{10}$-transformed values of CMs in a split-plot ANOVA[41] with individuals nested within aviaries and single portions of droppings nested within individuals. Because gender varies among individuals, it is tested against the residual variation among individuals within aviaries. All other factors, including interactions, vary among portions of droppings and are therefore tested at the lowest level of the experiment, that is, using the residual mean square, which quantifies the variation among portions of droppings nested within individuals. To test the significance of particular group differences, we compared them with the least significant difference at an α level of 0.05. Analysis was conducted using the GenStat package.[42]

RESULTS

Radiometabolism Study

Excretion of ^{3}H-corticosterone started immediately because above-background radioactivity was measured in the first droppings collected 1 h after injection. All four birds had one main peak, followed by several smaller peaks. The radioactivity returned to background levels about 33 h after injection. Three individuals showed a broad peak between 1 and 4 h after injection (FIG. 1). Female B differed from the other three birds by showing one sharp peak of radioactivity 1 h and a second lower and broader peak 3–4 h after injection. This female was quite nervous and was the only bird with more liquid droppings (see MATERIALS AND METHODS).

The RP-HPLC separations of the droppings with the highest radioactivity revealed the presence of three to four major radioactive peaks (mainly between fractions 30 and 50) and many smaller peaks, indicating that several CMs were excreted

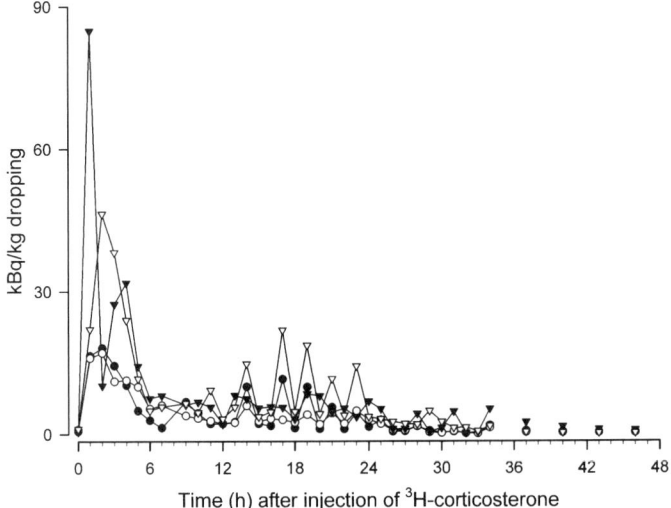

FIGURE 1. Time course of excretion of ^{3}H-corticosterone in droppings of 2 male and 2 female Capercaillies. The *circles* represent the two males, the *open triangles* female A and the *filled triangles* the results of female B.

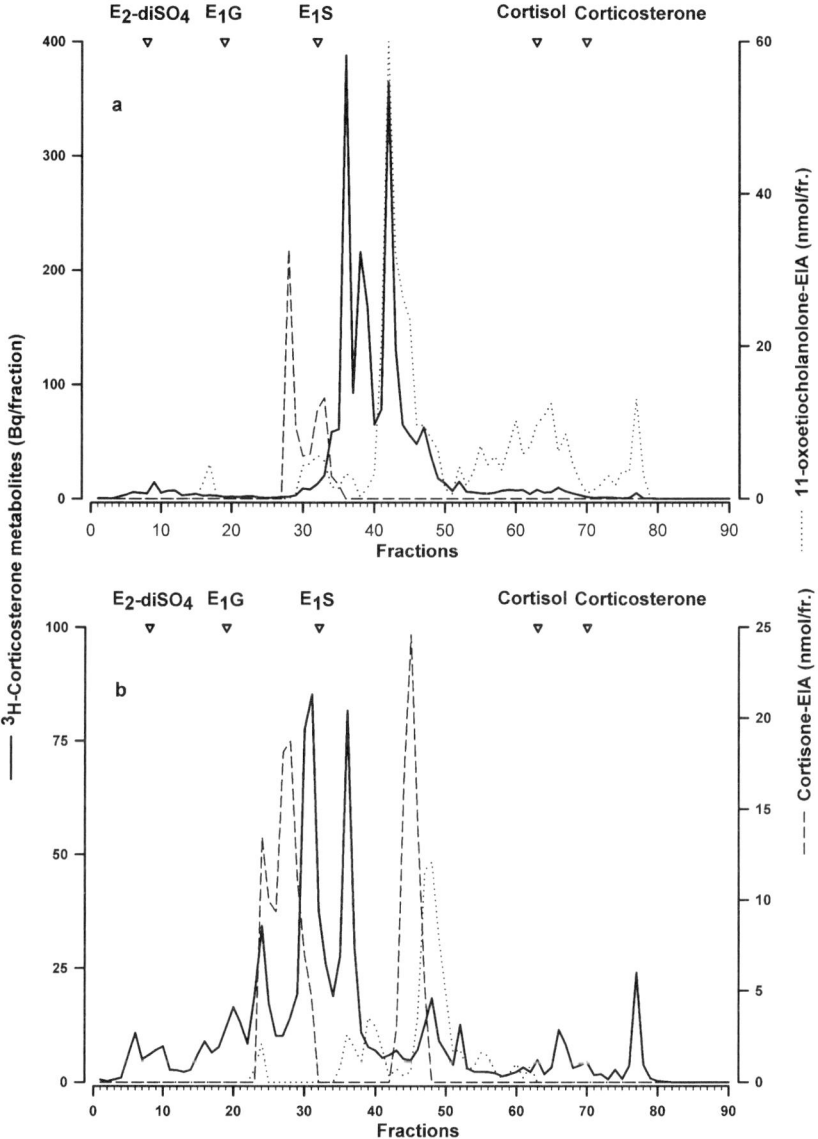

FIGURE 2. RP-HPLC separation of ^{3}H-CM in droppings of (**a**) a male and (**b**) a female Capercaillie. Radioactivity of each fraction was determined by liquid scintillation counting. The immunoreactivity of ^{3}H-CM with two different EIAs is given in nmol per fraction. *Open triangles* mark the approximate elution position of respective standards: E_2-diSO_4 = estradiol-17β-disulphate; E_1G = estroneglucuronide; E_1S = estronesulphate, cortisol, and corticosterone.

(FIG. 2). Only negligible amounts of radioactivity (if any) eluted at the position of corticosterone. In the HPLC fractions, the highest amounts of immunoreactivity were detected with the cortisone EIA, measuring metabolites with a common 3,11-dione structure, and the 11-oxoetiocholanolone assay, measuring CM with a 5-3-11-one structure (FIG. 2a and 2b).

Physiological Validation

To assess the biological relevance of the method, we injected ACTH in five birds to stimulate adrenocortical activity. The cortisone assay measured the highest concentrations in the samples. After injection, the concentration of CMs in the droppings increased sharply, peaked after 1–3 h, and returned to basal values between 4 and 13 h after injection. The increase in CMs above basal levels varied individually between a factor of about 5 up to a factor of about 60 (FIG. 3). In contrast, the 11-oxoetiocholanolone EIA detected lower concentrations and no clear response to the stimulation of the HPA axis (data not shown).

Concentrations of CMs excreted during the 28 h before injection of ACTH were regarded as basal values. Although some variation of the concentrations occurred, the pattern did not resemble a distinct diurnal rhythm (mean ± SD of the three individuals: 212 ± 70; 228 ± 178; 391 ± 192 nmol/kg of droppings).

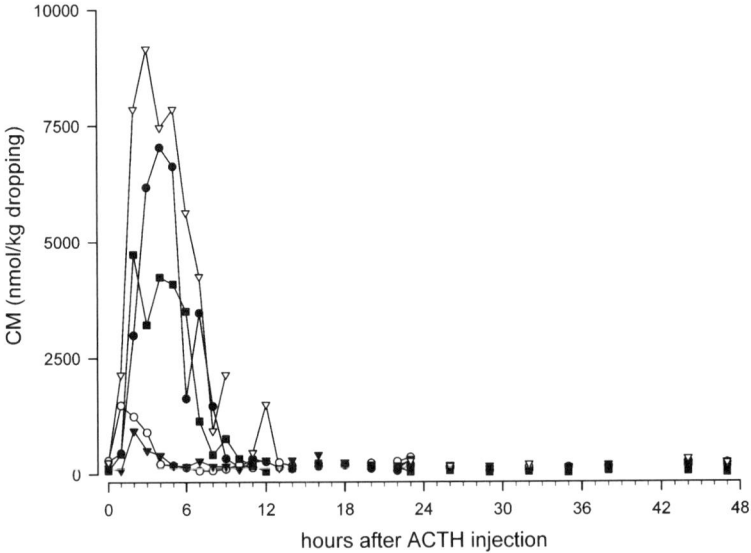

FIGURE 3. Concentrations of CM (3,11-dioxo-CM, nmol/kg droppings) after administration of ACTH in droppings of 3 male and 2 female Capercaillies.

TABLE 1. Effect of gender, ambient temperature, and storage time on the concentration of CM in droppings

Source of variation	d.f.	m.s.	F	P
Aviary stratum	2	0.084	0.28	
Individual per aviary stratum				
Gender	1	1.726	5.85	.137
Residual	2	0.295	12.86	
Units per individual per aviary				
Temperature	2	0.92605	40.37	<.001
Time	3	0.39155	17.07	<.001
Gender*temperature	2	0.22437	9.78	<.001
Gender*time	3	0.02807	1.22	.302
Temperature*time	1	0.32875	14.33	<.001
Gender*temperature*time	1	0.29662	12.93	<.001
Residual	251	0.02294		
Total	268			

NOTE: Nested ANOVA with ln[CM] as dependent variable and aviary, gender, temperatre, storage time, and interactions as independent variables. For details of the model, see the text. Units of the analysis consist of a single portion of the homogenized droppings.

ABBREVIATIONS: d.f. = degrees of freedom; m.s. = mean squares; F-value and significance P are given.

Effects of Storage

The average concentration of CMs in the samples frozen within 1 day after voidance was 537 ± 79 nmol/kg for males and 339 ± 50 nmol/kg of droppings for females. There were statistically significant effects of temperature and storage time on the concentration of CMs as well as of the gender–temperature, temperature–time, and gender–temperature–time interactions (TABLE 1). The main effects of gender and the gender–time interaction were not statistically significant.

Both temperature and storage time changed the concentration of CMs in the droppings compared with that in droppings frozen within 24 h after voidance (FIG. 4). In samples incubated at 8°C, CMs were not significantly different from those in frozen samples, but in samples exposed to a temperature of 21°C for 21 days, concentrations were significantly higher in both genders.

DISCUSSION

Radiometabolism Study

The radiometabolism study aimed to track the temporal excretion pattern of CMs and to select the droppings with the highest radioactivity to characterize the excreted CMs. The excretion of ^{3}H-corticosterone started immediately after injection. In three birds, high concentrations of radioactivity were measured during 1–4 h, ap-

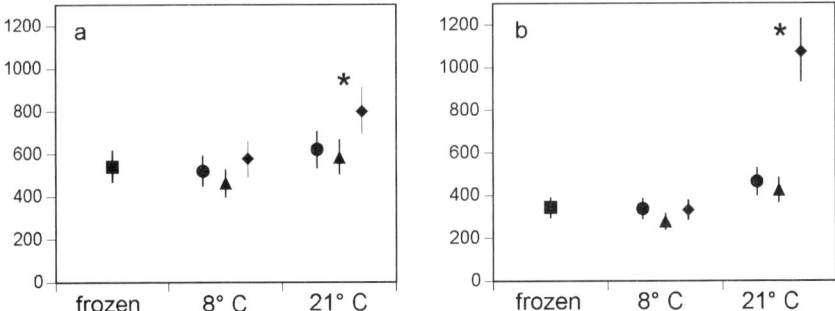

FIGURE 4. Concentrations (mean ± SE; $n = 15$) of 3,11-dioxo-CM (nmol/kg droppings) after storage at 8°C and 21°C for 1 (*circle*), 7 (*triangle*), and 21 (*rhombus*) days, respectively, for (**a**) males and (**b**) females. Control samples ($n = 45$) were frozen at −23°C within 24 hours after collection. Conditions marked with an asterisk differ significantly ($P < .05$) from control samples.

pearing as one broad peak. Only in one female could two peaks of ^{3}H-CM be observed, 1 h and 3–4 h after injection. This biphasic excretion pattern of female B agrees with that described for domestic chicken.[28] In the chicken, the first peak could be assigned to the metabolites excreted via the urine and the second peak to the excretion of the metabolites via the feces.[28] Because birds excrete urine and feces together via the cloaca, the two components cannot readily be separated, as is possible in mammals.[34,37,43]

The unimodal temporal excretion pattern found in the three normally excreting individuals might be explained by the particular diet and specific characteristics of digestion of capercaillies. During winter, capercaillies feed mainly on conifer needles (>90%). This fiber-rich diet quickly passes through the intestine, except the nutrient-rich liquid component, which enters the ceca. This diet is poor in water and protein;[8,36] thus, only small amounts of urine are excreted. Furthermore, urine is transported from the cloaca back to the ceca by reverse peristalsis for reabsorption of water.[44] Hence, urine is mixed with feces by back-and-forth movement in the rectum. Consequently, the CMs secreted by the kidney and the bile are mixed and excreted together, which results in a broad peak appearing relatively quickly. The fact that the single hen with the biphasic pattern excreted liquid (and therefore less mixed) droppings rather than the normal dry ones supports this idea.

Glucocorticoids are heavily metabolized before excretion.[34] In the RP-HPLC analysis of the droppings with the highest concentrations of radiolabeled CMs, corticosterone itself could be found only in very small amounts, if at all, in the birds. This finding is in accordance with other studies, which also found no corticosterone in droppings (black grouse *Tetrao tetrix*,[39] great tit *Parus major*,[27] European stonechats *Saxicola torquata*,[25] and domestic chicken[28]). The RP-HPLC analysis indicated that metabolites more polar than corticosterone were excreted. Because the three to four most prominent fractions were polar, conjugated or polar unconjugated metabolites were most abundant. These results agree with those in other bird species, in which radiometabolism studies were performed (chicken, geese, and black grouse), all showing mainly polar metabolites.[22,28,39]

Physiological Validation

Stimulation of the HPA axis by injection of ACTH promotes synthesis and secretion of glucocorticoids,[45] which finally results in an increase of the CM levels in the droppings.[28] Following ACTH injection, the capercaillies' CM concentrations increased after the first hour and peaked after about 3–4 h before they slowly returned to pretreatment levels. This finding agrees with the temporal pattern of excretion in the other grouse species, the black grouse.[39] In chickens[28] and spotted owls (*Strix occidentalis caurina*[26]), peak concentrations of CMs appeared 2 h after stimulation. The delayed excretion of peak concentrations in tetranoids is probably due to urine's being transported back from the cloaca to the ceca and therefore mixed with the feces, as discussed before.

The cortisone EIA, which detects metabolites with a 3,11-dioxo structure, and the 11-oxoetiocholanolone assay, measuring CM with a 5β-3α-11-one structure, showed a high immunoreactivity in the HPLC fractions. However, only the cortisone assay showed an expressed increase of concentrations following administration of ACTH. This result demonstrates the importance of a physiological validation. Because this group-specific antibody also proved best suited for chicken and black grouse,[28,39] it seems to be the most adequate for the determination of the CMs excreted by Galliforms.

Effects of Storage

The concentration of CMs in the feces can vary with storage conditions and time.[31,32] Because the exact time since voidance of capercaillie droppings found in the field is unknown, we measured CM concentrations in droppings that were exposed for different time spans and at different temperatures. This approach may help establish a sampling protocol suited for field conditions.

The results of the storage experiment showed that the concentration of CMs varied with gender, storage time, and temperature—but only at room temperature. In both genders the concentrations increased significantly after 21 days of storage at room temperature. In comparison with the frozen samples, no significant change could be observed in droppings stored at 8°C up to 21 days. Therefore, in studies conducted in the mountains during the winter with ambient temperatures near or below the freezing point, changes in the concentration of CMs after voidance will most probably be insignificant.

An increase in glucocorticoid metabolite concentrations was also observed in mammals, when feces were incubated at room temperature. These studies measured increased concentrations after a storage time of 4 h in cattle and 24 h in horses. Most probably, naturally occurring bacteria metabolize steroids with their enzymes.[31,32]

Gender differences in pattern and amounts of excreted fecal CMs were described previously for chickens.[28] Because of the various difficulties of raising and keeping capercaillies, only a few animals could be used in our study. Therefore, statistically significant gender differences could not be proved. However, gender should be taken into account when interpreting the results of studies of fecal samples. For the purposes of our sampling protocol in the field, this gender difference poses no problem, because the droppings of male capercaillies from roosting trees can be distinguished from those of females by their larger size.

Diurnal Pattern

The samples collected during 28 h revealed no diurnal rhythm. This finding agrees with those in the black grouse[39] but is in contrast to several studies measuring plasma corticosterone levels in other birds,[46,47] in which corticosterone concentration rises in the early morning before activity starts. Grouse species during winter typically have a particular activity pattern. They feed only twice per day, in the morning and afternoon, while they roost in snow burrows, below, or on trees during the rest of the day. This bimodal activity and feeding pattern is maintained in captivity. Hence, it is not surprising that their daily corticosterone pattern differs from that of other bird species kept in captivity.

As for the sampling protocol in the field, it seems that there is no need to observe time of day. However, because there are differences in corticosterone metabolite concentrations between droppings,[39] for example, possibly related to a pulsed excretion of corticosterone by the bile (Klasing, personal communication), several droppings from below a roost tree or a burrow should be collected and homogenized. With this method, a mean concentration of CMs excreted over a longer time span will be obtained.

CONCLUSION

This study demonstrates that the concentration of CMs can be reliably measured in droppings collected from capercaillies under field conditions during winter. CMs excreted in droppings are best determined with the cortisone EIA (measuring 3,11-dioxo CMs), as shown by the radiometabolism study and the ACTH validation experiment. Genders probably differ in the concentration of CMs in droppings, but in most cases droppings can be separated according to gender in the field. Droppings should be collected within about 20 days and should not be exposed to temperatures exceeding 8°C. The maximum age of the droppings can usually be determined when observing the last snowfall, and ambient temperatures exceeding 8°C are rare during winter in the habitat of the capercaillie. Several droppings should be analyzed together to avoid variations between droppings and to obtain a mean value over a longer period. Hence, the method to estimate whether capercaillies are physiologically stressed on the basis of contents of droppings seems to be promising.

ACKNOWLEDGMENTS

We thank Peter Berthold, Max Planck Institute for Ornithology, Radolfzell, for permission to do the experiments with his capercaillies; Alexandra Kuchar and Ruedi Meier for laboratory assistance; Marc Kéry for statistical advice; and Niklaus Zbinden and Lukas Jenni for valuable comments.

Birds were held and all experiments were carried out under license from the Regierungspräsidium Freiburg, G-03/21.

REFERENCES

1. HAGEMEIJER, W.J. & M.J. BLAIR. 1997. The EBCC Atlas of European Breeding Birds: Their Distribution and Abundance. Poyser. London.
2. STORCH, I. (compiler) 2000. Grouse status survey and conservation action plan 2000–2004. WPA/BirdLife/SSC Grouse Specialist Group. IUCN, Gland. Switzerland and Cambridge, UK and the World Pheasant Association, Reading, UK.
3. KELLER, V., N. ZBINDEN, H. SCHMID & B. VOLET. 2001. Rote Liste der gefährdeten Brutvogelarten der Schweiz. BUWAL-Reihe Vollzug Umwelt. Bundesamt für Umwelt, Wald & Landschaft. Bern & Schweizerische Vogelwarte, Sempach.
4. BAUER, H.-G. et al. 2002. Rote Liste der Brutvögel Deutschlands. Vogelschutz **39**: 13–60.
5. GEPP, J. 1994. Rote Listen gefährdeter Tiere österreichs. Grüne Reihe des Bundesministeriums für Umwelt, Jugend und Familie. Band 2. Styra, Graz.
6. MOLLET, P. et al. 2003. Verbreitung und Bestand des Auerhuhns Tetrao urogallus in der Schweiz 2001 und ihre Veränderungen im 19 und 20. Jahrhundert. Orn. Beob. **100**: 67–86.
7. ZEITLER, A. 1995. Skilauf und Rauhfusshühner. Orn. Beob. **92**: 227–230.
8. KLAUS, S., et al. 1989. Die Auerhühner. Die Neue Brehm-Bücherei. A. Ziemsen Verlag. Wittenberg Lutherstadt, Germany.
9. FRID, A. & L. DILL. 2002. Human caused disturbance stimuli as a form of predation risk. Conserv. Ecol. **6**: 11. [online].
10. MÜLLER, F. 1974. Territorialverhalten und Siedlungsstruktur einer mitteleuropäischen Population des Auerhuhns (Tetrao urogallus major). Ph.D. thesis, Marburg/Lahn, Germany.
11. BÉLANGER, L. & J. BÉDARD. 1990. Energetic cost of man-induced disturbance to staging Snow Geese. J. Wildl. Manage. **54**: 36–41.
12. GAVRIN, V.F. 1973. Die Wirkung des Angstfaktors auf die Produktivität von Federwildpopulationen. In Proc. XI. Int. Congr. Game Biol.: 401–403. Stockholm.
13. WINGFIELD, J.C. & L.M. ROMERO. 1999. Adrenocortical responses to stress and their modulation in free-living vertebrates. In Handbook of Physiology—Coping with the Environment. B. McEwen, Ed.: 211–239. Oxford University Press. New York.
14. VOLK, H. 1983. Winterport und Biotopschutz. Hat das Auerhuhn in Skilanglaufgebieten eine Chance? Nat. Landschaft **58**: 454–459.
15. SAPOLSKY, R.M. 1987. Stress, social status, and reproductive physiology in free-living baboons. In Psychobiology of Reproductive Behavior: An Evolutionary Perspective. D. Crews, Ed.: 291–392. Prentice Hall. Englewood Cliffs, NJ.
16. MUNCK A., P.M. GUYRE & N.J. HOLBROOK. 1984. Physiological functions of glucocorticoids in stress and their relation to pharmacological actions. Endocr. Rev. **5**: 25–44.
17. VON HOLST, D. 1998.The concept of stress and its relevance for animal behavior. Adv. Study Behav. **27**: 1–131.
18. SAPOLSKY, R.M. 1992. Stress, the Aging Brain, and the Mechanisms of Neuron Death. MIT Press. Cambridge, MA.
19. HOFER, H. & M.L. EAST. 1998. Biological conservation and stress. Adv. Study Behav. **27**: 405–525.
20. CREEL, S. et al. 2002. Snowmobile activity and glucocorticoid stress responses in wolves and elk. Conserv. Biol. **16**: 809–814.
21. PALME, R. et al. 2005. Stress hormones in mammals and birds: comparative aspects regarding metabolism, excretion and noninvasive measurement in fecal samples. Ann. N.Y. Acad. Sci. **1040**: 162–171.
22. KOTRSCHAL, K., K. HIRSCHENHAUSER & E. MÖSTL. 1998. The relationship between social stress and dominance is seasonal in greylag geese. Anim. Behav. **55**: 171–176.
23. NAKAGAWA, S., E. MÖSTL & J.E. WAAS. 2003. Validation of an enzyme immunoassay to measure faecal glucocorticoid metabolites from Adélie penguins (Pygoscelis adeliae): a non-invasive tool for estimating stress? Polar Biol. **62**: 491–493.

24. QUILLFELDT, P. & E. MÖSTL. 2003. Resource allocation in Wilson's storm-petrels *Oceanites oceanicus* determined by measurement of glucocorticoid excretion. Acta Ethol. **5:** 115–122.
25. GOYMANN, W., E. MÖSTL & E. GWINNER. 2002. Corticosterone metabolites can be measured noninvasively in excreta of European stonechats (*Saxicola torquata rubicola*). Auk **119:** 1167–1173.
26. WASSER, S.K. *et al.* 2000. A generalized fecal glucocorticoid assay for use in a diverse array of nondomestic mammalian and avian species. Gen. Comp. Endocrinol. **120:** 260–275.
27. CARERE, C. *et al.* 2003. Fecal corticosteroids in a territorial bird selected for different personalities: daily rhythm and response to social stress. Horm. Behav. **43:** 540–548.
28. RETTENBACHER, S. *et al.* 2004. Measurement of corticosterone metabolites in chicken droppings. Br. Poult. Sci. **45:** 704–711.
29. MÖSTL, E. & R. PALME. 2002. Hormones as indicators of stress. Dom. Anim. Endocrinol. **23:** 67–74.
30. TOUMA, C. & R. PALME. 2005. Measuring fecal glucocorticoid metabolites in mammals and birds: the importance of validation. Ann. N.Y. Acad. Sci. **1046:** 54–74.
31. MÖSTL, E. *et al.* 1999. Measurement of glucocorticoid metabolite concentrations in faeces of domestic livestock. Zentralbl. Veterinarmed. A **46:** 621–631.
32. HUNT, K.E. & S.K. WASSER. 2003. Effect of long-term preservation methods on fecal glucocorticoid concentrations of grizzly bear and African elephant. Physiol. Biochem. Zool. **76:** 918–928.
33. TOUMA, C., N. SACHSER, E. MÖSTL & R. PALME. 2003. Effects of sex and time of day on metabolism and excretion of corticosterone in urine and feces of mice. Gen. Comp. Endocrinol. **130:** 267–278.
34. PALME, R. 2005. Measuring fecal steroids: guidelines for practical application. Ann. N.Y. Acad. Sci. **1046:** 75–80.
35. GREMMELS, H.-D. 1986. Das Verdauungssystem der Rauhfusshühner—Eine Übersicht zur Physiologie und Mikroanatomie dieses Organsystems. Z. Jagdwiss. **32:** 96–104.
36. LIESER, M. 1996. Zur Nahrungswahl des Auerhuhns *Tetrao urogallus* im Schwarzwald. Orn. Beob. **93:** 47–58.
37. SCHATZ, S. & R. PALME. 2001. Measurement of faecal cortisol metabolites in cats and dogs: a non-invasive method for evaluating adrenocortical function. Vet. Res. Commun. **25:** 271–287.
38. MÖSTL, E. *et al.* 2002. Measurement of cortisol metabolites in faeces of ruminants. Vet. Res. Commun. **26:** 127–139.
39. BALTI, M., S. JENNI-EIERMANN, R. ARLETTAZ & R. PALME. 2005. A noninvasive technique to evaluate human-generated stress in the black grouse. **1046:** 81–95.
40. PALME, R. & E. MÖSTL. 1997. Measurement of cortisol metabolites in faeces of sheep as a parameter of cortisol concentration in blood. Int. J. Mammal. Biol. **62**(Suppl II): 192–197.
41. SNEDECOR, W.G. & W.G. COCHRAN. 1989. Statistical Methods. 8th ed. Iowa State University Press, Ames.
42. PAYNE, R.W. *et al.* 1993. Genstat 5, Release 3. Reference Manual. Clarendon Press. Oxford.
43. PALME, R., P. FISCHER, H. SCHILDORFER & M.N. ISMAIL. 1996. Excretion of infused ^{14}C-steroid hormones via faeces and urine in domestic livestock. Anim. Reprod. Sci. **43:** 43–63.
44. MOSS, R. & I. HANSSEN. 1980. Grouse nutrition. Nutr. Abstr. Rev. B **50:** 555–567.
45. WINGFIELD, J.C. & M. RAMENOFSKY. 1999. Hormones and the behavioral ecology of stress. *In* Stress Physiology in Animals. P.H.M. Balm, Ed.: 1–51. Academic Press. Sheffield, UK.
46. SMOAK, K.D. & G.P. BIRRENKOTT. 1986. Daily variation of corticosterone and thyroid hormones in the broiler cockerels. Poult. Sci. **65:** 197–198.
47. BREUNER, C., J.C. WINGFIELD & L.M. ROMERO. 1999. Diel rhythms of basal and stress-induced corticosterone in a wild, seasonal vertebrate, Gambel's white-crowned sparrow. J. Exp. Zool. **284:** 334–342.

Noninvasive Measures of Reproductive Function and Disturbance in the Barred Owl, Great Horned Owl, and Northern Spotted Owl

SAMUEL K. WASSER AND KATHLEEN E. HUNT

Center for Conservation Biology, Department of Biology, University of Washington, Seattle, Washington 98195-1800, USA

ABSTRACT: There is an urgent need for noninvasive methods to study reproduction and environmental stress in at-risk species such as the northern spotted owl (*Strix occidentalis caurina*). Two related owl species (barred owl and great horned owl) were used as surrogates to validate hormone assays for fecal metabolites of progesterone, 17β-estradiol, testosterone, and corticosterone. Infusions of radiolabeled hormones showed that the owls excreted most hormone within 6 h. Feces and urine contained roughly equal amounts of hormone, and most fecal hormone metabolites were quite polar. The testosterone and corticosterone assays in this study bound to the major excreted metabolites of these hormones, but two progesterone assays did not appreciably bind to the major progesterone metabolites. All assays showed excellent parallelism with hydrolyzed and unhydrolyzed samples and with previously dried or undried fecal samples. Thus, samples do not require hydrolysis or prior drying. Samples from a female barred owl had significantly higher fecal estrogen, lower fecal testosterone, and higher fecal estrogen/testosterone ratio than samples from two male barred owls. The fecal estrogen/testosterone ratio was the most accurate predictor of owl gender, particularly if two or more samples are available from the same individual. Fecal corticosterone metabolites also demonstrated considerable utility for wild northern spotted owls. Fecal glucocorticoid levels varied by gender and breeding stage, being highest in male northern spotted owls early in the breeding season and highest in females when nestlings were fledging. Collectively, these studies show that noninvasive fecal hormone measurements show great promise for noninvasive assessment of reproduction and stress in wild owls.

KEYWORDS: noninvasive methods; assay validations; HPLC; radiolabel infusion; fecal hormones; owls; stress; reproduction; conservation biology

INTRODUCTION

Northern spotted owl (NSO; *Strix occidentalis caurina*) populations in the Pacific Northwest have been in a serious decline since the late 1980s.[1] Considerable atten-

Address for correspondence: Samuel K. Wasser, Center for Conservation Biology, Department of Biology, Box 351800, University of Washington, Seattle, WA 98195-1800. Voice: 260-543-1669; fax: 206-616-2011.
 wassers@u.washington.edu

Ann. N.Y. Acad. Sci. 1046: 109–137 (2005). © 2005 New York Academy of Sciences.
doi: 10.1196/annals.1343.010

tion was drawn to this issue during the 1992 forest summit in Portland, Oregon, attended by both the President and Vice President of the United States. Despite this attention, the most recent 5-year demographic study indicated that the NSO continues to decline over most of its range and may soon go extinct in British Columbia, with populations in Washington State following closely.[1] Some of the most serious pressures facing the NSO are suspected to be loss of habitat from timber harvest, land conversion and uncontrolled fires, and the invasion of the barred owl from the eastern United States. However, the relative contributions of these pressures remain quite controversial despite hundreds of thousands of research hours and millions of dollars in funding. This controversy poses serious problems for the politically controversial management of this species because scientists and managers are unable to agree upon what factor(s) actually needs to be mitigated. Noninvasive physiological measures of stress and reproduction could be of great help. Such measures are likely to be especially valuable for partitioning stressors on a species such as the NSO that tends to show minimal behavioral response to external pressures despite undergoing a chronically precipitous decline. Given the right indices, physiological responses to external pressures are much harder to mask and less subject to interpretation than are behavioral responses.[2–4] Such measures are also relatively easy to noninvasively collect in a manner that does not require disturbing the owl.[5] For these reasons, noninvasive physiological measures can provide a useful tool for partitioning environmental impacts as well as tracking effectiveness of mitigation efforts. At least one application of these methods on the NSO found that fecal corticosterone metabolite measures were doubled in individuals whose home ranges were within 0.5 km of a major road or timber-management activities, compared with owls nesting farther away.[2] This study, as well as a related study[5] using such measures on California spotted owls (*Strix occidentalis occidentalis*), also showed that stress levels are likely to vary over the breeding season. Thus, fecal hormone measures may be able to not only partition relative impacts of stressors but also address the periods when such stressors are likely to have their greatest impacts. Such information would be enormously valuable to management endeavors.

Noninvasive hormone measures must be carefully validated if they are to be effective research or management tools. Validity is especially important for species at risk such as the NSO, whose proper management can have serious economic impacts. Fecal hormone measures have been extensively validated in mammals.[6–23] However, relatively few studies have validated such measures for avian species[10,24–29] (for a detailed list see Touma and Palme[43]). Such validations are especially critical in birds, in which feces, urine, and urates are all excreted together. Excretion lag times are likely to differ in avian species, as are bacterial actions on steroids, enterohepatic recirculation, and, ultimately, the metabolites being shed.[30] This study thoroughly validates fecal glucocorticoid and reproductive steroid hormone measures for the NSO, using the barred owl (*Strix varia*) and the great horned owl (*Bubo virginianus*) as surrogate species. We conducted the more invasive procedures on the barred and great horned owls, presuming that if these methods work well for both the closely related barred and distantly related great horned owls, they should also apply to the NSO. We provide evidence to this effect.

We used radiolabel infusion studies to determine the relative amounts of fecal corticosterone, estradiol, progesterone, and testosterone metabolites excreted in feces, urine, and urates; their excretion lag times; and the metabolic forms of their excretion. We also used these metabolites to determine the extraction method that resulted in the highest recovery of fecal metabolites from feces, the effectiveness of various antibodies used for the measurement of these hormones in radioimmunoassay, the relative polarity of the major fecal metabolites, the usefulness of hydrolysis to remove conjugate groups, and the necessity of drying samples prior to extraction and expressing results in per gram of dry weight, as is common practice for mammalian feces.[7] Finally, we assess the practical utility of these hormone assays: For the reproductive hormones, we investigated whether fecal estrogen and fecal testosterone can be used to noninvasively identify gender in barred owls. We also test whether fecal corticosterone metabolites vary with gender or breeding stage among wild NSOs.

METHODS

Animals and Housing

Six birds from the Virginia Wildlife Center were used for the initial validation studies: two female barred owls (#01 and #0045), three male barred owls (#02, #137, and #287), and one male great horned owl (GHO). A female (01) and male (02) barred owl and the male GHO were used for the radiolabel infusion study after being transferred to the Conservation and Research Center of the National Zoological Park. All three birds were unreleasable due to injuries. The remaining three barred owls (male #287, male #137, and female #0045) were sampled serially for measurement of fecal estradiol and testosterone metabolite levels. All six birds were wild-origin adults of unknown age.

All experiments took place in May–July 1993. Subjects were housed in single cages and fed a diet of mice. All experiments were conducted in accordance with all applicable local, state, and federal animal welfare regulations.

Validation comparisons using an NSO were also conducted at the Woodland Park Zoo in Seattle and on free-ranging NSOs in Oregon and Washington State. These data have been previously described,[2,10] but are presented with new analyses here.

Radiolabeled Hormone Infusions

In the first infusion experiment, female barred owl 01 and the GHO were each injected at 9:00 A.M. on May 18, 1993 with 5 µCi of ^{14}C-labeled testosterone and 50 µCi of ^{3}H-labeled corticosterone. In the second infusion experiment, female barred owl 01, male barred owl 02, and the GHO were each injected with 5 µCi of ^{14}C-labeled progesterone and 50 µCi of ^{3}H-labeled 17β-estradiol, at 9:00 A.M. on June 1, 1993. In each infusion, the two hormones were combined in 1 mL of saline and injected intramuscularly by a wildlife veterinarian. We used an order of magnitude less ^{14}C than ^{3}H steroids because of the higher specific activity of ^{3}H versus ^{14}C steroids. This approach also reduced dual counting error resulting from overlap in particle energy detection windows of these two isotopes.

After each infusion, all feces and urine were collected from individuals for a minimum of 45 h postinjection. Removable trays under each cage were inspected hourly for the presence of fecal and urine samples. Fecal samples were separated from urine/urates at the time of collection by sliding the membrane-encapsulated fecal pellets outward with a clean wooden stir stick. Fecal samples were then briefly rinsed with distilled water to remove any residual urine and urates and transferred to storage vials; urine/urates were pipetted off the collection tray into storage vials. The urine/urate mixture was separated by centrifugation (15 min at $1500 \times g$). All sample parts were then weighed, mixed well, and subsampled for measurement of radioactivity. Subsamples were placed in a scintillation vial with 5 mL of scintillation fluid and allowed to equilibrate for several hours to minimize quenching of counting efficiency. The samples were then counted for 3 min in Ultima Gold scintillation fluid (Perkin-Elmer, Wellesley, MA), using a Beckman liquid scintillation counter with a dual-label quench-corrected program for simultaneously counting ^3H and ^{14}C in the same sample.

Steroid Extraction Methods

Extraction recovery of metabolites from the four radiolabeled steroids was measured using the peak-radioactivity excretion samples for each steroid from the barred owl female 01 and the GHO. Peak samples were divided into three subsamples and extraction recoveries compared using 100% ethanol, 90% ethanol (10% distilled water), or 80% ethanol (20% distilled water). (Further subsampling was not possible due to the small size of owl fecal samples.) Extraction recovery should be highest using pure ethanol if the metabolites were excreted as nonpolar metabolites, whereas extraction recovery should progressively increase by increasing the percentage of water if the metabolites were more polar.

All subsamples were extracted using the method described by Wasser *et al.*[2] In brief, weighed subsamples were boiled with 10 mL of 100%, 90%, or 80% ethanol, for 20 min at 70°C and spun down; supernatant was collected. Pellets were vortexed with an additional 5 mL of the appropriate percentage of ethanol and centrifuged, and the supernatants were combined. Supernatants were then dried, resuspended in 1 mL of methanol, and then diluted in assay buffer. For the recovery measures, a subsample was transferred to scintillation vials containing 5 mL of scintillation fluid and compared with an identical amount of that sample counted prior to extraction, as previously described.

HPLC

High-performance liquid chromatography (HPLC) analyses were used to determine the relative polarity and assay affinities for metabolites of each of the four radiolabeled steroids in the peak excretion samples from female barred owl 01 (and male barred owl 02 for progesterone only). Prior to injection into the HPLC column, samples were passed through a 2-μm-pore-size filter followed by a C18 matrix column (Spice Cartridges; Rainin, Woburn, MA) and eluted off the column with 5 mL of 80% methanol, as described by Shackleton.[31] Fifty microliters of each purified sample was then injected onto a reverse-phase HPLC column (Microsorb C18, Rainin) and eluted at 1 mL/min over 80 min. For corticosterone and testosterone, we used a linear gradient from 20% methanol (80% water) to 40% methanol (60% water) over 80 min. For estradiol and progesterone, we used a linear gradient from 20%

methanol to 50% methanol over 80 min. These gradients were selected to maximize metabolite separation after initial testing showed that all major fecal metabolites from barred and GHOs were polar. (Pure steroid hormones would elute after 80 min in all four gradients.)

Radioactivity of HPLC fractions was measured in a Beckman scintillation counter, as previously described.

To assess immunoreactivity, we analyzed three more fecal samples collected from the same female barred owl before the infusion study (i.e., containing only endogenous, nonradioactive steroids). These samples were run with the HPLC gradients described earlier for progesterone, estradiol, testosterone, and corticosterone, and the resulting fractions were assayed for immunoreactivity with the corresponding assays (see below).

Hormone Assays

We tested two radioimmunoassays (RIAs) each for progestins, estrogens, androgens, and glucocorticoids. In each case, the two assays were compared for parallelism and accuracy in dried fecal samples, as well as for hydrolyzed versus unhydrolyzed fecal samples (see below). All comparisons were conducted on both barred owl and GHO feces. For each hormone, the best-performing assay was then used for all other analyses—for example, assays of HPLC fractions and comparison of dried and undried samples. We also tested assays for dihydrotestosterone, androstanediol glucuronide, pregnanediol glucuronide, and estrone sulfate; these assays did not appear useful for these owl species and results are not described here.

Progesterone

We tested a ^{3}H RIA and a ^{125}I RIA kit for progesterone metabolite measurement. The ^{3}H assay was an in-house assay in phosphate-buffered saline gel (PBSG; pH = 7). Bound from unbound hormone was separated using a 15-min incubation with 1.25% dextran-coated charcoal, followed by centrifugation (20 min at 1500 g), and a 3-min count of the supernatant in 5 mL of Ultima Gold scintillation fluid (Perkin-Elmer) in a Beckman LS6500 liquid scintillation counter. This assay used 10 pairs of standards spanning 7.5 to 480 ng/mL; see Wasser et al.[8] for further details and cross-reactivities. The ^{125}I progesterone assay was a coated-tube RIA kit (Coat-A-Count #TKPG-3; Diagnostic Products, Los Angeles, CA). The manufacturer's protocol was used with the 1-h incubation in a 37°C water bath. Cross-reactivities were 9.0% for 5α-pregnan-3,20-dione, 3.4% for 17α-hydroxyprogesterone, 3.2% for 5β-pregnan 3,20 dione, 2.2% for 11-deoxycorticosterone, and less than 1% for all other tested steroids.

Estrogens

We tested a ^{3}H 17β-estradiol RIA and a ^{125}I 17β-estradiol RIA kit for measurement of 17β-estradiol metabolites. The ^{3}H estradiol assay was an in-house assay using the same general protocol as the ^{3}H progesterone assay; see Risler et al.[32] for further details and cross-reactivities. The ^{125}I assay was a coated-tube RIA (Coat-A-Count kit #TKE2-4; Diagnostic Products). The manufacturer's protocol was used with a 1-h incubation in a 37°C water bath. Cross-reactivities were 10% for estrone,

4.4% for D-equilenin, 1.8% for ethinyl estradiol, 1.8% for estrone-β-D-glucuronide, and less than 1% for all other tested steroids.

Testosterone

We tested a ^{3}H RIA and a ^{125}I RIA kit for measurement of testosterone metabolites. The ^{3}H assay was an in-house assay using the same protocol as the other ^{3}H assays (see earlier). The testosterone antibody for this assay was #250 from the G. Niswender Laboratory (Colorado State University), diluted to 1:10,000 in PBSG. Cross-reactivities were 69% dihydrotestosterone, 22% for 3β-androstanediol, 14% for 3α-androstanediol 1% for androst-4-ene-3,17-dione, and <1.0% for all other tested steroids. (For further assay details, see Velloso *et al.*[33]) The ^{125}I testosterone assay was a double-antibody kit (Catalog #07-189102; MP Biomedicals [formerly ICN Biomedicals], Costa Mesa, CA). The manufacturer's protocol was used. (See the manufacturer's protocol for cross-reactivities.)

Corticosterone/cortisol

Two glucocorticoid assays were tested: a corticosterone ^{125}I RIA kit and a cortisol ^{125}I RIA kit. The corticosterone assay was a ^{125}I double-antibody radioimmunoassay kit (#07-120103; MP Biomedicals); see Wasser *et al.*[10] for details and cross-reactivities. The cortisol kit was a coated-tube RIA (GammaCoat Cortisol, Catalog #CA1529; DiaSorin [formerly IncStar], Stillwater, MN). The manufacturer's protocol was used. Cross-reactivities were 77% for prednisolone, 43% for 6-methylprednisolone, 6.3% for 11-deoxycortisol, 1.2% for 17-hydroxyprogesterone, and less than 1.0% for all other tested steroids.

Hydrolyzed versus Unhydrolyzed Samples

The HPLC and extraction experiments suggested that all metabolites were quite polar (see RESULTS), and hence likely to be excreted as conjugates. We therefore conducted parallelism studies on both hydrolyzed and nonhydrolyzed samples to determine whether samples needed to be hydrolyzed prior to extraction to improve assay efficiency. Hydrolysis was conducted using the acid solvolysis method described by Hodges *et al.*,[34] with parallelism tested with sample extracts serially diluted from 1:1 to 1:64. These tests were done using one large fecal sample selected from each radiolabel infusion from each bird.

Dried versus Undried Samples

Studies in mammals suggest the importance of freeze-drying samples prior to extraction to improve extraction efficiency and control for diet-related changes in hormone excretion.[7,35] However, the small size of owl fecal mass and rapid steroid excretion rates may make their concentrations less susceptible to diet-related changes. Very small samples have also been reported to overestimate hormone concentrations,[2,5] presumably due to underestimation of initial fecal weights (see also DISCUSSION). Drying the samples prior to analysis would further decrease starting sample weights, potentially exacerbating such impacts. We therefore compared results from parallelism studies of dried and undried samples for progesterone and corticosterone. Samples were dried within 24 h in a centrifugal evaporator (Thermo Savant [formerly Savant], Holbrook, NY). Hormone concentrations of dried and undried extracts of the same samples collected over 42 days from three barred owls (female 0045, male 137, and male 287) were also compared for both progesterone and corticosterone metabolites.

Sexing with Fecal Estrogen/Testosterone Ratios

To assess the utility of fecal estrogen (E) and fecal testosterone (T) metabolites for noninvasive assessment of owl reproductive status, we examined the accuracy of fecal E, T, and the E/T ratio to discriminate male from female samples. Fifty-one samples collected over 45 h from one female and two male barred owls (female 0045, male 137, and male 287; $n = 17$ samples from each bird) were assayed for E and T metabolites. Fecal E/T ratios were calculated for each sample, and results were compared with fecal estrogen and fecal testosterone concentrations separately as predictors of owl gender. Data were log-transformed and compared across the three birds using an ANOVA.

Seasonal Profiles of Fecal Corticosterone in Wild NSOs

The biological validity of the ^{125}I corticosterone assay for NSOs has been reported elsewhere.[2,10] These analyses are supplemented here by a new analysis of data from wild NSOs, previously presented in the work of Wasser *et al.*[2] A general linear model analysis was used to examine gender differences in stress profiles across the breeding season of more than 200 fecal samples collected throughout Washington and Oregon in 1994 during owl censuring by crews from the Washington Department of Fish and Wildlife, U.S. Fish and Wildlife Service, U.S. Forest Service, Washington Department of Natural Resources, and private industry scientists. Fecal pellets were removed as previously described (see METHODS) and placed in a 7-mL vial containing 2.5 mL of 90% ethanol as a preservative. Samples were then transported at ambient temperature within 2–5 h and stored frozen until lyophilized for analysis.[36] Gender-specific temporal patterns were further delineated by simple regression analyses conducted separately on each sex.

RESULTS

Radiolabeled Hormone Infusions

For all four hormone infusions, barred owls typically excreted peak levels of radioactivity in both feces and urine within 6 h after injection, generally in the first or second sample, declining to baseline within approximately 24 h (FIGS. 1–4). The GHO showed a similar pattern, but with considerable enterohepatic recycling of metabolites reappearing between 10 and 13 h. When data are expressed on a per-gram or per-milliliter basis, two to three times more radioactivity was excreted per unit of feces versus urine. However, the absolute level of radioactivity excreted in urine was often equivalent to or higher than that excreted in feces (due to a greater mass of urine than feces). Negligible amounts of hormone were recovered from urates, presumably due to residual amounts of urine still mixed with the urates.

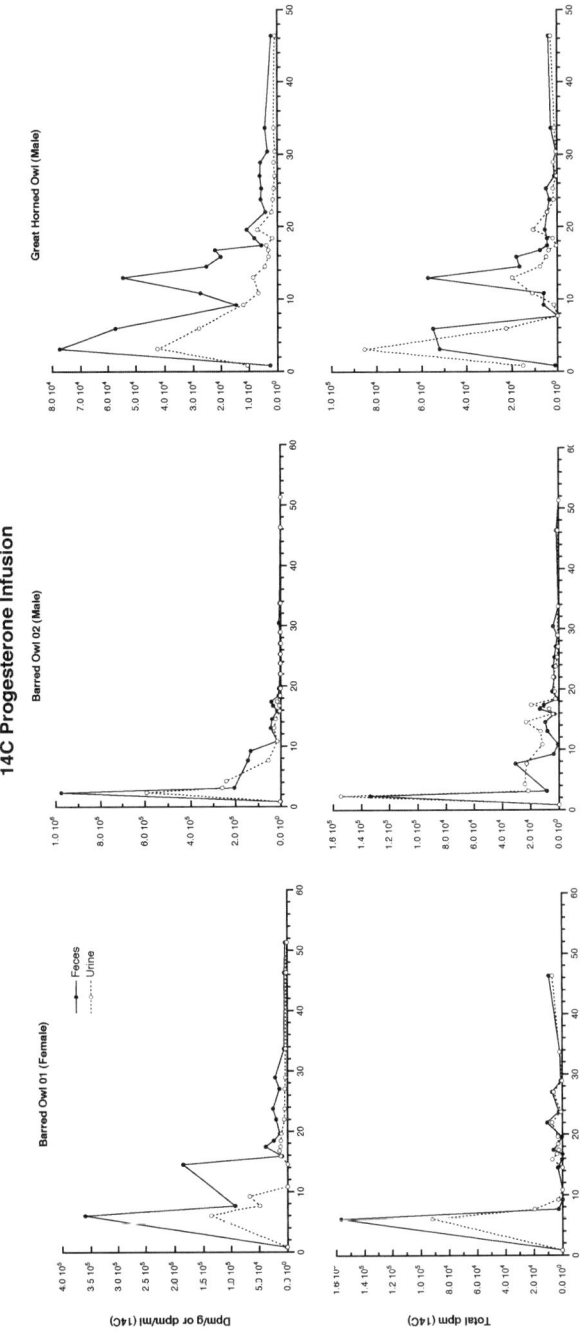

FIGURE 1. Excretion profiles of ^{14}C-labeled progesterone infused into two barred owls and one GHO. Data are shown for feces and urine, expressed either as dpm/g or dpm/mL (**top panel**) or as total dpm per fecal or urine sample (**bottom panel**).

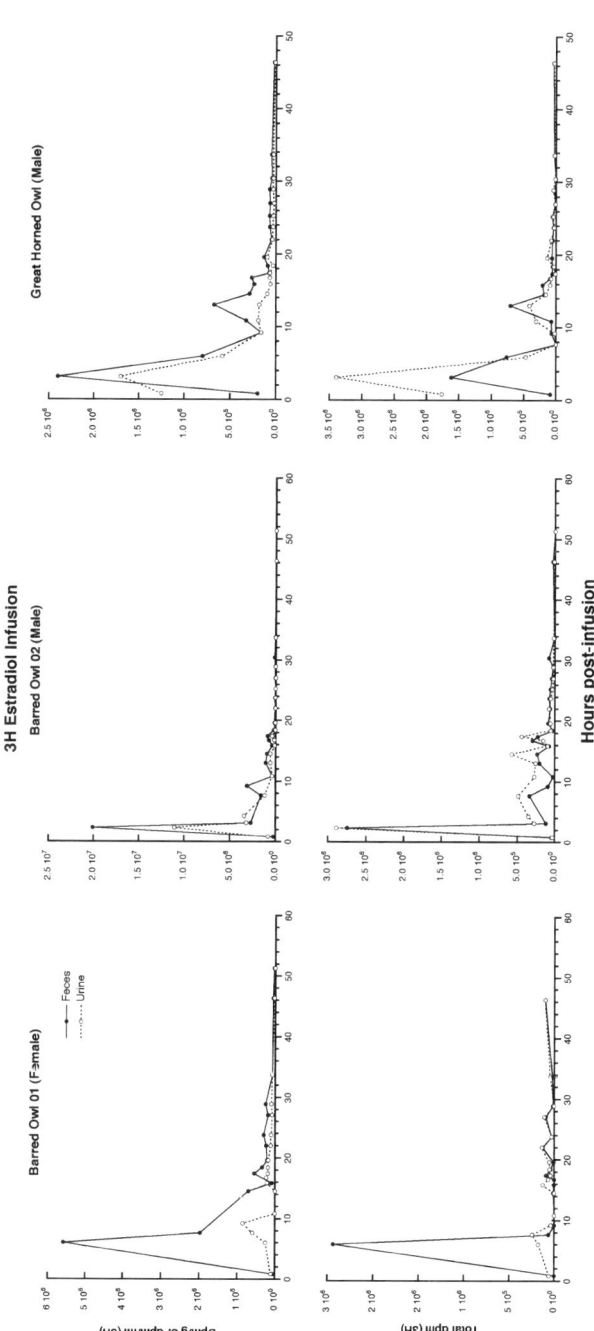

FIGURE 2. Excretion profiles of 3H-labeled 17β-estradiol infused into two barred owls and one GHO. Data are shown for feces and urine, expressed either as dpm/g or dpm/mL (**top panel**) or as total dpm per fecal or urine sample (**bottom panel**).

Progesterone

In all three owls, peak excretion of ^{14}C progesterone fecal and urinary metabolites, when expressed on a per-gram or per-milliliter basis, occurred in the second sample (6.0 h postinfusion for the female barred owl, 2.3 h for the male barred owl, and 3.1 h for the GHO; FIG. 1). The GHO also exhibited a prominent secondary peak at 13 h postinfusion in both feces and urine, likely due to enterohepatic recirculation of hormones. In feces, this second peak had slightly higher total disintegrations per minute (dpm) than the 3.1-h or 5.9-h samples of the earlier GHO peak. The female barred owl also exhibited a secondary peak in fecal dpm per gram at 14.5 h, but this was not apparent in total dpm excretion.

Estradiol

Peak excretion of ^3H estradiol metabolites occurred in the second fecal samples from the male barred owl (2.3 h postinfusion) and the GHO (3.1 h; FIG. 2). In the female barred owl, the fecal ^3H peak was also in the second sample (6 h), but the urinary ^3H peak was in the third sample (7.7 h). The GHO again exhibited a secondary excretion peak in both feces and urine at 13 h.

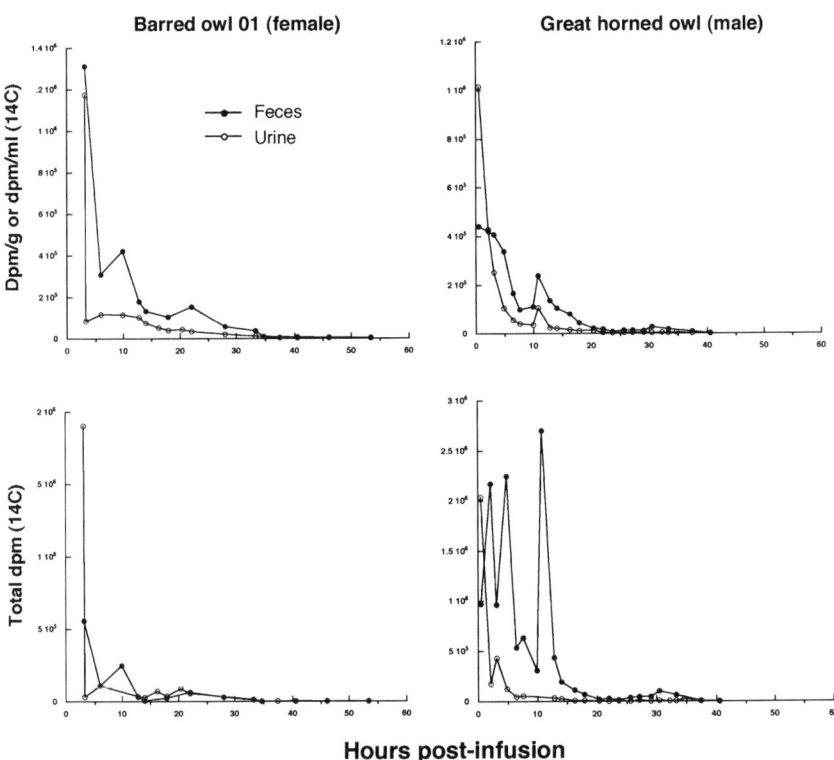

FIGURE 3. Excretion profiles of ^{14}C-labeled testosterone infused into a barred owl and a GHO. Data are shown for feces and urine, expressed either as dpm/g or dpm/mL (**top panel**) or as total dpm per fecal or urine sample (**bottom panel**).

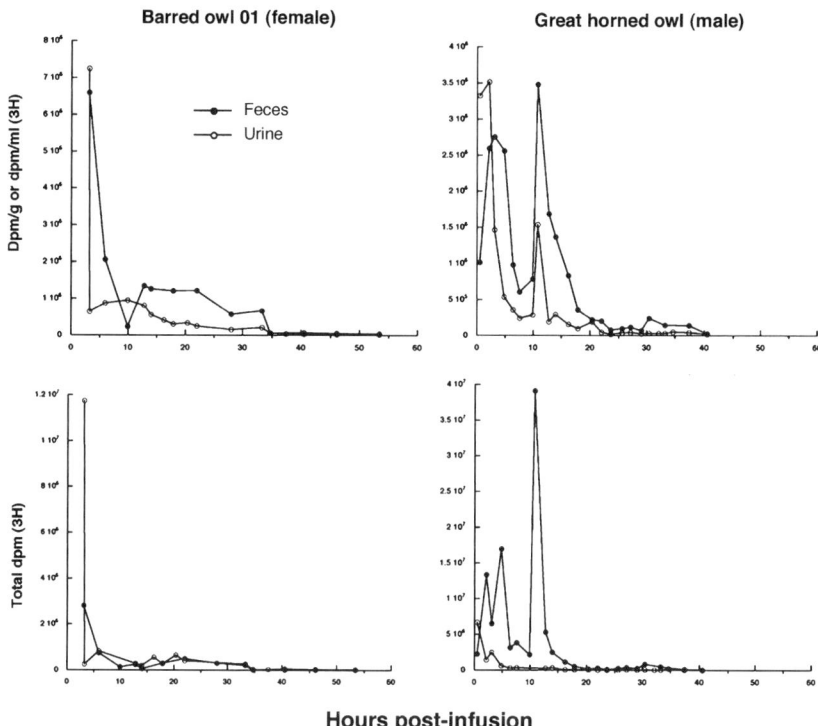

FIGURE 4. Excretion profiles of ^{3}H-labeled corticosterone infused into a barred owl and a GHO. Data are shown for feces and urine, expressed either as dpm/g or dpm/mL (**top panel**) or as total dpm per fecal or urine sample (**bottom panel**).

Testosterone

Peak excretion of ^{14}C testosterone metabolites occurred in the first fecal sample for the female barred owl (3.1 h; F$_{IG}$. 3), with a small secondary peak at 9.9 h. Both owls exhibited small secondary peaks in dpm per gram of fecal sample. For the GHO, total dpm in feces varied widely over the first 12 h postinfusion, peaking at 10.8 h. Urinary excretion for the GHO peaked in the first sample (0.5 h).

Corticosterone

The female barred owl excreted peak levels of ^{3}H corticosterone metabolites in the first fecal and first urine samples (3.1 h; F$_{IG}$. 4). The GHO again exhibited two prominent peaks in fecal excretion, at 3.1 h and 10.75 h, with peak total dpm excretion in the second peak (10.75 h). In urine, the GHO's peak total dpm excretion was at 3.1 h.

Steroid Extraction Methods

Recovery of radioactive hormone metabolites averaged 73% for 100% ethanol, 92% for 90% ethanol, and 98% for 80% ethanol. Although only a single sample from

TABLE 1. Percentage of radioactive hormone metabolites recovered from feces of barred owl and great horned owl following radiolabeled hormone infusions

	Ethanol percentages		
	100%	90%	80%
Barred owl			
Corticosterone	82	98	99
Testosterone	88	98	99
Estradiol	82	97	99
Progesterone	67	92	98
Great horned owl			
Corticosterone	68	90	98
Testosterone	71	85	97
Estradiol	60	82	97
Progesterone	66	93	99

NOTE: Steroids were extracted with 20-min boiling in 100% ethanol, 90% ethanol (10% distilled water), or 80% ethanol (20% distilled water). Total radioactivity in the ethanol extracts was compared to radioactivity originally present in an unextracted, well-mixed subsample of the same fecal sample.

each owl was tested for each hormone, all four hormones gave similar results for both species (TABLE 1). We therefore used 80% ethanol to extract samples for all other analyses.

Initial Assay Validations

All assays except the cortisol assay showed good parallelism and accuracy for fecal hormone metabolites of barred owls and GHOs. On the basis of these results, we selected the ^{3}H progesterone assay, the ^{125}I estradiol assay, the ^{3}H testosterone assay, and the ^{125}I corticosterone assay for all further analyses (see below for parallelism of selected assays).

HPLC

Progesterone

HPLC fractions showed at least six prominent progesterone metabolites based on ^{14}C radioactivity per fraction of the peak-radioactivity, postinfusion fecal sample from the female barred owl (FIG. 5). Five of these peaks were also seen in a postinfusion fecal sample from the male barred owl (FIG. 5, lower panel). The progesterone assay detected several prominent immunoreactive peaks that did not correspond to these ^{14}C peaks, indicating that the progesterone antibody may not have bound well to the most abundant progesterone metabolites (FIG. 5, top panel). HPLC fraction slippage between the two HPLC runs may also have obscured matches in peaks; however, the major peaks do not align well even when attempts are made to adjust for possible slippage.

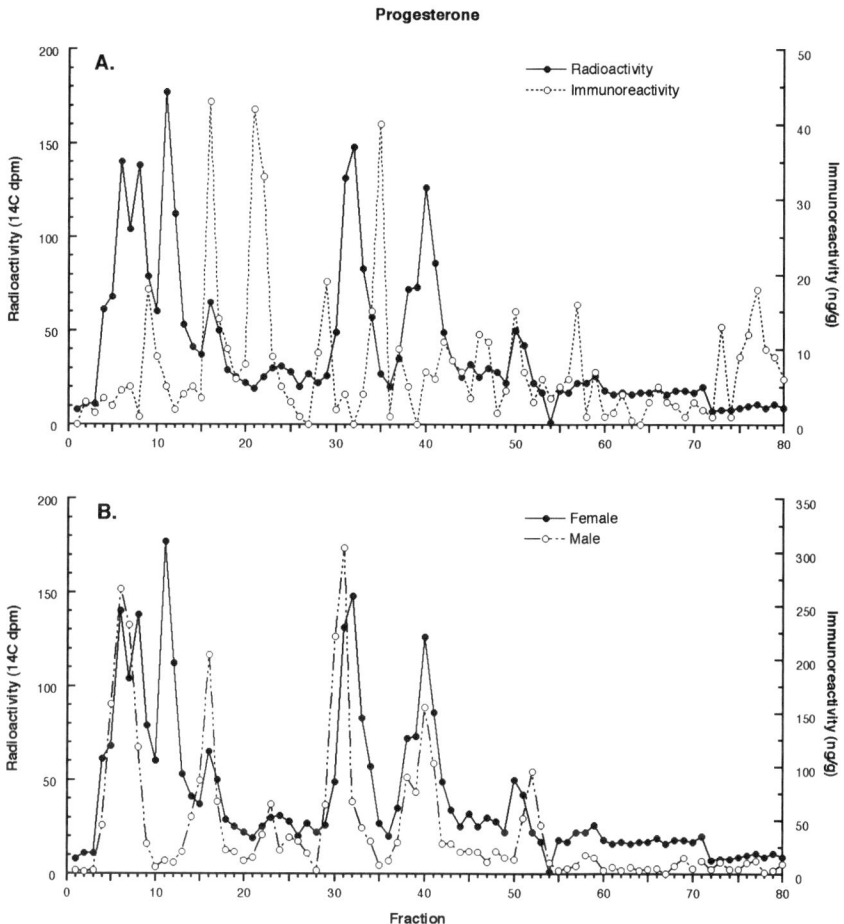

FIGURE 5. (**A**) HPLC elution profiles of radioactive metabolites (*solid lines*) and their antibody-specific immunoreactivities (*dashed line; top panel* only) following infusion of radiolabeled progesterone in a female barred owl. (**B**) Comparison of the female's radioactive metabolite elution profile (*solid lines*) with those from a similarly infused male barred owl (*broken line*).

Estradiol

HPLC fractions from the peak-radioactivity fecal sample after ^{3}H-estradiol infusion had two major and two minor peaks in ^{3}H radioactivity (FIG. 6, top panel). All metabolites in the owl samples were quite polar. However, the very early peak in fraction 5 is most likely the result of a solvent front containing ^{3}H that separated from the labeled molecule. The few radioactive metabolites we found are also consistent with radio- and immunoreactive studies of estrogens in other species.[8,37,38] The small number of estradiol metabolites is presumed to occur because estradiol is at the end of the metabolite pathway for all related steroids. (We attempted to assay

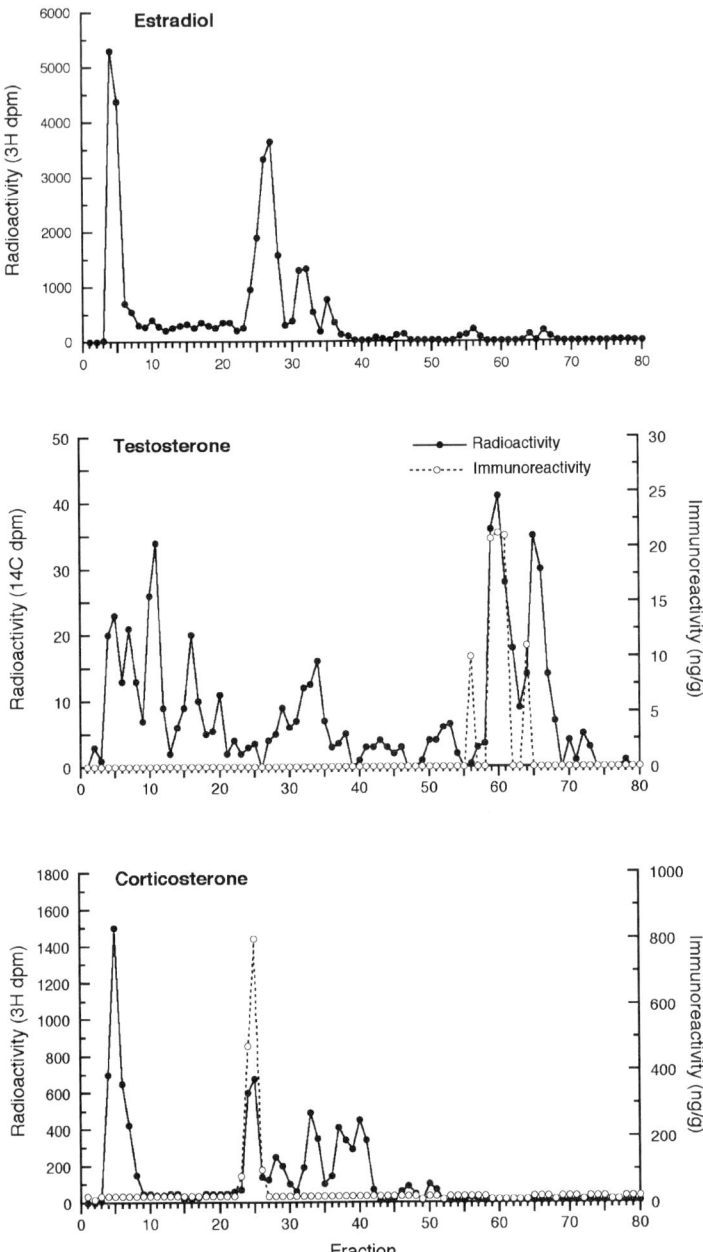

FIGURE 6. HPLC elution profiles of radiolabeled metabolites (*solid lines*) and their antibody-specific immunoreactivity (*dashed lines*) excreted in feces following radiolabel hormone infusions of ^{3}H 17β-estradiol (**top panel**), ^{14}C testosterone (**middle panel**), and ^{3}H corticosterone (**bottom panel**) in a female barred owl. *Solid lines* show radioactivity of HPLC fractions; *dashed lines* show immunoreactivity.

these fractions for 17β-estradiol immunoreactivity, but results were below assay sensitivity threshold.)

Testosterone

HPLC fractions from a fecal sample after ^{14}C-testosterone infusion show two major and relatively nonpolar peaks in ^{14}C radioactivity, and several smaller, more polar peaks (FIG. 6, middle panel). The ^{3}H testosterone assay revealed one major and two minor peaks in immunoreactivity, all quite nonpolar. The major immunoreactive peak aligns perfectly with the major radioactive peak, indicating that this assay detects the dominant testosterone fecal metabolite in barred owls.

Corticosterone

HPLC fractions from a fecal sample after ^{3}H-corticosterone infusion show a large, highly polar ^{3}H peak in fraction 5, which is probably a ^{3}H-containing solvent front, as described for estradiol. This was followed by a set of smaller, less polar ^{3}H

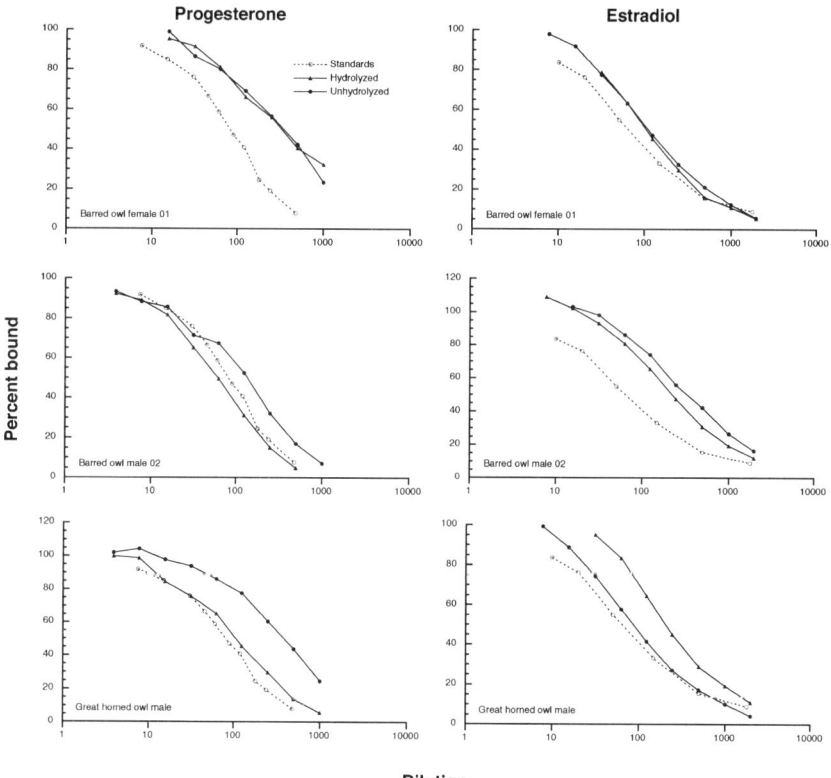

FIGURE 7. Assay parallelism in hydrolyzed versus nonhydrolyzed fecal samples for progesterone and 17β-estradiol assays (see text).

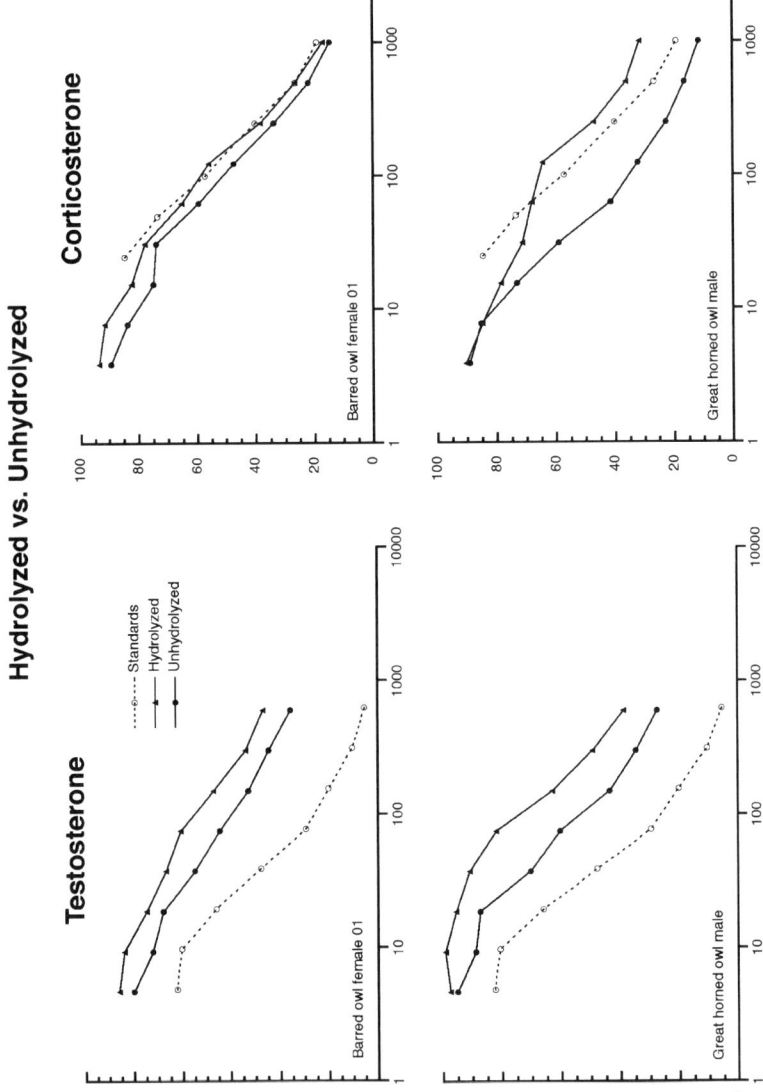

FIGURE 8. Assay parallelism in hydrolyzed versus nonhydrolyzed fecal samples for testosterone and corticosterone assays (see text).

peaks (FIG. 6, bottom panel). The ^{125}I corticosterone antibody binds well to the first and largest of these radioactive peaks.

Hydrolyzed versus Unhydrolyzed Samples

Overall, parallelism was generally good for both hydrolyzed and unhydrolyzed fecal metabolites of all four hormones from both species (FIGS. 7 and 8). Hydrolysis did not markedly improve parallelism for any of the four hormone assays and appeared to decrease parallelism in at least one case (corticosterone in the GHO; FIG. 8).

Dried versus Undried Samples

Generally, dried and undried fecal samples both had good parallelism for all four hormone assays (FIGS. 9 and 10), indicating that either method is acceptable. Profiles of dried versus undried assay results were nearly identical for progesterone as well as for corticosterone metabolites for the three barred owls (FIG. 11). However, occasionally the dried samples produced extremely high values not seen in wet samples or in serum. These are presumed to be outliers resulting from inaccurate (underestimated) weights of dried samples (see DISCUSSION).

Sexing with Fecal Estrogen/Testosterone Ratios

Fecal E/T ratios were significantly higher in the female barred owl than in the two male barred owls (FIGS. 12 and 13; ANOVA, $F_{2,48} = 25.362$; $P < .0001$; post hoc test with Fisher's PLSD, $P < .0001$ for female vs. each male). The two males were not significantly different from each other (Fisher's PLSD, $P = .9158$). The two males had significantly higher fecal testosterone levels than did the female (ANOVA, $F_{2,48} = 13.604$, $P < .0001$). Post hoc tests (Fisher's PLSD) revealed that male 287 had significantly higher fecal testosterone than male 137 ($P = .0101$) and female 0045 ($P < .0001$), and male 137 also had higher fecal testosterone than the female ($P = .0145$). The three birds also had significantly different fecal estrogen levels (ANOVA, $F_{2,48} = 4.961$, $P = .0110$). Post hoc tests (Fisher's PLSD) revealed that the female had significantly higher estrogen than male 137 ($P = .0029$), but not higher than male 287 ($P = .0736$), and the two males did not have different fecal estrogen levels ($P = .1976$).

Although all three measures (fecal T, fecal E, and fecal E/T ratio) show promise for sexing unknown birds, in our data set the fecal E/T ratio performed better than fecal E or fecal T alone. The fecal E/T ratio performed best at a threshold E/T ratio of 300, above which all samples were female samples, and below which there were seven female samples misidentified as male (13.7% inaccurate calls). Fecal T had a broad area of overlap in the lower range of values, with many male samples having low T; best results were obtained using a threshold value of 15 ng/g, above which one female sample was misidentified as a male and below which 19 male samples were misidentified as females (39% inaccurate calls). Fecal estrogen also showed very broad overlap, with a male sample having the highest fecal estrogen level and a female sample having the lowest estrogen level. Thus, it was not possible to identify either sex with complete accuracy using fecal estrogen. The best sex discrimination for fecal estrogen was at a threshold value of 2 ng/g, below which 6 female

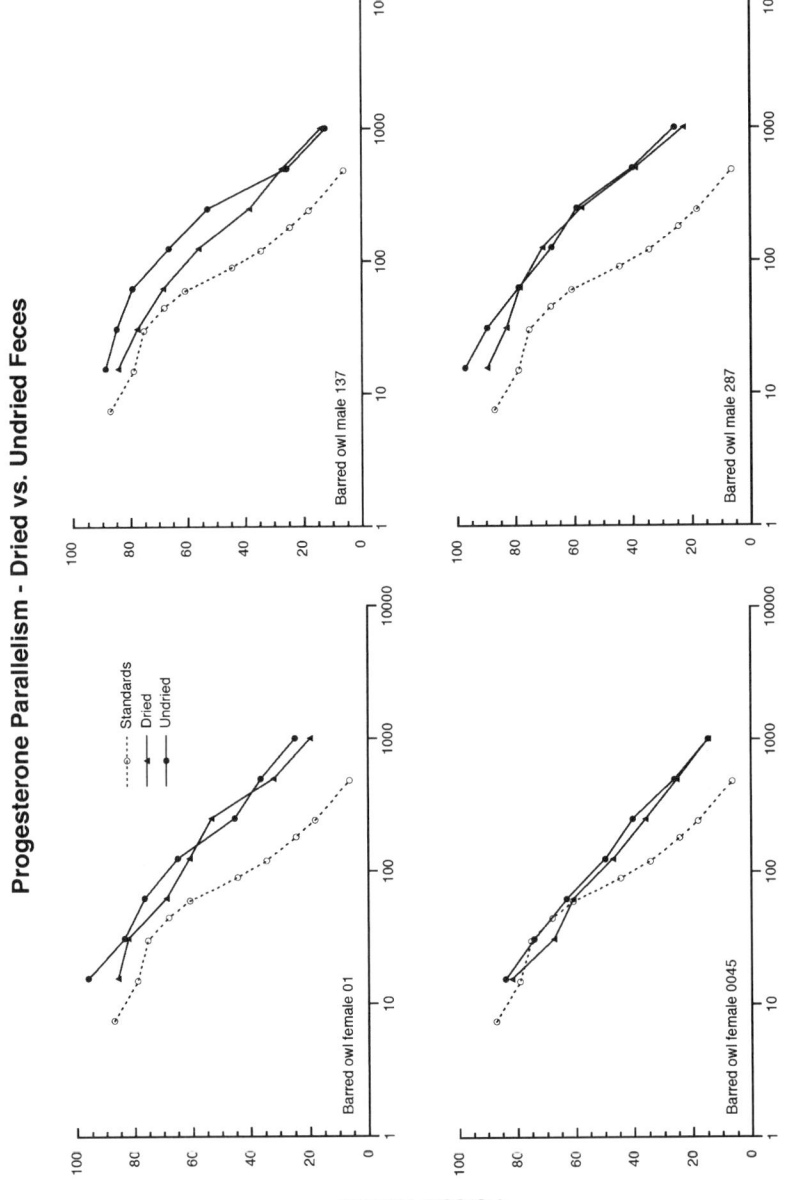

FIGURE 9. Progesterone assay parallelism in dried versus undried fecal samples from two female and two male barred owls.

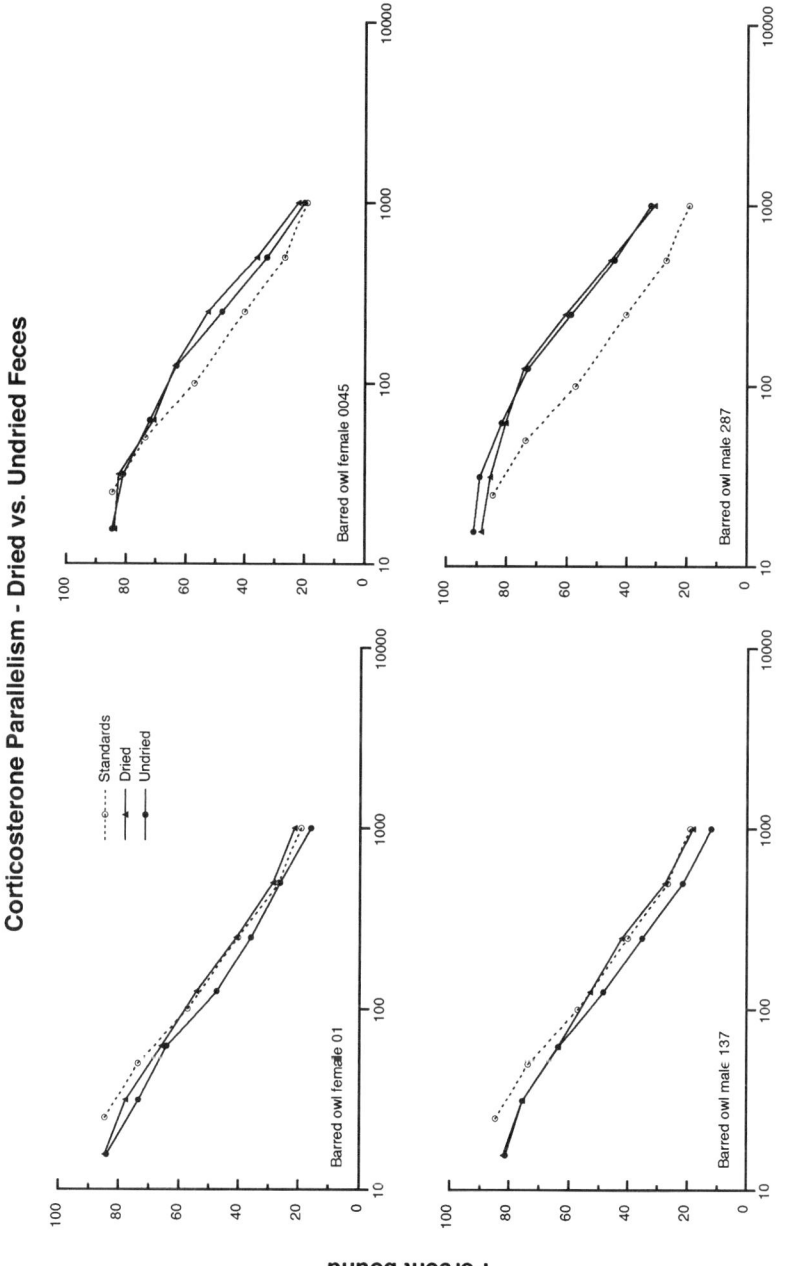

FIGURE 10. Corticosterone assay parallelism in dried versus undried fecal samples from two female and two male barred owls.

128 ANNALS NEW YORK ACADEMY OF SCIENCES

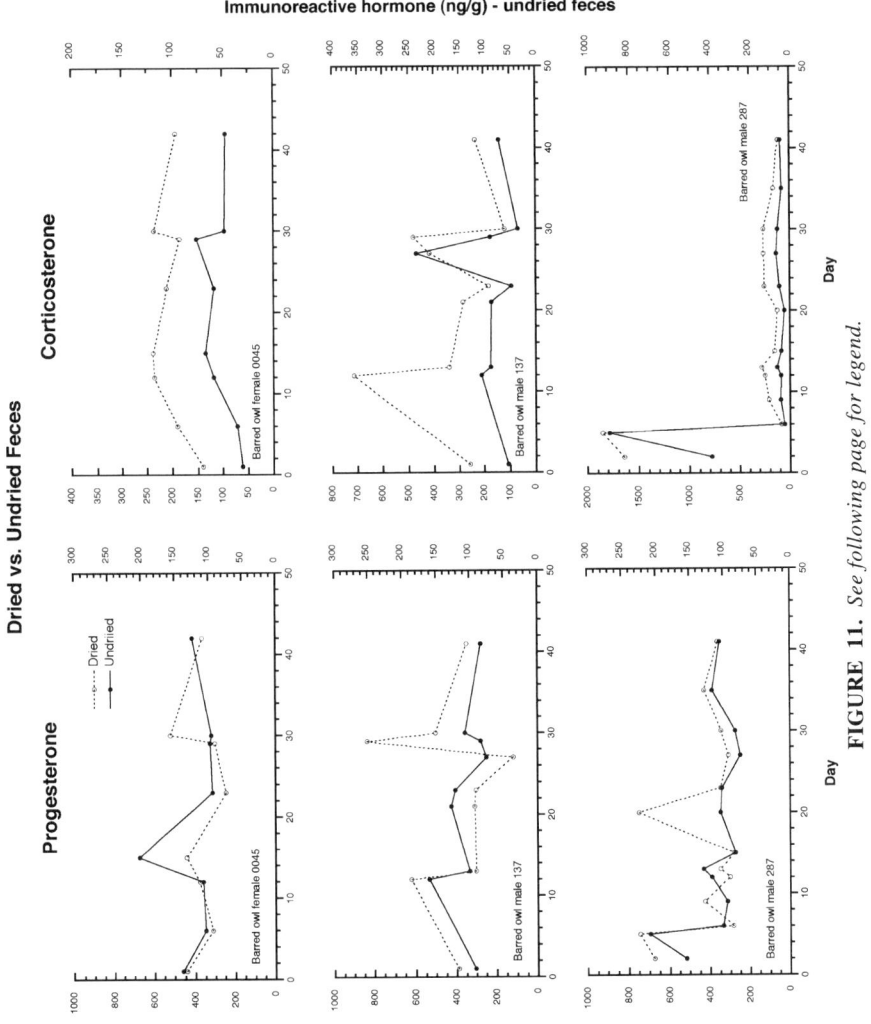

FIGURE 11. *See following page for legend.*

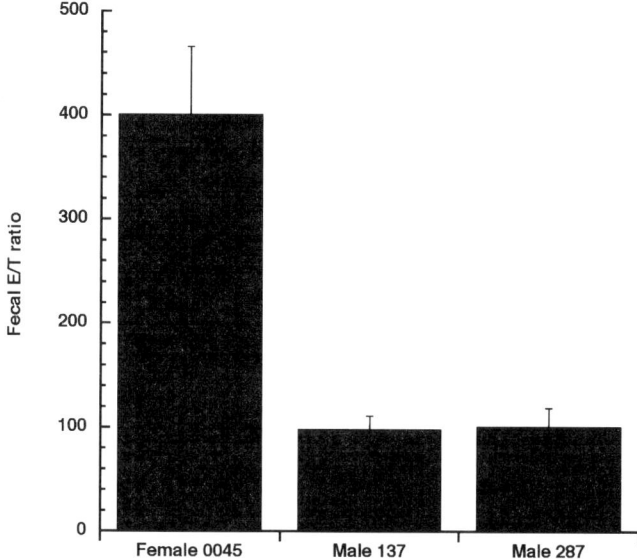

FIGURE 12. Average fecal estrogen/testosterone metabolite ratios (±SEM) from serially collected fecal samples of one female and two male barred owls.

samples were misidentified as male and above which 7 male samples were misidentified as female (25.5% inaccuracy).

In principle, sex identification using E/T ratios could be markedly improved if two or three serial samples from individuals were available, rather than a single sample (FIG. 13). In our data set, three serial samples would result in correct identification of the bird's gender for all but one of our 51 samples (i.e., <2% error rate). This data set is, of course, limited to one female bird and two males, but indicates that the fecal E/T ratio shows promise as a noninvasive method for determining gender of owls.

Seasonal Profiles of Fecal Corticosterone in Wild NSOs

FIGURE 14 shows fecal glucocorticoid (GC) levels in wild male and female NSOs collected across the breeding season throughout Washington and Oregon. Analyses using a general linear model showed significant effects of gender ($F = 7.2$, $P = .008$) and a significant interaction between gender and month ($F = 2.74$, $P = .03$) on fecal GC concentrations. Gender-specific simple regression analyses showed that male fecal GC levels significantly declined over time (in days) from March through July ($r =$

FIGURE 11. Comparison of progesterone (**left panel**) and corticosterone (**right panel**) hormone metabolites from the same serially collected samples extracted and expressed as dried (*dashed line*) versus undried (*solid line*) feces collected over 42 days from one female and two male barred owls.

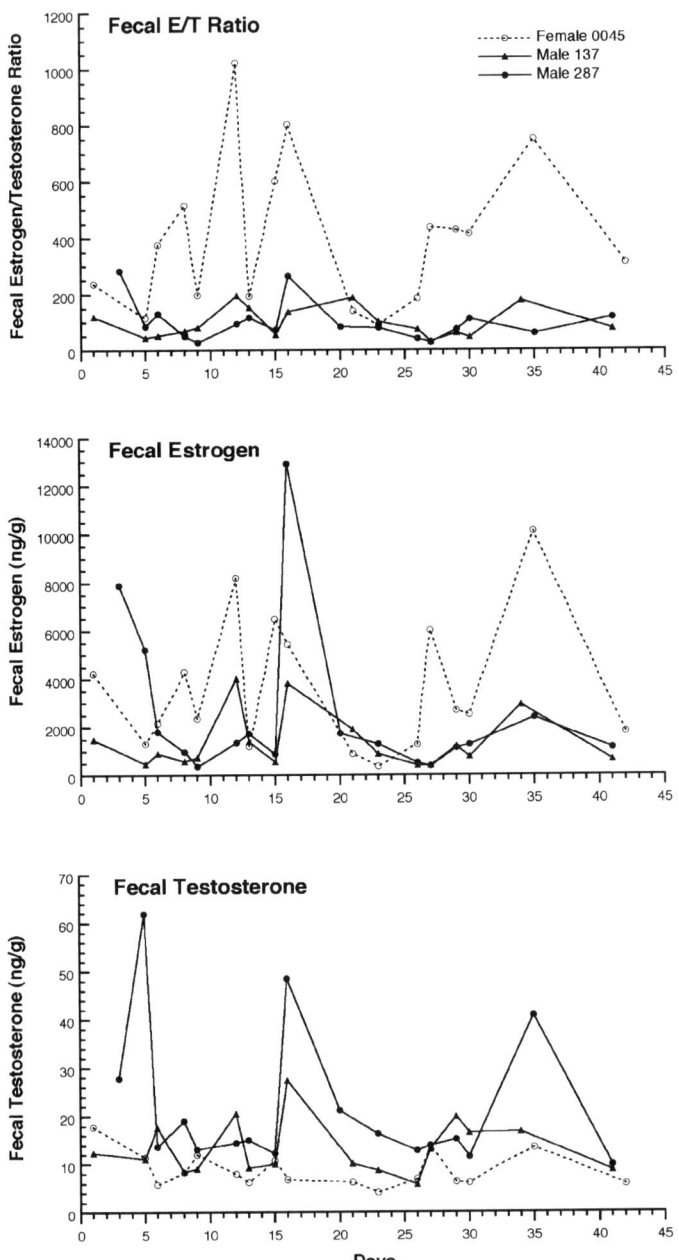

FIGURE 13. Longitudinal profiles of fecal estrogen/testosterone metabolite ratios (**top panel**) in one female and two male barred owls, compared with fecal estrogen alone (**middle panel**) or fecal testosterone alone (**bottom panel**). Data same as from FIGURE 12.

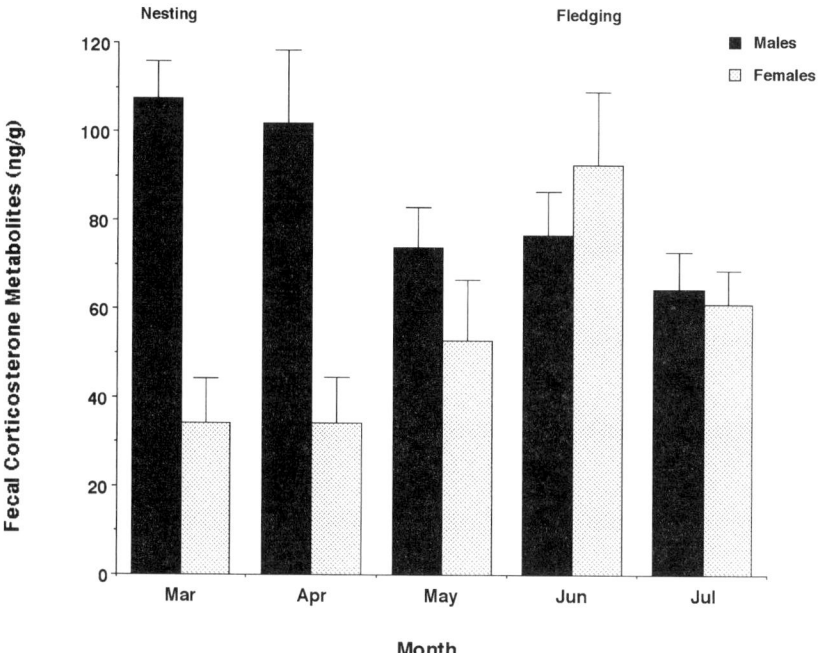

FIGURE 14. Fecal corticosterone metabolites in male and female wild NSOs sampled across the breeding season in Washington State and Oregon by state and federal managers during 1994.

.28, $P = .012$), whereas fecal GC levels significantly increased in females from March to June ($r = .4$, $P = .006$). Given the fall in GC levels among females in July, an ANOVA was also conducted on female GC levels using month (March–July) as the unit of analysis ($F = 2.64$, $P = .04$), with significant differences found between March and June ($P = .02$), April and June ($P = .015$), and June and July ($P = .05$). Essentially, stress levels were highest in male owls early in the breeding season (during nesting and mating), progressively declining thereafter. By contrast, glucocorticoid levels were relatively low among females during the early nesting period and then increased progressively up to and during the fledgling stage in June, falling off thereafter.

DISCUSSION

Fecal hormone measures provide a promising tool for conservation biology, as well as physiological and behavioral ecology studies in avian species. In birds, the short lag time from secretion until excretion and the small fecal mass appear to minimize many of the problems of mammalian fecal hormones (e.g., effects of diet and

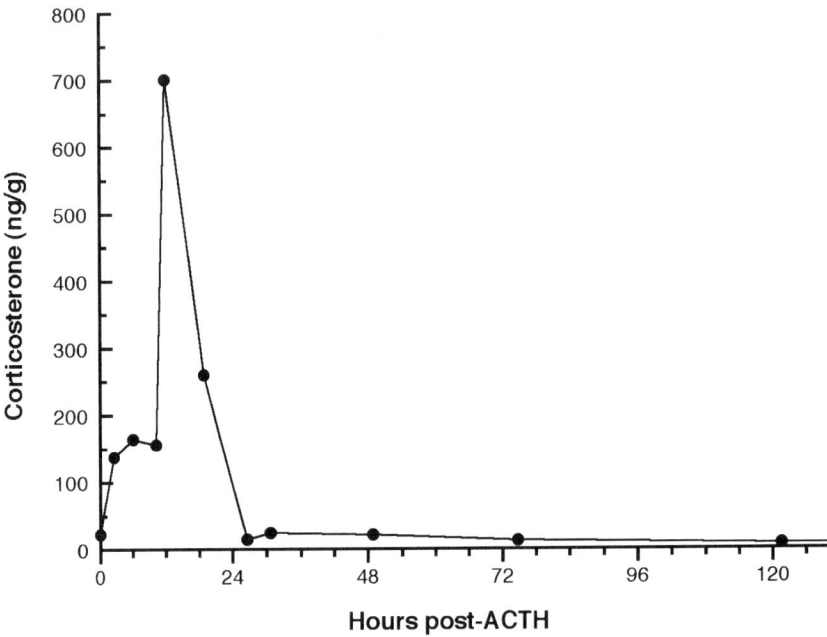

FIGURE 15. Immunoreactive fecal corticosterone profile after ACTH challenge of a barred owl. (Data from Wasser et al.[10])

bacterial metabolism[7]). Results were generally similar for glucocorticoids and reproductive hormones, for both males and females, as well as between owl species (barred owls, NSOs, and GHOs). Steroids were rapidly excreted (generally within the first two samples, 1–4 h postinfusion, in both urine and feces of barred owls and GHOs. This same time course was also found in response to an ACTH challenge in the NSO (FIG. 15).[10] More hormone was typically excreted in feces than in urine when expressed on a per-gram-of-feces or per-milliliter-of-urine basis. Total dpm (i.e., per excrement) for feces compared with urine was generally more variable, because total urine volume is so variable. For these reasons, hormones expressed per gram of feces should be far more reliable than those expressed per milliliter of urine or per volume of total excrement. Problems with expressing urinary hormones are further exacerbated by the tendency for urine to be soaked into the ground and the need to separate out nonhormone-containing urates. For all these reasons, fecal hormones are likely to be more useful than urinary hormones for field studies of owls and probably other birds.

True extraction efficiency, based on recovery of radiolabeled metabolites, was quite high (mean = 98% for all hormones across species using 80% ethanol). Extraction recovery increased with the amount of water, suggesting that the hormone metabolites were fairly polar. This finding was confirmed by HPLC of the various radiolabeled hormone metabolites. HPLC comparisons of radioactive versus immunoreactive metabolites in the feces further showed that the primary antibodies examined were measuring one or more of the primary metabolites. One exception to this trend was for progesterone, where poor correspondence between radioactive and immunoreactive metabolite peaks suggests that the progesterone antibodies tested may have been binding to nonprogesterone-derived metabolites. Other progesterone-based antibodies should be examined for these species.

High polarity of metabolites suggested that at least some metabolites might be conjugated, leading us to examine whether hydrolysis was a necessary component of the extraction prior to measurement using RIA. Parallelism was excellent for both hydrolyzed and nonhydrolyzed samples, suggesting either that the polar metabolites were not conjugated or that the metabolite affinities to the antibodies we used in our RIAs were not affected by the conjugates (i.e., the antibodies were binding metabolites at sites different from where the conjugates were attached). Thus, hydrolysis is not a necessary component of these extractions. The excellent parallelism results seen with the progesterone assay are in contrast to the poor overlap seen in radioactive and immunoreactive HPLC data for progesterone metabolites. This finding emphasizes the importance of testing antibody suitability in multiple complementary ways.

Parallelism studies were also used to test whether it was better to dry samples prior to extraction and express concentrations per gram of dry weight. The latter is often preferred for mammalian hormone data, largely because it controls for diet-related variation in steroid excretion rates.[7] Parallelism studies suggested that reliability of hormone measures in the owls did not change appreciably for samples expressed as wet versus dry weight. If anything, drying samples resulted in slightly but not significantly poorer parallelism. We further examined this trend by plotting progesterone and corticosterone metabolite concentrations in serially collected samples that were analyzed both dried and undried. The correspondence was quite close for all but a few samples in these comparisons. The few discordant samples most often had higher (relative to other samples) concentrations in dried versus undried subsamples. We suspect that this is simply the result of measurement error when weighing the sample. Underestimated weights of very small samples can significantly inflate the hormone concentration expressed per unit volume (Wasser et al.[2]; Tempel and Gutiérrez[5]). Small samples would weigh even less after water removal, potentially exacerbating such problems. As samples get smaller, there is also an increased likelihood of hormones being unevenly distributed in the small mass analyzed. This source of error may also favor inflated values because blank samples are more likely to be ignored or reextracted.

These studies clearly illustrate the importance of carefully validating any new assay. This general theme is echoed by other studies in this volume, in which investigators are extending fecal hormone analysis to a whole new class of vertebrates. Thorough validations are particularly important in cases such as the threatened species in this study, for which results could have far-reaching conservation and economic implications.

This study also examined biological activity of the hormones being measured in two important ways. For the reproductive steroids, we examined how well the reproductive hormones predicted gender. The E/T ratio was 86% reliable at identifying gender from any given sample. Moreover, no individual sample with a high ratio was misclassified; all samples classified as "female" were indeed from the female. Those female samples that fell within the male samples' range were almost always followed by samples with "female" concentrations within the next one to two samples. Using two to three consecutive samples thus increased gender assignment accuracies to more than 98%.

Biological activity for the glucocorticoids was previously assessed by an ACTH challenge study on a captive NSO (FIG. 15), as well as wild studies examining gender differences in fecal glucocorticoid concentrations resulting from environmental disturbance and gender-specific changes in stress levels over the breeding season of NSOs.[2,10] The ACTH challenge showed a 4-fold response at 2 h and a 15-fold response at 16 h. The 2-h response is consistent with the time course of radiolabeled metabolites found in the barred owl corticosterone infusion study. The 16-h response is more surprising. We suspect that the 16-h response was exacerbated by stress resulting from repeated sample collections immediately following capture and ACTH injection. Thus, accumulated stress continued to elevate glucocorticoid levels. This stress increased GCs by 5-fold more than that resulting from ACTH challenge alone. This result has also been reported in other species (e.g., chronic chair experiments of hypothalamic activity[4]), suggesting that ACTH may not necessarily maximally stimulate the adrenal gland. Presumably, this occurs because the stress response has many more synergistic components than simply that triggered by a rise in ACTH.

Fecal GC measures applied to wild owls also appear to provide a sensitive measure of disturbance impacts in wild owls that is reflected by gender differences in the stress response.[2] As shown in FIGURE 14, fecal GC levels collected from wild owls across the breeding season in Washington and Oregon showed that stress levels were highest in male owls early in the breeding season (during nesting and mating), progressively declining thereafter. By contrast, stress levels were relatively low among females during the early nesting period and then increased progressively up to and during the fledging stage in June, falling off thereafter. This gender difference in the stress response could have important implications for the timing of timber harvests and other controlled disturbances, simultaneously showing the potential utility of easily collected fecal hormone measures for wildlife conservation and management. For example, a management plan of restricting logging early in the nesting period (to avoid interfering with breeding), but then allowing logging at increasing intensity for the remainder of the owls' breeding season, could result in unforeseen problems. The above gender differences in fecal glucocorticoid concentrations over time suggest that such timber practices might have disproportionately negative impacts on females. Unrestricted timber harvest later in the year, continuing into the fledging period, would thus be high when female stress levels are already at their highest. Such management practices should be given close attention given the pivotal role of the female NSO in successfully rearing young.

The threatened NSO is an excellent example of a species whose management could greatly benefit from the kinds of analyses validated in this study. NSOs are declining rapidly from disturbance in the Pacific Northwest, presumably from a variety of pressures that are difficult to partition and therefore mitigate. Yet, this species tends to demonstrate very little behavioral response to disturbance. More sensitive measures are needed whose responses can be readily interpreted. Noninvasive physiological measures of stress and reproductive health hold great promise here because, unlike behavior, these physiological responses should occur whenever the bird is disturbed, providing a more quantitative, less subjective measure of disturbance impacts. This makes noninvasive physiological indices particularly useful for monitoring the relative impacts of disturbance, delineating the pressures that most need mitigation as well as tracking the effectiveness of such mitigation efforts. Moreover, acquiring such measures, accurately but noninvasively, will reduce any confounders created by incidental disturbance associated with invasive sampling such as blood draw, while also permitting a higher sample size.

Fortunately, there is a treasure trove of information in fecal samples. Our approach capitalizes on this diversity of measures, aiming to monitor a host of physiological responses from a single sample. This method is important in the field when it may be difficult to acquire repeated samples from the same individual and/or there may be advantages to sampling as many individuals as possible over the landscape. Monitoring the relation of diverse endocrine measures as well as other physiological (e.g., immunoglobulin A[39]) and even genetic measures[36,40–42] enables one to approach these problems much like a physician conducting a health panel on a patient to maximize information from a single visit. However, like any good measure, reliable application of these tools requires that their quantification be adequately validated. This study successfully validated a variety of fecal stress and reproductive hormones to this end.

ACKNOWLEDGMENTS

Support for this work was provided by the U.S. Fish and Wildlife Service and the Bureau of Indian Affairs. We thank Nicole Presley, Kathy Cooper, Steve Monfort, and Lisa Ware for assistance with all phases of this project. Subjects and sample collection assistance were provided by the Virginia Wildlife Center and the Woodland Park Zoo. Sample collection from wild spotted owls were provided by Eric Hanson, Ken Bevis, Gina King, E. Forsman, C. Meslow, L. Young, K. Radeke, D. Herter, L. Hicks, and their crews.

REFERENCES

1. COURTNEY, S.P. *et al.* 2004. Scientific Evaluation of the Status of the Northern Spotted Owl. Report by Sustainable Ecosystems Institute, Portland, OR.
2. WASSER, S.K., K. BEVIS, G. KING & E. HANSON. 1997. Noninvasive physiological measures of disturbance in the northern spotted owl. Conserv. Biol. **11:** 1019–1022.
3. MILLSPAUGH, J.J. *et al.* 2001. Fecal glucocorticoid assays and the physiological stress response in elk. Wildl. Soc. Bull. **29:** 899–907.
4. RUYS, J.D., S.P. MENDOZA, J.P. CAPITANIO & W.A. MASON. 2004. Behavioral and physiological adaptation to repeated chair restraint in rhesus macaques. Physiol. Behav. **82:** 205–213.
5. TEMPEL, D.J. & R.J. GUTIÉRREZ. 2003. Factors related to fecal corticosterone levels in California spotted owls: implications for assessing chronic stress. Conserv. Biol. **18:** 538–547.
6. WASSER, S.K., S.L. MONFORT & D.E. WILDT. 1991. Rapid extraction of faecal steroids for measuring reproductive cyclicity and early pregnancy in free-ranging yellow baboons (*Papio cynocephalus cynocephalus*). J. Reprod. Fertil. **92:** 415–423.
7. WASSER, S.K. *et al.* 1993. Effects of dietary fibre on faecal steroid measurements in baboons (*Papio cynocephalus cynocephalus*). J. Reprod. Fertil. **97:** 569–574.
8. WASSER, S.K., S.L. MONFORT, J. SOUTHERS & D.E. WILDT. 1994. Excretion rates and metabolites of oestradiol and progesterone in baboon (*Papio cynocephalus cynocephalus*) faeces. J. Reprod. Fertil. **101:** 213–220.

9. WASSER, S.K., S. PAPAGEORGE, C. FOLEY & J.L. BROWN. 1996. Excretory fate of estradiol and progesterone in the African elephant (*Loxodonta africana*) and patterns of fecal steroid concentrations throughout the estrous cycle. Gen. Comp. Endocrinol. **102:** 255–262.
10. WASSER, S.K. *et al.* 2000. A generalized fecal glucocorticoid assay for use in a diverse array of nondomestic mammalian and avian species. Gen. Comp. Endocrinol. **120:** 260–275.
11. HUNT, K.E. & S.K. WASSER. 2004. Effect of long-term preservation methods on fecal glucocorticoid concentrations of grizzly bear and African elephant. Physiol. Biochem. Zool. **76:** 918–928.
12. HUBER, S., R. PALME & W. ARNOLD. 2003. Effects of season, sex, and sample collection on concentrations of fecal cortisol metabolites in red deer (*Cervus elaphus*). Gen. Comp. Endocrinol. **130:** 48–54.
13. PALME, R., P. FISCHER & M.N. ISMAIL. 1993. Excretion of infused ^{14}C-oestrone and ^{14}C–progesterone via faeces and urine of ewes. 2nd Sci. Congr. of the Egyptian Society for Cattle Diseases. Cairo.
14. PALME, R., P. FISCHER, H. SCHILDORFER & M.N. ISMAIL. 1996. Excretion of infused ^{14}C-steroid hormones via faeces and urine in domestic livestock. Anim. Reprod. Sci. **43:** 43–63.
15. PALME, R. *et al.* 1999. Measurement of faecal cortisol metabolites in ruminants: a non-invasive parameter of adrenocortical function. Wien. Tieraerztl. Mschr. **86:** 237–241.
16. MÖHLE, U., M. HEISTERMANN, R. PALME & J.K. HODGES. 2002. Characterization of urinary and fecal metabolites of testosterone and their measurement for assessing gonadal endocrine function in male nonhuman primates. Gen. Comp. Endocrinol. **129:** 135–145.
17. MÖSTL, E. *et al.* 2002. Measurement of cortisol metabolites in faeces of ruminants. Vet. Res. Commun. **26:** 127–139.
18. SCHWARZENBERGER, F., E. MÖSTL, R. PALME & E. BAMBERG. 1996. Faecal steroid analysis for non-invasive monitoring of reproductive status in farm, wild and zoo animals. Anim. Reprod. Sci. **42:** 515–526.
19. LYNCH, J.W. *et al.* 2003. Concentrations of four fecal steroids in wild baboons: short-term storage conditions and consequences for data interpretation. Gen. Comp. Endocrinol. **132:** 264–271.
20. TOUMA, C., N. SACHSER E. MÖSTL & R. PALME. 2003. Effects of sex and time of day on metabolism and excretion of corticosterone in urine and feces of mice. Gen. Comp. Endocrinol. **130:** 267–278.
21. BROWN, J.L. *et al.* 1997. Faecal steroid analysis for monitoring ovarian and testicular function in diverse wild carnivore, primate and ungulate species. 1st Int. Symp. on Physiologie and Ethology of Wild and Zoo Animals. Suppl. II: 27–31. Berlin.
22. LASLEY, B.L. & J.F. KIRKPATRICK. 1991. Monitoring ovarian function in captive and free-ranging wildlife by means of urinary and fecal steroids. J. Zoo. Wildl. Med. **22:** 23–31.
23. ZIEGLER, T.E. 2001. Effective use of fecal and urinary cortisol measurements for determining health conditions in wild and captive nonhuman primates (Abstract #90). Am. J. Primatol. **54**(Suppl. 1): 44 .
24. DEHNHARD, M. *et al.* 2003. Measurement of plasma corticosterone and fecal glucocorticoid metabolites in the chicken (*Gallus domesticus*), the great cormorant (*Phalacrocorax carbo*), and the goshawk (*Accipiter gentilis*). Gen. Comp. Endocrinol. **131:** 345–352.
25. FRIGERIO, D., E. MÖSTL & K. KOTRSCHAL. 2001. Excreted metabolites of gonadal steroid hormones and corticosterone in greylag geese (*Anser anser*) from hatching to fledging. Gen. Comp. Endocrinol. **124:** 246–255.
26. GOYMANN, W., E. MÖSTL & E. GWINNER. 2002. Non-invasive methods to measure androgen metabolites in excrements of European stonechats, *Saxicola torquata rubicola*. Gen. Comp. Endocrinol. **129:** 80–87.
27. KELEMEN, K., P. PECZELY, Z. SZOKE & V. LADJANSZKY. 2003. A comparative methodical study of the faecal steroid analysis on birds: looking for a valid method of testosterone determination. Acta Biol. Hung.. **54:** 285–298.

28. NAKAGAWA, S., E. MÖSTL & J.R. WAAS. 2003. Validation of an enzyme immunoassay to measure faecal glucocorticoid metabolites from Adelie penguins (*Pygoscelis adeliae*): a non-invasive tool for estimating stress? Polar Biol. **26:** 491–493.
29. TELL, L.A. 1997. Excretion and metabolic fate of radiolabeled estradiol and testosterone in the cockatiel (*Nymphicus hollandicus*). Zoo Biol. **16:** 505–518.
30. KLASING, K.C. 1998. Comparative Avian Nutrition. CABI Publishing. Cambridge, MA.
31. SHACKLETON, C.H. 1986. Profiling steroid hormones and urinary steroids. J. Chromatogr. **379:** 91–156.
32. RISLER, L., S.K. WASSER & G.P. SACKETT. 1987. Measurement of excreted steroids in Macaca nemestrina. Am. J. Primatol. **12:** 91–100.
33. VELLOSO, A.L., S.K. WASSER, S.L. MONFORT & J.M. DIETZ. 1998. Longitudinal fecal steroid secretion in maned wolves (*Chrysocyon brachyurus*). Gen. Comp. Endocrinol. **112:** 96–107.
34. HODGES, J.K., R. TARARA, J.P. HEARN & J.G. ELSE. 1986. The detection of ovulation and early pregnancy in the baboon by direct measurement of conjugated steroids in urine. Am. J. Primatol. **10:** 329–338.
35. WASSER, S.K., L. RISLER & R.A. STEINER. 1988. Excreted steroids in primate feces over the menstrual cycle and pregnancy. Biol. Reprod. **39:** 862–872.
36. WASSER, S.K. *et al.* 1997. Techniques for application of faecal DNA methods to field studies of Ursids. Mol. Ecol. **6:** 1091–1097.
37. ZIEGLER, T.E. *et al.* 1989. Excretion of estrone, estradiol, and progesterone in the urine and feces of the female cotton-top tamarin (*Saguinus oedipus oedipus*). Am. J. Primatol. **17:** 185–195.
38. HEISTERMANN, M., M. AGIL, A. BUTHE & J.K. HODGES. 1998. Metabolism and excretion of oestradiol-17beta and progesterone in the Sumatran rhinoceros (*Dicerorhinus sumatrensis*). Anim. Reprod. Sci. **53:** 157–172.
39. ERIKSSON, E. *et al.* 2004. Effect of metabolic cage housing on immunoglobulin A and corticosterone excretion in faeces and urine of young male rats. Exp. Physiol. **89:** 427–433.
40. WASSER, S.K. *et al.* 2004. Assigning African elephant DNA to geographic region of origin: applications to the ivory trade. Proc. Natl. Acad. Sci. USA. **101:** 14847–14852.
41. WASSER, S.K. *et al.* 2004. Scat detection dogs in wildlife research and management: application to grizzly and black bears in the Yellowhead Ecosystem, Alberta, Canada. Can. J. Zool. **82:** 475–492.
42. LUDDERS, J.W., J.A. LANGENBERG, N.M. CZEKALA & H.N. ERB. 2001. Fecal corticosterone reflects serum corticosterone in Florida Sandhill Cranes. J. Wildl. Dis. **37:** 646–652.
43. TOUMA, C. & R. PALME. 2005. Measuring fecal glucocorticoid metabolites in mammals and birds: the importance of validation. Ann. N.Y. Acad. Sci. **1046:** 54–74.

Synthesis of Measuring Steroid Metabolites in Goose Feces

KATHARINA HIRSCHENHAUSER,[a] KURT KOTRSCHAL,[a] AND ERICH MÖSTL[b]

[a]*Konrad Lorenz Forschungsstelle, A-4645 Grünau, Austria, and University of Vienna, Vienna, Austria*

[b]*Veterinary University, Vienna, Austria*

> ABSTRACT: The reliability of noninvasively measuring steroid hormones from feces in greylag geese (*Anser anser*) and domestic geese (*A. domesticus*), both qualitatively and quantitatively, was tested experimentally. Geese are mainly herbivorous birds with a short gut-passage time (2–3 h). Groups of eight outdoor-housed male domestic geese were subjected to two different experiments, injection of either GnRH or ACTH, which were replicated in three different seasons (spring, summer, and fall). GnRH stimulation resulted in significant increases of response fecal testosterone metabolites (TM; 17β-OH-androgens) in spring and fall, but not during the summer photorefractoriness. Testosterone response patterns obtained from plasma samples paralleled those from feces; however, no direct correlation between individual immunoreactive plasma and feces contents was observed. To improve the sample handling during extraction and the assay sensitivity, we promote the use of a group-specific antibody against 17-oxoandrogens that does not require deconjugation prior to the analysis. ACTH robustly increased fecal corticosterone in all seasons. The polar nature of glucocorticoids, however, seems to make a distinction between conjugated and nonconjugated types difficult, and the available avian literature on this topic is discussed.
>
> KEYWORDS: *Anser anser; Anser domesticus;* fecal hormones; steroid excretion; testosterone; corticosterone; progesterone; estradiol; enzyme immunoassay; fecal sample storage; GnRH; ACTH; seasonal patterns; diurnal patterns

INTRODUCTION

Hormones are produced by endocrine glands and are transported via the bloodstream. They are rapidly metabolized and excreted. Measuring immunoreactive steroid metabolites from excrements carries several advantages over more traditional blood sampling. Collection of fecal samples enables continuous monitoring from the same individual, even over long periods, with minimal disturbance of its activities and social environment. Handling of the animal is not required, which is especially valuable in studies with a focus on adrenal stress hormones (i.e., glucocorticoids).[1]

Address for correspondence: Katharina Hirschenhauser, Konrad Lorenz Forschungsstelle, A-4645 Grünau 11, Vienna, Austria. Voice: +43-7616-8510; fax: +43-7616-85104.
 k.hirschenhauser@aon.at

However, gonadal steroid secretion may also be affected by adrenal stress responses (e.g., due to the sampling procedure), probably by an intrahypothalamic neuroendocrine circuit through which stress may inhibit the release of GnRH.[2] This factor particularly delays egg laying in female birds.[3,4]

Fecal measures of steroids are cumulative levels over the gut-passage time, and therefore reflect environmental effects over time.[5] In contrast, steroid measures from blood represent single-condition snapshots in time. The drawbacks of fecal hormone measurements are that care must be taken to prevent misidentification of samples when animals group closely together or when the vegetation level is high. Optimal storage of samples in the field may be problematic.[6-8] The separation of urinary and fecal compounds may be unfeasible,[9] further confounding the results. In birds, as in mammals, renal excretion via urine is faster than excretion via feces. However, fecal and urinary glucocorticoid metabolite concentrations were highly correlated in Wilson's storm petrels (*Oceanites oceanicus*[10]) and chicken (*Gallus domesticus*[11]). In the case of geese, however, the uric acid compound is placed as a cap at one end of the dropping (8% of the dropping in barnacle geese *Branta leucopsis*[12]), and so at least collection of the white urea may easily be avoided.

Increasing evidence indicates that there are large differences between species and that, so far, no single method for noninvasive assessment of steroid hormones can be used for cross-species application, which necessitates customized approaches.[5-6,13-15] The optimization of the mode of extraction and validation of the assay system are time-consuming and expensive. Validation and the specificity of the assay should always be stringent requests when fecal steroid data are presented.[16-19]

The free-ranging nonmigratory population of monogamous and biparental greylag geese (*Anser anser*) in Grünau is an exceptionally suitable subject for the use of fecal steroid measurements. All birds are marked with colored leg bands and habituated to the close presence of human observers; the individual life histories have been recorded since 1973.[20-21] As a starting point for hormonal studies in the greylag geese, the diurnal[22] and ontogenetic[23] patterns have been examined for integration into sampling paradigms. From individual fecal samples of adult male and female geese, we monitored the seasonal patterns of immunoreactive corticosterone,[24] testosterone,[25,26] estradiol, and progesterone[27] metabolites. We focused on the behavioral modulation of steroid hormones by comparing the seasonal patterns of different social categories, such as paired or unpaired, parental or nonparental. To do this, a large set of fecal samples was collected throughout a complete annual cycle. Fecal steroid data were grouped by means of seasonal phase; the phases were individually arranged by the date a female laid her first egg. Several issues had to be resolved beforehand, such as the collection and storage of the samples and the qualitative and quantitative clarifications of what the assay basically detected in the goose excrement. The fundamental indication for the method's applicability was the biological relevance of a match between hypotheses and first results. For example, the annual cycles of fecal sex steroids in greylag geese[25,27] generally matched those of plasma in barheaded geese (*A. indicus*),[28] and the seasonal patterns of fecal corticosterone were associated with social status and behavioral interactions.[24,26] Also, androgen metabolite levels in males were affected by the winning of interactions and pair-bond challenges during the breeding season,[26] as predicted by the "challenge hypothesis."[29] And finally, implants of exogenous testosterone resulted in higher levels of androgen metabolites in the feces of adult males.[30] Meanwhile, we have done more detailed analyses of short-term hormonal responses to behavior and changes in the social environment.[31–34] Here, we aim to present a synthesis of the technical aspects of measuring steroids (particularly androgens and glucocorticoids) in goose feces.

GUT-PASSAGE TIME AND REFLECTANCE OF CIRCULATING LEVELS

The lag between a plasma hormone peak and its appearance in the feces depends on gut-passage time.[13] The time course of excretion varies widely between species and is faster via urine than via feces,[5,10,11] in which it may be influenced by dietary fiber content.[16,35] Steroid hormones reach the gut after passing through the liver and the bile.[36,37] Peak excretion of both androgens and glucocorticoids is in the range of days in large mammals,[36,38–41] and in the range of minutes to hours in birds.[13,40,42–48]

To determine the basic gut-passage time of geese, we attempted two different methods. For greylag geese, as a marker we used a silver mica powder nucleus set inside a piece of bread (W. Haberl, personal communication), which after continuous observation was excreted after precisely 2 h.[49] In the second method we used charcoal dust–marked food, which revealed a gut-passage time of 2–5 h in domestic geese (*A. domesticus*[13]). Both approaches corresponded with the average throughput time reported for domestic geese (between 44 and 137 min[50]). Therefore, we assume that steroid levels obtained from goose feces reflect cumulative secretion over the 2 h preceding defecation. However, defecation rates of geese may vary between seasons,[12] even if provided with standardized food,[13] which indicates a potential overlap of the periods reflected in a dropping.

The central assumption that fecal levels of hormone metabolites reflect circulating levels of biologically active compounds is inevitable. However, the relationship between fecal and plasma steroid concentrations is not always a direct one. High plasma–feces correlations for testosterone and estradiol were reported in domestic fowl using the droppings during the 3 h before the plasma was sampled[43] or, alternatively, using the fecal samples at a lag of 4 h after the plasma sampling.[48] Correlations of corticosterone were reported in the Northern spotted owl (*Strix occidentalis caurina*), with blood samples taken on three mornings just after a fresh fecal sample was obtained.[45] However, the pulsatile and episodic pattern of glucocorticoid secretion, for example, may complicate this validation approach. In geese, we observed no short-term correlations between individual plasma and fecal testosterone measures, although plasma and feces resulted in parallel androgen response patterns to GnRH challenges[13] (FIG. 1, and see below). It seems that the amount of "noise" in fecal steroid data[51] suggests that they cannot be used to distinguish between animals that have only small differences in plasma steroid concentrations. Rather, fecal steroids are useful to distinguish between animals with large plasma differences (J. Cockrem, personal communication). In fact, for most questions the use of integrative cumulative hormone measurements from feces may even provide a better picture of hormone patterns than data derived from blood sampling.[5]

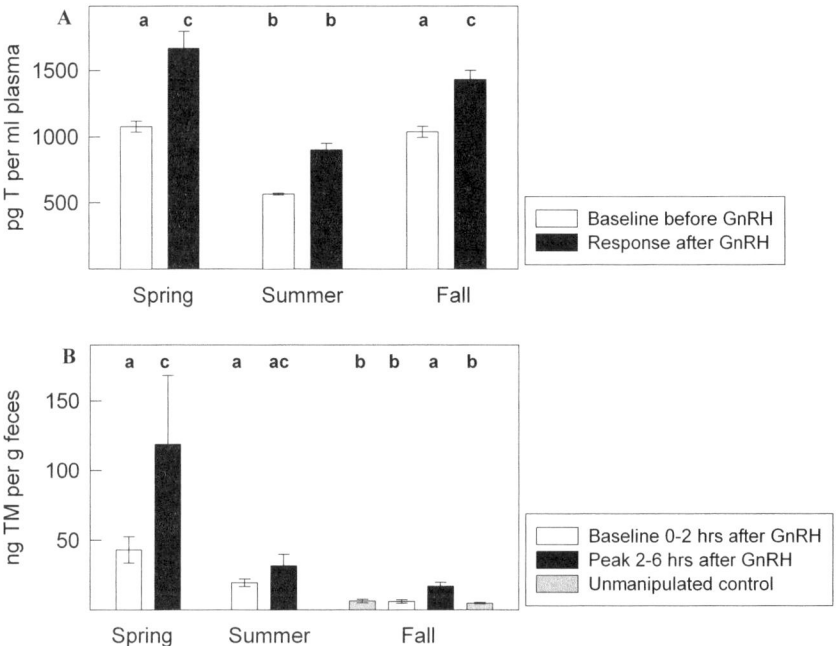

FIGURE 1. Seasonal comparison (mean ± SEM) of baseline and androgen response in domestic ganders. (**A**) Baseline testosterone levels (T) in plasma were sampled before GnRH stimulation and responses 90 min after injection. (**B**) In feces, baseline testosterone metabolites (TM) were sampled within 0–2 h after stimulation and peak response from samples within 2–6 h after stimulation. An unmanipulated control group was available for feces sampling only in fall. Two groups of eight ganders were available per season. Sample sizes vary because individuals were included only when both baseline and response values were available (depending mainly on irregular defecation). Significant differences between seasons, as well as between baseline and response levels, are indicated by different letters, a–c. In the (**A**) plasma, both baseline and response T differed significantly between seasons (one-way ANOVA: baseline, $F = 24.32$, $df = 2$, $P < .0001$; response, $F = 9.14$, $df = 2$, $P = .0012$); spring and fall levels significantly exceeded summer levels (Bonferroni *post hoc* test). Within seasons, response T significantly exceeded baseline T in spring and fall (Wilcoxon: spring, $Z = -2.80$, $n = 10$, $P = .005$; fall, $Z = -3.06$, $n = 12$, $P = .002$) but nonsignificantly so in summer ($Z = -1.83$, $n = 4$, $P = .068$). In the (**B**) feces, baseline TM significantly varied between seasons (one-way ANOVA: $F = 10.46$, $df = 2$, $P = .0006$); spring baseline was significantly higher than fall baseline (Bonferroni *post hoc* test). Seasonal differences of peak TM responses were only marginally significant ($F = 3.13$, $df = 2$, $P = .063$). Peak TM levels significantly exceeded baseline levels in spring (Wilcoxon: $Z = -2.40$, $n = 10$, $P = .017$) and in fall ($Z = -3.06$, $n = 12$, $P = .002$), but not in summer ($Z = -0.73$, $n = 4$, $P = .465$). (Figure modified from Hirschenhauser *et al.*,[13] with permission from Elsevier.)

SAMPLE HANDLING

The stability of the immunoreactive content in fecal samples due to handling and storage is an essential topic, particularly for field work, and may potentially be one source of noise in fecal steroid data.[8] The time between defecation and freezing the droppings is critical, because oxidation and bacterial metabolism can alter the steroid contents in the feces within hours after deposition.[6,52] In unpreserved baboon (*Papio cynocephalus*) fecal samples kept at room temperature, Wasser *et al.*[53] showed that changes in fecal estrogen and progesterone levels occurred within 6 h. In chicken droppings, no effect of room temperature on testosterone metabolites (TMs) or estrogen concentrations was observed for up to 48 h.[43] This may be another species-specific issue. It is probably essential to collect fresh droppings immediately after deposition and to preserve them as quickly as possible, optimally by freezing at −20°C. If freezing is not available, it is possible to store aliquots of the sample in alcohol.[6,54] However, caution is suggested because long-term preservation of elephant (*Loxodonta africana*) feces in ethanol resulted in elevated immunoreactive glucocorticoid concentrations of up to 300% of those in lyophilized frozen control samples.[55] Furthermore, repeated sample thawing and the exposition of the droppings to rainfall increased fecal glucocorticoid metabolite measurements.[7] A study on fecal glucocorticoid and estrogen metabolite content of baboon feces stored in 95% ethanol at room temperature or at −20°C, suggested keeping the samples no longer than a month at either temperature and if possible to extract and deconjugate the fecal sample as soon as possible.[6]

Stability of Testosterone in Fecal Samples: From Collection to Storage

We stored the fecal samples from geese by freezing unpreserved dry matter at −20°C within 2 h of collecting fresh feces. Care was also taken to keep them frozen, or at least cool, during transport. In a small data set, we aimed to define a sampling–freezing field protocol for fecal androgen measurements from goose feces taken during the summer season. We were also interested in whether the collection of "older" feces—from incubating females at their nests, for example—would be feasible.[49] To determine the latest possible time before freezing the fecal samples from geese, we divided large droppings into four portions, one of which was stored at −20°C immediately after deposit, while the remaining three portions were kept at room temperature (20°C) and then frozen after 1 h, 2 h, and 3 h ($n = 3$ females and 4 males). The samples were then extracted and analyzed for levels of testosterone (TM),[25] progesterone (PM), and estradiol (EM)[27] metabolites. Sample sizes were too small to statistically test the resulting patterns; however, variation within individuals was especially large with regard to TM in males (>200% of the initial TM content; FIG. 2B). In all cases, freezing of the fecal samples as soon as possible after collection is highly recommended.

To compare the storage of fecal samples as untreated dry matter with that stored in alcohol, we divided the available three to four feces portions of one female and one male into halves. One half was frozen at −20°C with no preservatives, and we stored 0.5 g of the other half in a 6-mL aqueous methanol dilution, and then stored it at −20°C immediately, after 1 h, 2 h, and 3 h at room temperature. The immunoreactive contents of TM, PM, and EM in the male samples were hovering around 100% of the initial value (FIG. 2B, 2D, 2F), and the same result was observed for the EM contents in female goose feces stored in alcohol (FIG. 2E). However, PM levels in female fecal samples frozen in alcohol were considerably higher than the initial values of dry frozen matter (FIG. 2C), whereas in the samples from males this pattern was less pronounced and, if any, reversed (FIG. 2D). TM in the female fecal samples preserved in alcohol started with an identical 100% of the initial value in the unpreserved frozen sample, but then increased to more than 200% (FIG. 2A).

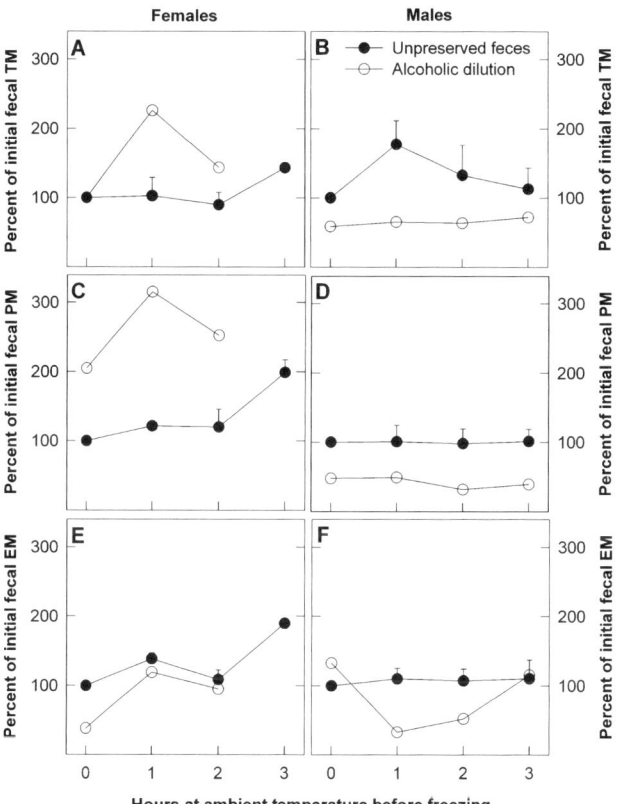

FIGURE 2. Test series of the immunoreactive steroid contents [(**A**, **B**) TM: fecal 17β-OH-androgen metabolites; (**C**, **D**) PM: fecal progesterone metabolites; (**E**, **F**) EM: fecal estrogen metabolites] in greylag goose feces frozen as dry matter or in alcohol at different times at ambient temperature. Up to four portions of the same fecal samples were frozen as unpreserved fecal matter (*filled symbols*) immediately after collection (0 h), and after 1 h, 2 h, and 3 h at room temperature (20°C). For comparison, 0.5-g aliquots of the fecal sample portions of one female and one male were preserved in aqueous alcoholic dilution before freezing (*open circles*). Immunoreactive steroid contents are expressed as percent of the initial value (frozen unpreserved immediately after deposition). Left-side panels show the fecal steroid patterns from females ($N = 3$), right-side panels from males ($N = 4$).

QUANTITATIVE TESTS OF STEROID EXCRETION IN GEESE

Testosterone

To test the basic assumption that TM levels in feces are biologically meaningful and proportionally reflect plasma levels, we compared basal and GnRH-stimulated (10 g Ovurelin per individual; Reana, Budapest, Hungary) plasma T concentrations with fecal TM in two groups of eight domestic ganders.[13] The experiment was repeated during three seasons: the spring peak of reproductive activity, the summer photorefractoriness, and the fall sexual reactivation. Plasma was sampled 90 min following the treatment; feces were collected continuously from all individuals throughout the day of the experiment.

Plasma T-level increases in response to GnRH were followed by significantly elevated fecal TM 2–6 h after the GnRH treatment in spring and fall (FIG. 1). During the summer refractoriness, however, T responses to GnRH remained nonsignificant in both plasma and feces. The high correspondence between general androgen response patterns measured from plasma and feces emphasizes the reliability of using fecal androgen measures for pattern comparisons.[13] On the other hand, direct correlations between individual plasma T and fecal TM levels remained nonsignificant ($P > .05$), which indicates the lack of a short-term relationship between plasma and excreted androgen metabolites. Episodic fluctuations seem to be dampened rather than reflected by the fecal metabolite concentration.[48]

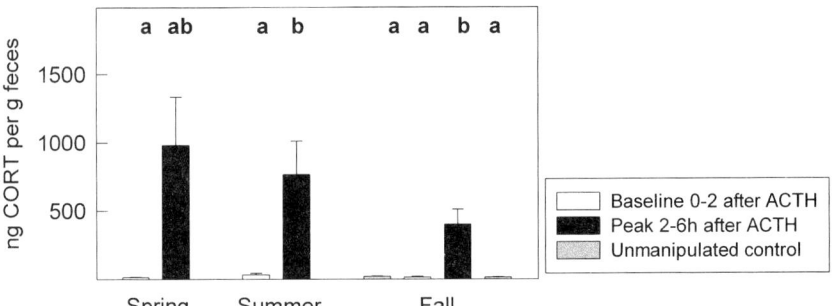

FIGURE 3. Seasonal comparison (mean ± SEM) of fecal baseline corticosterone metabolites (CORT) sampled 0–2 h after ACTH treatment and peak responses within 2–6 h after ACTH injection in domestic geese. An unmanipulated control group was available only in fall. Two groups of eight ganders were available per season. Different letters (a, b) indicate significant differences between baseline and response levels, as well as between seasons. Neither baseline CORT levels nor peak response levels differed significantly between seasons (one-way ANOVA: baselines $F = 0.68$, $df = 2$, $P = .523$; peaks $F = 1.09$, $df = 2$, $P = .363$). However, in all seasons, peak CORT levels were significantly elevated in response to the ACTH treatment (Wilcoxon baseline vs. peak response: spring $Z = -1.825$, $n = 4$, $P = .068$; summer $Z = -2.52$, $n = 8$, $P = .012$; fall $Z = -2.02$, $n = 5$, $P = .043$). Sample sizes vary because individuals were included only when both baseline and response values were available (mainly depending on defecation, which varied between seasons). (From Kotrschal *et al.*[46] Reprinted with permission from Springer.)

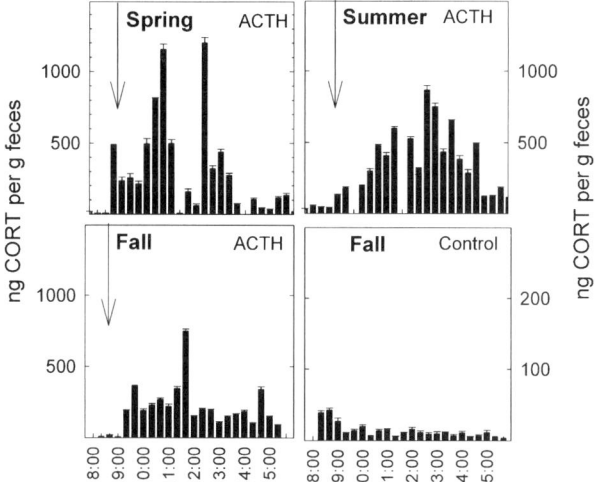

FIGURE 4. Diurnal excretion patterns of mean fecal corticosterone metabolites (CORT) of eight domestic ganders ± SEM within 20-min sampling periods following ACTH injection (indicated by the *arrows*) in spring (**Upper-Left Panel**), summer (**Upper-Right Panel**), fall (**Lower-Left Panel**), and an unmanipulated control group in fall (**Lower-Right Panel**). Because individual defecation rates were irregular, the number of samples integrated per bar varies between one and eight. In bars lacking SEM, $n = 1$. (From Kotrschal et al.[46] Reprinted with permission from Springer.)

Corticosterone

Two groups of eight domestic ganders were also treated with ACTH in spring, summer, and fall to stimulate corticosterone (CORT) responses. Individual fecal samples were collected throughout the treatment day.[46] ACTH clearly increased peak fecal CORT levels in all seasons (FIG. 3), with two to three peaks over the day (one after approximately 2 h, a second one after 4 h, and possibly a third one after 6 h; FIG. 4). In the unmanipulated fall control group we observed the typical small CORT peaks in the morning,[22,56] followed by consistently low levels throughout the day. We therefore consider the oscillatory excretion pattern of CORT throughout the day as specific responses to the treatment rather than an effect of an endogenous circadian excretion regimen.

QUALITATIVE TESTS OF STEROID CONTENT IN GOOSE EXCREMENT

Androgens

To determine the excretory testosterone metabolites in domestic geese, we injected intravenously two ganders with 370 kBq of ^{14}C-testosterone (NEC-101; New England Nuclear, Vienna). All droppings were collected, and 0.5 g of each sample was

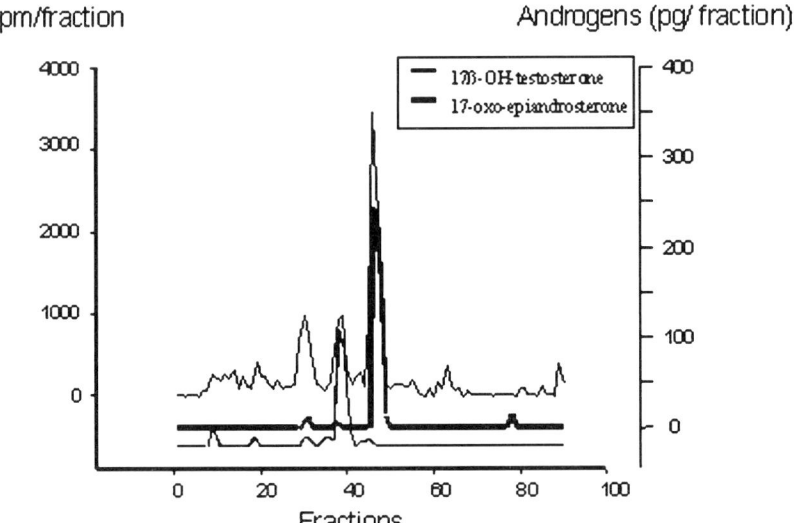

FIGURE 5. HPLC immunogram after intravenous application of ^{14}C-testosterone into a domestic gander. The fecal sample with the peak radioactivity was separated via a SepPac C_{18} column, followed by reverse-phase HPLC. The 17β-OH-androgen assay bound to a few minor metabolites, whereas the 17-oxoandrogen assay detected the major metabolite peak. (Modified from Hirschenhauser et al.,[13] with permission from Elsvier.)

extracted with methanol, centrifuged, and the supernatant measured by liquid scintillation counting. The samples with highest radioactivity were then assayed with and without prior deconjugation (enzymatic hydrolysis with β-glucuronidase/ arylsufatase; Merck 4114[25,27]) using enzyme immunoassays. We compared two group-specific antibodies,[52,57] one against 17β-OH-androgens (4-androstene-17β-ol-3-on-carboxymethyloxine-albumine-CMO from rabbits), which we had used hitherto,[13,25–27,46] with an epiandrosterone antibody against 17-oxoandrogens. The 17-oxoandrogen assay detected more immunoreactive metabolites, and the biological sensitivity of the assay therefore increased. Using the 17β-OH-androgen assay, immunoreactive metabolites were measured mostly after deconjugation. When using the assay for the 17-oxoandrogens, however, hydrolysis diminished the immunoreactivity.[13] An HPLC immunogram (high-pressure liquid chromatography separation) indicated that all three major radioactive peaks showed immunoreactivity (FIG. 5). Both assays reacted with testosterone metabolites in the feces, but the dominating radioactive peak was detected by the 17-oxoandrogen assay without prior hydrolysis. Thus, the 17-oxoandrogen antibody allows a direct assay without the necessity of deconjugation. Because this antibody binds to another position of the androgen molecules, one can measure both unconjugated and conjugated fractions in one procedure, which is probably a solution comparable to the steroid–conjugate assays proposed by Bishop and Hall.[42]

To test the robustness of our previous results, we reanalyzed selected samples from domestic ganders (three individuals in spring, summer, and fall) with high TM

TABLE 1. Summary of the literature on measuring immunoreactive steroids from avian feces with or without deconjugation prior to the analysis

Order	Species	Focal immunoreactive steroid	Hydrolysis	Reference
Galliformes	Japanese quail (*Coturnix coturnix japonica*)	Testosterone	Yes	42
Galliformes	Japanese quail	17-oxoandrogens	No[a]	Own unpublished data
Galliformes	Chinese painted quail (*Excalfactoria chinensis*)	17-oxoandrogens	No[a]	Own unpublished data
Galliformes	Chicken (*Gallus domesticus*)	Testosterone	No[b]	43
Anseriformes	Greylag goose (*Anser anser*)	17β-OH-androgens	Yes	13, 25, 27
	Domestic goose (*A. domesticus*)	17β-OH-androgens	Yes	13
Anseriformes	Domestic goose	17-oxoandrogens	No[a]	13
Piciformes	Downy woodpecker (*Picoides pubescens*)	Testosterone	No[b]	64
Psittaciformes	Cockatiel (*Nymphicus hollandicus*)	Testosterone	Yes	65
Passeriformes	Brown dipper (*Cinclus pallasii*)	Testosterone	No[b]	66
Passeriformes	European stonechat (*Saxicola torquata rubicola*)	Testosterone	Yes[b]	47
Galliformes	Chicken	Corticosterone	No	48
Galliformes	Chicken	Cortisone	No	11
Anseriformes	Greylag goose	Corticosterone (11β,21-diol-20-one)	Yes	24
	Domestic goose			46, 60
Anseriformes	Greylag goose	Corticosterone (11β-Hydroxyetiocholanolone)	Yes	62
Sphenisciformes	Adelie penguin (*Pygoscelis adeliae*)	Corticosterone (Tetrahydrocorticosterone)	No	63
Procellariformes	Wilson's stormpetrel (*Oceanites oceanicus*)	Corticosterone (Tetrahydrocorticosterone)	No	10
Apodiformes	Rufous hummingbird (*Selasphorus rufus*)	Corticosterone	No	67
Falconiformes	Goshawk (*Accipiter gentilis*)	Corticosterone	No[b]	48

TABLE 1. (*continued*) Summary of the literature on measuring immunoreactive steroids from avian feces with or without deconjugation prior to the analysis

Order	Species	Focal immunoreactive steroid	Hydrolysis	Reference
Strigiformes	Northern spotted owl (*Strix occidentalis caurina*)	Corticosterone	No	45
	Barred owl (*Strix varia*)			40
	Great horned owl (*Bubo virginianus*)			40
Passeriformes	European stonechat	Corticosterone	Yes	61
Passeriformes	Great tit (*Parus major*)	Corticosterone (11β-Hydroxyetiocholanolone)	Yes	56
Galliformes	Japanese quail	Estradiol	Yes[c]	42
Galliformes	Japanese quail	Pregnanediol	Yes[c]	42
Galliformes	Chicken	Estradiol	No[b]	43
Psittaciformes	Cockatiel	Estrone/Estradiol	Yes	65
Passeriformes	Brown dipper	Estradiol	No[b]	66

[a]Antibody also detects conjugated androgen molecules.
[b]The original article provided no information on whether testing had proved the necessity of using hydrolysis.
[c]Antisera used against steroid conjugates (i.e., glucuronides).

concentrations after GnRH stimulation[13] by using the 17-oxoandrogen assay. Results obtained with this assay were parallel to the results obtained by the 17β-OH-androgen assay ($r_s = 0.98$, $n = 9$, $P < .0001$[13]), which confirmed the results obtained by the earlier assay. All of these tests were restricted to male geese, and we are currently testing for potential sex-specific differences of immunoreactive androgen metabolites[16,58,59] in goose feces.

Early studies revealed a large proportion of unconjugated steroid metabolites in Japanese quail (*Coturnix coturnix japonica*) feces for both testosterone (31% of TM compared with only 6.5% in the starling's feces) and estradiol (15% of EM). However, in quail feces the dominant metabolites were also present as glucuronide conjugates.[42] Recent 17-oxoandrogen assays of Japanese and Chinese painted quail (*Excalfactoria chinensis*) feces resulted in biologically meaningful androgen patterns without deconjugation of the extracted fecal samples (unpublished data).

The fact that unconjugated metabolites were detectable in chicken androgens,[11,43,48] whereas in the goose model we observed a large proportion of conjugated metabolites,[60] suggests a systematic pattern of the degree of conjugation or polarity. However, a summary of the existing literature on this topic does not confirm this proposition, and a systematic pattern of the degree of excreted steroid metabolites does not seem obvious (TABLE 1).

Glucocorticoids

Until recently, fecal glucocorticoid metabolites in goose feces were assayed by enzyme immunoassay using an antibody against corticosterone-3-CMO:BSA and corticosterone-3-CMO-dioxaoctane-biotin as label.[24,46] We measured the glucocorticoid metabolites after infusions of radioactive corticosterone metabolites in domestic ganders. Most glucocorticoid metabolites were not diethylether extractable, but enzymatic hydrolysis of the methanol-extracted fecal samples with β-glucuronidase/arylsufatase (Merck 4114) increased the recovery of excreted radioactive corticosterone metabolites to 23% compared with almost 0% without prior hydrolysis.[46,60] This finding indicated that in geese, glucocorticoid metabolites were either excreted as predominantly conjugated or as polar unconjugated metabolites. Similar results were meanwhile obtained with great tits (*Parus major*[56]), European Stonechats (*Saxicola torquata rubicola*[61]), and quails (*C. c. japonica* and *E. chinensis*; unpublished data).

More recently, an assay for measuring a group of fecal glucocorticoid metabolites (3,11β-dihydroxyandrogens) has been developed and applied to goose feces[62] (hydroxyetiocholanolone was used as the standard and as label). This assay resulted in considerably higher glucocorticoid metabolite concentrations than did measuring the same samples with the earlier assay and, thus, in more fine-tuned patterns. Preparation and extraction of the fecal samples, however, had to be continued as before.[24,46]

The necessity to deconjugate prior to the assay is not always clear. Glucocorticoids may be excreted as tetra- and pentahydroxylated metabolites. These are highly polar molecules, which are not ether extractable. Therefore, a clear classification of glucocorticoids into conjugated or nonconjugated is problematic.

To find a systematic pattern in the degree of conjugation of excreted steroid metabolites, we compiled a summary of the existing literature (TABLE 1). This summary shows that specifically in the (purely carnivorous) birds of prey[40,45,48] and Adelie penguins (*Pygoscelis adeliae*[63]), deconjugation prior to the analysis curiously did not improve the measured content of immunoreactive CORT in the feces, although the metabolites were highly polar. Deconjugation was also not used to measure TM in woodpeckers (*Picoides pubescens*[64]), which feed exclusively on insects, and for TM[43] and CORT[11,48] in chicken which, if permitted, prefer a high-protein diet. The binding of steroid molecules to glucuronides and sulfates is regulated by enterobacteria in the gut and ceca. A high-protein diet changes the microflora in the ceca, whereas the fiber content of the diet decreases steroid resorption.[9] Whether the location of steroid conjugates may vary with diet or with excretion pathway (i.e., proportion of feces or urine excreted), or whether those are just highly polar substances, remains an open question. A further confounding variable is the use of different antibodies that detect different metabolites.

In summary, over the last decade, several studies in the wild greylag geese, as well as in domestic geese, have demonstrated that fecal steroid measurements from goose feces result in biologically meaningful patterns. This method makes a wide array of questions accessible for research. Nevertheless, researchers in this area of study should be aware of the drawbacks of the method, such as sufficient number of samples to overcome variation within individuals, as well as the potential pitfall of sex-specific differences in excreted metabolites.

ACKNOWLEDGMENTS

Funding was provided by the Austrian FWF (P10483-BIO and R30-B03), and permanent support came from the Verein der Förderer der KLF. Many people have contributed to the work. In particular, we appreciate the support of J. Dittami, B. Ferro, B. Föger, D. Frigerio, J. Hemetsberger, W. Kabinek, M. Krawany, A. Kuchar, P. Peczely, F. Schachinger, A. Schöbitz, and B. Wallner. We thank J. Cockrem and S. Wasser for discussions, and W. Goymann, I. Scheiber, and B. Weiss for critiques on an earlier draft of the manuscript.

REFERENCES

1. ROMERO, L.M. & R.C. ROMERO. 2002. Corticosterone responses in wild birds: the importance of rapid initial sampling. Condor **104**: 129–135.
2. CALOGERO, A.E., G. BAGDY & R. D'AGATA. 1998. Mechanisms of stress on reproduction. Ann. N.Y. Acad. Sci. **851**: 364–370.
3. HARVEY, S., J.G. PHILLIPS, A. REES & T.R. HALL. 1984. Stress and adrenal function. J. Exp. Zool. **232**: 633–645.
4. SALVANTE, K.G. & T.D. WILLIAMS. 2003. Effects of corticosterone on the proportion of breeding females, reproductive output and yolk precursor levels. Gen. Comp. Endocrinol. **130**: 205–214.
5. WHITTEN, P.L., D.K. BROCKMAN & R.C. STAVISKY. 1998. Recent advances in noninvasive techniques to monitor hormone-behavior interactions. Am. J. Phys. Anthropol. Suppl. **27**: 1–23.
6. KHAN, M.Z., J. ALTMANN, S.S. ISANI & J. YU. 2002. A matter of time: evaluating the storage of fecal samples for steroid analysis. Gen. Comp. Endocrinol. **128**: 57–64.
7. WASHBURN, B.E. & J.J. MILLSPAUGH. 2002. Effects of simulated environmental conditions on glucocorticoid metabolite measurements in white-tailed deer feces. Gen. Comp. Endocrinol. **127**: 217–222.
8. LYNCH, J.W. *et al.* 2003. Concentrations of four fecal steroids in wild baboons: short-term storage conditions and consequences for data interpretation. Gen. Comp. Endocrinol. **132**: 264–271.
9. KLASING, K.C. 2005. Potential impact of nutritional strategy on noninvasive measurements of hormones in birds. Ann. NY Acad. Sci. **1046**: 5–16.
10. QUILLFELDT, P. & E. MÖSTL. 2003. Resource allocation in Wilson's storm-petrels *Oceanites oceanicus* determined by measurement of glucocorticoid excretion. Acta Ethol. **5**: 115–122.
11. RETTENBACHER, S. *et al.* 2004. Measurement of corticosterone metabolites in chicken droppings. Br. Poult. Sci. **45**: 704–711.
12. PROP, J. & T. VULINK. 1992. Digestion by barnacle geese in the annual cycle: the interplay between retention time and food quality. Funct. Ecol. **6**: 180–189.
13. HIRSCHENHAUSER, K. *et al.* 2000. Seasonal relationships between plasma and fecal testosterone in response to GnRH in domestic ganders. Gen. Comp. Endocrinol. **118**: 262–272.
14. MÖHLE, U., M. HEISTERMANN, R. PALME & J.K. HODGES. 2002. Characterization of urinary and fecal metabolites of testosterone and their measurement for assessing gonadal endocrine function in male nonhuman primates. Gen. Comp. Endocrinol. **129**: 135–145.
15. BUCHANAN, K. & A.R. GOLDSMITH. 2004. Noninvasive endocrine data for behavioural studies: the importance of validation. Anim. Behav. **67**: 183–185.
16. GOYMANN, W. 2005. Noninvasive monitoring of hormones in bird droppings: biological validations, sampling, extraction, sex differences, and the influence of diet on hormone metabolite levels. Ann. N.Y. Acad. Sci. **1046**: 35–53.
17. MÖSTL, E., S. RETTENBACHER & R. PALME. 2005. Measurement of corticosterone metabolites in birds' droppings: an analytical approach. Ann. N.Y. Acad. Sci. **1046**: 17–34.

18. PALME, R. 2005. Measuring fecal steroids: guidelines for practical application. Ann. N.Y. Acad. Sci. **1046:** 75–80.
19. TOUMA, C. & R. PALME. 2005. Measuring fecal glucocorticoid metabolites in mammals and birds: the importance of a biological validation. Ann. N.Y. Acad. Sci. **1046:** 54–74.
20. LORENZ, K. 1988. Hier bin ich—wo bist Du? Ethologie der Graugans. Piper Verlag. München.
21. KOTRSCHAL, K. 1995. Im Egoismus vereint? Tiere und Menschentiere—das neue Weltbild der Verhaltensforschung. Piper Verlag. München.
22. SCHÜTZ, K., B. WALLNER & K. KOTRSCHAL. 1997. Diurnal patterns of steroid hormones from feces in greylag goslings (*Anser anser*). Adv. Ethol. **32:** 66.
23. FRIGERIO, D., E. MÖSTL & K. KOTRSCHAL. 2001. Excreted metabolites of gonadal steroid hormones and corticosterone in greylag geese (*Anser anser*) from hatching to fledging. Gen. Comp. Endocrinol. **124:** 246–255.
24. KOTRSCHAL, K., K. HIRSCHENHAUSER & E. MÖSTL. 1998. The relationship between social stress and dominance is seasonal in greylag geese. Anim. Behav. **55:** 171–176.
25. HIRSCHENHAUSER, K., E. MÖSTL & K. KOTRSCHAL. 1999. Within-pair testosterone covariation and reproductive output in greylag geese *Anser anser*. Ibis **141:** 577–586.
26. HIRSCHENHAUSER, K. *et al.* 2000. Endocrine and behavioural responses of male greylag geese (*Anser anser*) to pairbond challenges during the reproductive season. Ethology **106:** 63–77.
27. HIRSCHENHAUSER, K., E. MÖSTL & K. KOTRSCHAL. 1999. Seasonal patterns of sex steroids determined from feces in different social categories of greylag geese (*Anser anser*). Gen. Comp. Endocrinol. **114:** 67–97.
28. DITTAMI, J.P. 1981. Seasonal changes in behavior and plasma titers of various hormones in barheaded geese, *Anser indicus*. Z. Tierpsychol. **55:** 289–324.
29. WINGFIELD, J.C. *et al.* 1990. The 'challenge hypothesis': theoretical implications for patterns of testosterone secretion, mating systems and breeding strategies. Am. Nat. **136:** 829–846.
30. FRIGERIO, D. *et al.* 2004. Experimentally elevated testosterone increases status signaling in male greylag geese (*Anser anser*). Acta Ethol. **7:** 9–18.
31. PFEFFER, C., J. FRITZ & K. KOTRSCHAL. 2002. Hormonal correlates of being an innovative greylag goose, *Anser anser*. Anim. Behav. **63:** 687–695.
32. FRIGERIO, D. *et al.* 2003. Social allies modulate corticosterone excretion and increase success in agonistic interactions in juvenile hand-raised greylag geese (*Anser anser*). Can. J. Zool. **81:** 1746–1754.
33. Weiss, B.M. & K. Kotrschal. 2004. Effects of passive social support in juvenile greylag geese (*Anser anser*): a study from fledging to adulthood. Ethology **110:** 429–444.
34. SCHEIBER, I.B.R. *et al.* 2005. Active and passive social support in families of greylag geese (*Anser anser*). Behaviour. In press.
35. WASSER, S.K. *et al.* 1993. Effects of dietary fibre on faecal steroid measurements in baboons (*Papio cynocephalus cynocephalus*). J. Reprod. Fertil. **97:** 569–574.
36. TAILOR, W. 1971. The excretion of steroid hormone metabolites in bile and feces. Vitam. Horm. 29: 201–285.
37. HELTON, E.D. & W.N. HOLMES. 1973. The distribution and metabolism of labeled corticosteroids in the duck (*Anas platyrhynchos*). J. Endocrinol. **56:** 361–385.
38. MACDONALD, I.A. *et al.* 1983. Degradation of steroids in the human gut. J. Lipid Res. **24:** 675–700.
39. GOYMANN, W. *et al.* 1999. Non-invasive fecal monitoring of glucocorticoids in spotted hyenas, *Crocuta crocuta*. Gen. Comp. Endocrinol. **114:** 340–348.
40. WASSER, S.K. *et al.* 2000. A generalized fecal glucocorticoid assay for use in a diverse array of nondomestic mammalian and avian species. Gen. Comp. Endocrinol. **120:** 260–275.
41. DEHNHARD, M. *et al.* 2001. Noninvasive monitoring of adrenocortical activity in roe deer (*Capreolus capreolus*) by measurement of fecal cortisol metabolites. Gen. Comp. Endocrinol. **123:** 111–120.
42. BISHOP, C.M. & M.R. HALL. 1991. Non-invasive monitoring of avian reproduction by simplified faecal steroid analysis. J. Zool. London **224:** 649–668.

43. COCKREM, J.F. & J.R. ROUNCE. 1994. Faecal measurements of oestradiol and testosterone allow the non-invasive estimation of plasma steroid concentrations in the domestic fowl. Br. Poult. Sci. **35:** 433–443.
44. KIKUCHI, M., N. YAMAGUCHI, F. SATO & S. ISHII. 1994. Extraction methods for fecal hormone analysis in birds. J. Ornithol. **135:** 64.
45. WASSER, S.K. et al. 1997. Non-invasive physiological measures of disturbance in the northern spotted owl. Conserv. Biol. **11:** 1019–1022.
46. KOTRSCHAL, K. et al. 2000. Effects of physiological and social challenges in different seasons on fecal testosterone and corticosterone in male domestic geese (*Anser domesticus*). Acta Ethol. **2:** 115–122.
47. GOYMANN, W., E. MÖSTL & E. GWINNER. 2002. Non-invasive methods to measure androgen metabolites in excrements of European stonechats, *Saxicola torquata rubicola*. Gen. Comp. Endocrinol. **129:** 80–87.
48. DEHNHARD, M. et al. 2003. Measurement of plasma corticosterone and fecal glucocorticoid metabolites in the chicken (*Gallus domesticus*), the great cormorant (*Phalacrocorax carbo*), and the goshawk (*Accipiter gentiles*). Gen. Comp. Endocrinol. **131:** 345–352.
49. HIRSCHENHAUSER, K. (1995): Testosteron aus Kot und Verhalten: Ein Jahresgang an Graugänsen (*Anser anser*). Master's thesis, University of Vienna, Austria.
50. MATTOCKS, J.G. 1971. Goose feeding and cellulose digestion. Wildfowl **22:** 107–113.
51. SCHEIBER, I.B.R. et al. 2005. Sampling effort/frequency necessary to infer individual acute stress responses from fecal analysis in greylag geese (*Anser anser*). Ann. N.Y. Acad. Sci. **1046:** 154–162.
52. MÖSTL, E., S. RETTENBACHER & R. PALME. 2005. Measurement of corticosterone metabolites in birds' droppings: an analytical approach. Ann. N.Y. Acad. Sci. **1046:**
53. WASSER, S.K., L. RISLER & R.A. STEINER. 1988. Excreted steroids in primate feces over the menstrual cycle and pregnancy. Biol. Reprod. **39:** 862–872.
54. PETER, A.T., J.K. CRITSER & N. KAPUSTIN. 1996. Analysis of sex steroid metabolites excreted in the feces and urine of nondomesticated animals. The Compendium **18:** 781–791.
55. HUNT, K.E. & S.K. WASSER. 2003. Effect of long-term preservation methods on fecal glucocorticoid concentrations of grizzly bear and African elephant. Physiol. Biochem. Zool. **76:** 918–928.
56. Carere, C. et al. 2003. Fecal corticosteroids in a territorial bird selected for different personalities: daily rhythm and the response to social stress. Horm. Behav. **43:** 540–548.
57. MÖSTL, E. et al. 1987. Oestrogen determination in feces of mares by enzyme immunoassay on microtitre plates. Proc. Symp. Analysis Steroids, Sopron, Hungary, pp. 219–224.
58. TOUMA, C., N. SACHSER, E. MÖSTL & R. PALME. 2003. Effects of sex and time of day on metabolism and excretion of corticosterone in urine and feces of mice. Gen. Comp. Endocrinol. **130:** 267–278.
59. GOYMANN, W. & J.C. WINGFIELD. 2004. Competing females and caring males. Sex steroids in African black coucals, *Centropus grillii*. Anim. Behav. **68:** 733–740.
60. KRAWANY, M. 1996. Die Entwicklung einer nicht-invasiven Methode zum Nachweis von Steroidhormonmetaboliten im Kot von Gänsen. Ph.D. thesis, Veterinary University of Vienna, Austria.
61. GOYMANN, W., E. MÖSTL & E. GWINNER. 2002. Corticosterone metabolites can be measured noninvasively in excreta of European stonechats (*Saxicola torquata rubicola*). Auk **119:** 1167–1173.
62. FRIGERIO, D., J. DITTAMI, E. MÖSTL & K. KOTRSCHA. 2004. Excreted corticosterone metabolites co-vary with ambient temperature and air pressure in male greylag geese (*Anser anser*). Gen. Comp. Endocrinol. **137:** 29–36.
63. NAKAGAWA, S., E. MÖSTL & J.R. WAAS. 2003. Validation of an enzyme immunoassay to measure faecal glucocorticoid metabolites from Adélie penguins (*Pygoscelis adeliae*): a non-invasive tool for estimating stress? Polar Biol. **26:** 491–493.
64. KELLAM, J.S., J.C. WINGFIELD & J.R. LUCAS. 2004. Nonbreeding season pairing behavior and the annual cycle of testosterone in male and female downy woodpeckers, *Picoides pubescens*. Horm. Behav. **46:** 703–714.

65. TELL, L.A. 1997. Excretion and metabolic fate of radiolabeled estradiol and testosterone in the cockatiel (*Nymphicus hollandicus*). Zoo Biol. **16:** 505–518.
66. KOFUJI, H., M. KANDA & T. OISHI. 1993. Breeding cycles and fecal gonadal steroids in the brown dipper *Cinclus pallasii*. Gen. Comp. Endocrinol. **91:** 216–223.
67. HIEBERT, S.M. *et al.* 2000. Noninvasive methods for measuring and manipulating corticosterone in hummingbirds. Gen. Comp. Endocrinol. **120:** 235–247.

Sampling Effort/Frequency Necessary to Infer Individual Acute Stress Responses from Fecal Analysis in Greylag Geese (*Anser anser*)

ISABELLA B. R. SCHEIBER,[a] SIMONA KRALJ,[b] AND KURT KOTRSCHAL[a]

[a]*Konrad Lorenz Forschungsstelle für Ethologie, A-4645 Grünau 11, Austria*
[b]*University of Ljubljana, Biotechnical Faculty, Department of Biology, SI-1000 Ljubljana, Slovenia*

ABSTRACT: Measuring hormone metabolites from excreta is a powerful method to study hormone–behavior relationships. Currently, fecal corticosterone metabolite concentrations are used to estimate individual short-term stress responses. From the free-roaming, semitame flock of greylag geese (*Anser anser*), as many fecal samples as possible were collected over 3 h following a challenge (social density stress) or in a control situation. This time span corresponds to the gut passage time of geese. It was asked how many samples were necessary to determine differences in excreted corticosterone immunoreactive metabolites (CORTs) between control and social density stress and which parameters (means, maxima, range) reliably showed this difference. A large variation of CORT was found between consecutive samples. Still, means, maxima, and ranges of the samples in a fecal series consistently showed the response to a stressor both within and between individuals. Three samples sufficed if the maximum value of CORT was used, whereas four or more samples were necessary to work with the mean. It was concluded that by increasing the number of fecal samples collected, the course of CORT could be measured more precisely and an individual's acute stress response inferred more reliably.

KEYWORDS: corticosterone–immunoreactive metabolites; sampling effort; noninvasive sampling; individual variation; short-term stress response; greylag goose; *Anser anser*

INTRODUCTION

The method of invasive sampling of circulating hormones from blood is well established; however, it may also have disadvantages. For example, blood samples show concentrations that occur in a very narrow time frame.[1] Plasma glucocorticoids, for example, can vary with time of the day and have pulsatile secretory patterns;[2–4] therefore, timing of when to sample becomes an issue. Alternatively, hormone metabolites from excretion products may show integration over a certain period.

Address for correspondence: Isabella B.R. Scheiber, Konrad Lorenz Forschungsstelle für Ethologie, A-4645 Grünau 11, Austria. Voice: +43-(0)7616-8510; fax: +43-(0)7616-85104.
isabella_scheiber@t-online.de

Gut passage time and the dynamics of excretion determine the temporal resolution of the method. In geese, fecal samples are assumed to represent an integrated, proportional record of the plasma level within a frame of 2–4 h prior to defecation.[5] Short-term hormonal changes in reaction to specific situations were measured from feces,[6] as were long-term endocrine profiles due to seasonal variations[7–9] or patterns during stress-related disorders.[10] One potential disadvantage of fecal sampling is that the time course of excretion varies widely between species, ranging from a maximum of days in mammals[11] to a low of hours or minutes in passerine[12,13] and nonpasserine[14] birds. Another disadvantage is that the sensitivity of the method can vary between different assays, and results can differ between seasons[9] and weather conditions.[15]

Individuals differ greatly in their stress-induced glucocorticoid responses.[12,16,17] These interindividual differences might show an individual's ability to cope with potentially unfavorable environmental demands or challenging situations within the social context. The latter topic is of key interest in our study on our flock of free-ranging, semitame greylag geese *(Anser anser)*. Generally, we are interested in answering questions pertaining to individual costs and benefits of living socially. We intend to determine individual variation in response to a variety of stressors within the intact social environment as well as stress-reducing effects of social partners.[18] Because our geese are human acquainted, we opted to collect excreta, as has been done in the past.[7,8,19]

It was our goal to optimize the resolution of the noninvasive approach of fecal hormone collection through increasing our sampling efforts—that is, collection of as many fecal samples as possible and determining how many samples were necessary to consistently find a difference in excreted corticosterone immunoreactive metabolites (CORTs) between control and an experimental challenge. We also wanted to establish which parameter (mean or maximum) was most reliable in finding this difference. In addition, we asked whether CORT would show consistent interindividual variation.

We intended to record exact metabolite profiles of individual geese to see how CORT develops in a series of consecutive fecal samples within 3 h. Furthermore, we wanted to determine whether we could detect the short-term stress response and pinpoint if there was a peak excretion of CORT after a certain time after a challenge. We were also interested in determining how large the variation was in a fecal series collected over that time and whether there were comparable patterns within the series of one individual.

MATERIALS AND METHODS

Animals

A nonmigratory flock of greylag geese was introduced into the Upper Austrian valley of the Alm River by Konrad Lorenz in 1973,[20] and individual life history data and social backgrounds of all individuals have been continuously monitored since then.[21] Individuals are unrestrained and roam the valley between the Konrad Lorenz Forschungsstelle (KLF) and a lake approximately 10 km to the south, where they roost at night. Geese are supplied with pellets and grain twice daily in the meadows around the research station, with low quantities from spring to fall and with sustaining quantities during winter. The flock is subjected to natural selection, and losses to predators, the most common being red foxes *(Vulpes vulpes)*, white-tailed eagles *(Haliaeetus albicilla)*, and golden eagles *(Aquila chrysaetos)*, and may account for up to 10% of the flock loss per year.[22] The flock consisted of approximately 170 individuals at the time of data collection. All are marked with colored leg bands. In addition to being raised by hand, the geese readily breed in the valley, either at natural nest sites or breeding boxes provided by the KLF. All geese are habituated to the presence of humans,[6] and goose-raised flock members neither show avoidance if approached to a distance of 1 m, nor do they excrete elevated levels of corticosterone-immunoreactive metabolites following such situations (Frigerio, unpublished data). This finding indicates that human presence does not cause stress and probably does not negatively affect agonistic motivation even in goose-raised geese. More detailed information of flock demography is presented elsewhere.[23,24]

Data Collection and Analysis

Our main research goal was to quantify costs of behavior in individual greylag geese, and we therefore continuously collected data and fecal samples for extraction of hormone metabolites—in particular, CORT and testosterone–immunoreactive metabolites. In general, data collection was such that we observed the geese starting from the morning feeding for up to 1 h. Observations were performed either on a "control day" with no experimental stressor or on a "stress day" when some sort of stressor was presented. Data presented here were collected under a "social density stress" feeding situation. During this experiment, food was spread widely (~160 m^2) on a control day, whereas the same amount of food was spread over a much closer area (~40 m^2, social density stress). This situation has been previously shown to induce a competitive feeding situation[25] (Frigerio, unpublished data) and actually produces an increased excretion of corticosterone–immunoreactive metabolites (Frigerio and Scheiber, unpublished data).

From the beginning of the distribution of food until 3 h thereafter, we collected fecal samples for extraction of CORT. Because geese defecated up to 11 times (mean ± SE: 7.15 ± 0.212) within 3 h, we attempted to collect short series of samples of feces per individual per observation day. In geese, fecal samples represent an integrated, proportional record of the plasma level within a frame of 2–4 h prior to defecation,[5,14,26] as gut passage time is not constant. To avoid the effect of diurnal variation, we collected individual fecal samples only during morning hours. Our sampling started well after the early morning peak of endogenous corticosterone.[27] Because separation of urine and feces in geese is not entirely possible,[35] we collected and analyzed both together. Samples were frozen at −20°C within 1 h after collection.

We assayed fecal samples with an enzyme immunoassay (EIA).[8,28] Fecal samples (0.5 g) were extracted in methanol and hydrolyzed as described by Kotrschal *et al.*[8] Recently, a new group-specific antibody was developed (Möstl *et al.*, unpublished data) that recognizes groups of metabolites (i.e., 5b,3a,11b-diol glucocorticoid metabolites) other than those previously used, which recognized 11b,210H,20-oxo-corticosterone metabolites. To assess the resolution of the 5β,3α,11β-diol glucocorticoid metabolite assay, and to validate the procedure, we also collected fecal samples from four domestic geese *(Anser domesticus)* in collaboration with the Department of Veterinary Medicine, University of Vienna (Möstl *et al.*, unpublished data). In this case, we collected every dropping over the course of 16 h, that is, from dusk to dawn, from four geese: two males and two females (range: 27–49 droppings). The recently developed assay was validated both through an experiment infusing trace amounts of radioactive-labeled hormone and an ACTH challenge experiment (Möstl *et al.*, unpublished data). We found that the new assay is considerably more sensitive in the biological sense, resulting in higher peak values in the same sample.[15] Therefore, all data presented were analyzed using the $5\beta,3\alpha,11\beta$-diol glucocorticoid metabolite assay.

Concentration limits for reliable measurements ranged from 0.15 ng/g to 250.5 ng/g. Intra- and interassay coefficients of variation were determined from homogenized pool samples. The mean intraassay coefficient of variation was 7.4%, and the interassay coefficient of variation was 13.2%.

For this study, we combined data from individuals that defecated either six ($N = 18$), seven ($N = 11$), eight ($N = 6$), or nine ($N = 3$) times in 3 h, which yielded 38 individual fecal series. To avoid pseudoreplication, each individual was used only once per situation ($N_{control} = 9$; $N_{social\ density\ stress} = 11$). However, some were used once in control and once in stress ($N_{control/social\ density\ stress} = 9$). If more than one series of a particular individual was available, we picked the longest series.

Data Analysis

Data were analyzed using the SPSS statistical package.[29] Results of all tests are two-tailed, and the level of significance was set to 0.05. Whether data were normally distributed was tested with Shapiro–Wilk tests, and appropriate parametric or nonparametric tests were then applied accordingly. Fecal samples were collected from male and female geese both in control and social density stress feeding situations. We analyzed possible differences between the sexes with Student's *t*-tests.

We determined whether differences existed between control and social density stress situations within the following parameters: (a) defecation intervals, (b) CORT means of all samples per series, (c) CORT maxima of the individual series, (d) CORT range of the samples in the series, and (e) coefficients of variance. Defecation intervals were defined as the elapsed time between the first and last sample divided by the length of the series, that is, the number of samples in the series. The range of the series was defined as the difference of the sample with the lowest concentrations of CORT within a series subtracted from the highest concentration of CORT within the same series. To compare the relative amount of variation of CORT within the series, we calculated the coefficient of variance, which determines the spread of the standard deviation from the mean as a percentage. Statistical comparisons were performed with parametric Student's *t*-tests.

Next we compared the differences in control and social density stress feeding situations in the nine individuals for which we had data in both situations: We compared individual means, maxima, ranges, and the variance coefficients with paired Student's *t*-tests.

To determine whether there was an excretion peak in CORT following a challenge after a certain time, we conducted Kruskal–Wallis ANOVAs, comparing the occurrence of maxima over ten 15-min intervals, starting from 30 min after the beginning of the feeding situation and continuing until 3 h after the feeding started.

The change of concentration between two consecutive samples over time was defined as the slope of CORT changes over time. Because data were not normally distributed, we calculated differences between the possible combinations with nonparametric Wilcoxon Signed Rank tests.

Often it is impossible to collect every dropping of an individual goose over a certain period, because, for example, defecation might occur in the water. Therefore, we decided to determine the number of samples necessary to predictably find the difference in CORT between control and social density stress, as well as which parameter is most likely to find this difference. In an attempt to mimic the collection of an incomplete series, we reduced all original six-sample series by one randomly chosen sample to obtain a five-sample series. This procedure was repeated five times. If we still found significant differences, we continued to reduce the original six-sample series by two, etc. We then calculated whether we could still find the difference in CORT between control and social density stress, using either the mean or the maximum. Because test results were equivocal when we reduced the original series by two, three, or four samples—some test results within these reduced series were significant, whereas others were not—we wanted to pinpoint the exact number of samples necessary to ensure that significant differences did not just occur by chance. We therefore repeated the reduction procedure as previously described five more times to get a total of 10 repetitions. We then used binomial tests to determine whether the difference in either means or maxima was statistically different between the four- and three-sample series or between the three- and two-sample series.

RESULTS

Within a series, concentrations can vary widely from one dropping to the next, both in control (FIG. 1a–1c) and experimental stress conditions (FIG. 1d–1f). Fluctuations in CORT over the course of the day are shown in FIGURE 2.

CORT values (expressed in nanograms per gram) were normally distributed both in control and social density stress in males and females. There was no significant difference of mean CORT between the genders either in control (females: $N = 10$; $\overline{X} \pm SE$ 32.25 ± 4.00; males: $N = 7$; $\overline{X} \pm SE$ 36.96 ± 8.27; Student's t-test corrected for unequal variances: $T = = -0.512$; df = 8.827; $P = .627$) or during social density stress (females: $N = 13$; $\overline{X} \pm SE$ 49.13 ± 3.77; males: $N = 8$; $\overline{X} \pm SE$ 56.39 ± 6.61; Student's t-test: $T = -1.030$; df = 19; $P = .316$). Therefore, we combined data form both genders in further analyses.

Geese defecation intervals were shorter during social density stress feedings ($N = 21$) than during control feedings ($N = 17$) (control: $\overline{X} \pm SE$ 16.62 [min] ± 1.01; social density stress: $\overline{X} \pm SE$ 13.28 ± 0.93 [min] ± 0.81; Mann–Whitney U test: $U = 88$, $P = .016$), indicating enhanced systemic CORT.

FIGURE 3 reveals that means of CORT concentration maxima ($T = -4.256$, df = 36; $P < .001$), series means ($T = -3.369$, df = 36; $P = .002$), and series ranges ($T = -3.588$, df = 36; $P = .001$) differed significantly between control ($N = 17$) and social density stress ($N = 21$), whereas the coefficients of variance did not ($T = -1.165$, df = 36; $P = .252$). Therefore, we did not combine data from control and social density stress.

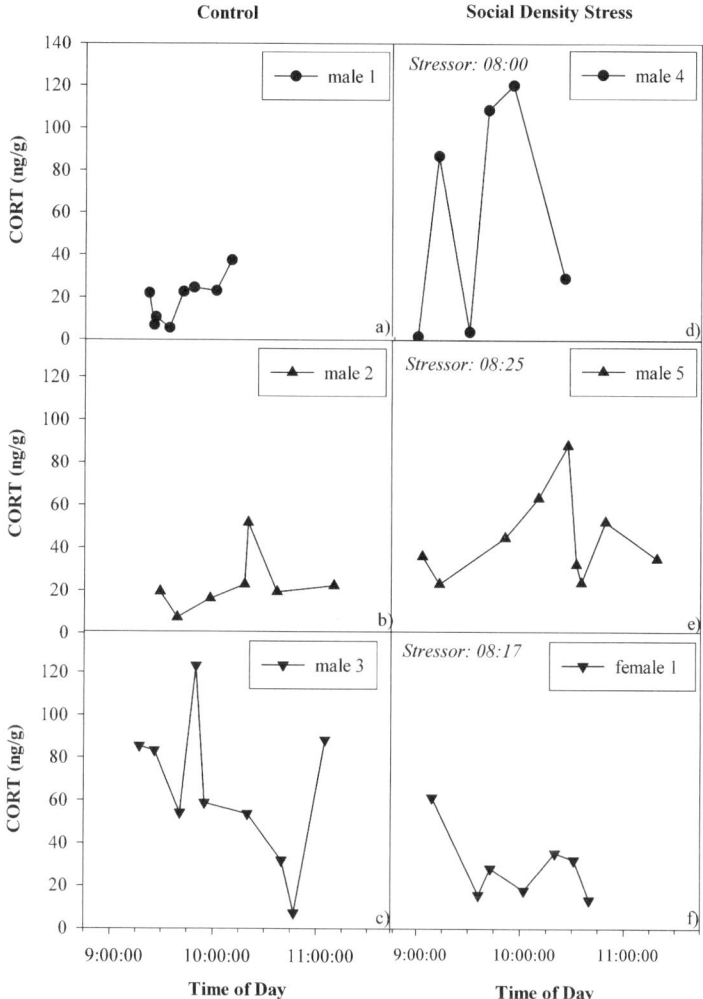

FIGURE 1. Examples of CORT excretion profiles over time during control (**a**)–(**c**) and social density stress feeding (**d**)–(**f**) in nanograms of CORT per gram of feces. Beginning of the social density stress feeding situation is depicted in the upper right corner of panels (d)–(f).

Similar results were obtained when we compared the nine individuals, for which we had data both during control and social density stress feeding situations (TABLE 1). Maxima and means of CORT concentration differed within individuals, and ranges of the series tended to differ, whereas the coefficients of variance did not differ between control and social density stress (TABLE 1).

The point when peak excretion occurred did not differ between control and social density stress (Mann–Whitney U test: $U = -159$; $P = .581$). Within either control series or social density stress series, there was no apparent excretion peak at a certain

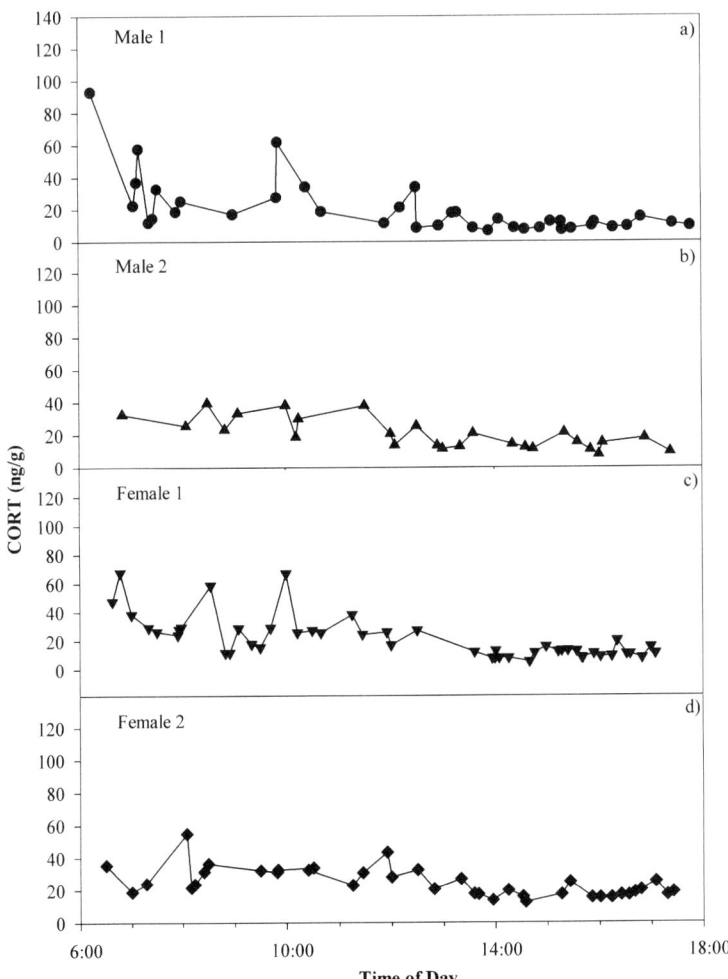

FIGURE 2. CORT excretion profiles of four domestic geese over time during control feeding in nanograms of CORT per gram of feces. Series are of different lengthes: (**a**) *filled circle*, 38-sample series; (**b**) *filled triange*, 49-sample series; (**c**) *inverted, filled triange*, 27-sample series; and (**d**) *filled diamond*, 38-sample series. Individuals in (a) and (b) are male; those in (c) and (d) are female.

point in time during the control period (Kruskal–Wallis ANOVA: $H = 10.5$, df = 7; $P = .62$). However, peak excretion occurred at approximately 120 min after a stressor was given (Kruskal–Wallis ANOVA: $H = 19.064$, df = 6; $P = .004$; FIG. 4).

We then calculated the concentration change over time between two consecutive samples (FIG. 5). All comparisons of control and social density stress values were not significantly different from one another and fluctuated around zero (Wilcoxon Signed Rank tests: results not significant).

TABLE 1. Statistical results of various parameters from CORT series of nine individual greylag geese used in control as well as social density stress (SDS)

	Mean ± S.E. control (ng/g)	Mean ± S.E. SDS (ng/g)	T	df	P
Maxima	39.56 ± 6.39	67.20 ± 9,30	−2.810	8	.023
Means	22.99 ± 5.02	40.34 ± 6.37	−2.390	8	.044
Ranges	25.13 ± 3.06	43.05 ± 9.20	−1.957	8	.086
Coefficients of variance	49.97 ± 6.93	47.15 ± 6.51	0.296	8	.775

NOTE: Paired Student's T-tests were applied.

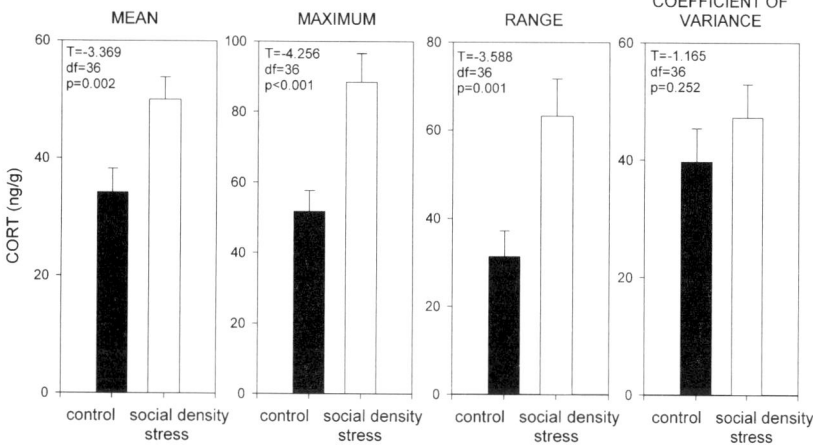

FIGURE 3. Differences in CORT (ng/g) means, maxima, ranges, and coefficients of variances. *Black bars* indicate control feeding situations; *white bars* indicate social density stress feeding situations. Standard errors are given. Note different scaling of the y-axis.

To determine how many samples were necessary and which parameter to use, we focused on series with six samples, for which we had the most examples ($N = 18$). Data for these series were normally distributed. Means and maxima differed significantly between control and social density stress situation (means: Student's t-test: $T = -2.591$, df = 16, $P = .020$; maxima: Student's t-test: $T = -3.673$, df = 16, $P = .002$).

When reducing the original six-sample series randomly by one sample to obtain five-sample series, the significant differences in control and social density stress between maxima and means were retained in all but one case (TABLE 2). This, however, was not the case if the original six-sample series were reduced by two samples at random to four-sample series. To calculate the number of samples necessary to reveal the difference between control and stress, we repeated the reduction procedure of the original six-sample series to four, three, and two samples, for a total of 10 repeats in means and maxima (TABLE 2). Four- versus three-sample series were significantly different in means (significant repeats four-sample series: $N = 8$, three-sample series: $N = 4$; bino-

TABLE 2. Random reduction of the original 6-sample series by one to four CORT samples repeated either five (1-sample reduction) or 10 times (2- to 4-sample reductions)

6-Sample series reduced by	Repeat 1/6	Repeat 2/7	Repeat 3/8	Repeat 4/9	Repeat 5/10
Mean					
1 ($\Rightarrow$ 5 samples used)	T = **–2.540** P = **.022**	T = **–2.177** P = **.045**	T = **–2.475** P = **.025**	T = –2.064 P = .056	T = **–2.804** P = **.013**
2 ($\Rightarrow$ 4 samples used)	T = –1.747 P = .124	T = **–2.956** P = **.021**	T = **–2.420** P = **.046**	T = **–2.933** P = **.022**	T = **–2.590** P = **.036**
	T = **–2.615** P = **.035**	T = –1.985 P = .088	T = **–2.905** P = **.023**	T = **–3.147** P = **.016**	T = **–2.947** P = **.022**
3 ($\Rightarrow$ 3 samples used)	T = –2.041 P = .081	T = –1.386 P = .208	T = **–2.941** P = **.022**	T = **–2.720** P = **.030**	T = –1.796 P = .115
	T = **–2.590** P = **.036**	T = –1.923 P = .096	T = –1.544 P = .166	T = **–2.420** P = **.046**	T = –1.455 P = .189
4 ($\Rightarrow$ 2 samples used)	T = –1.342 P = .221	T = –2.000 P = .086	T = –2.044 P = .080	T = –0.264 P = .800	T = –0.507 P = .335
	Z = –1.122 P = .262	T = –0.990 P = .355	Z = –1.828 P = .076	Z = –1.958 P = .052	T = –1.983 P = .088
Maximum					
1 ($\Rightarrow$ 5 samples used)	T = **–4.514** P = **.003**	T = **–2.664** P = **.032**	T = **–4.921** P = **.002**	T = **–2.475** P = **.025**	T = **–4.614** P = **.003**
2 ($\Rightarrow$ 4 samples used)	T = **–2.956** P = **.021**	T = **–4.095** P = **.005**	T = **–3.230** P = **.014**	T = **–3.668** P = **.008**	T = **–2.933** P = **.022**
	T = **–3.147** P = .016	T = **–2.047** P = **.022**	T = **–2.905** P = **.023**	T = **–3.987** P = **.005**	T = **–2.615** P = **.035**
3 ($\Rightarrow$ 3 samples used)	T = –2.041 P = .081	T = **–2.941** P = **.022**	Z = **–2.067** P = **.040**	T = **–3.237** P = **.014**	Z = –1.680 P = .093
	T = **–4.247** P = **.004**	T = **–3.685** P = **.008**	Z = –1.122 P = .262	T = –1.893 P = .100	T = **–2.353** P = **.050**
4 ($\Rightarrow$ 2 samples used)	Z = **–2.380** P = **.017**	Z = –1.820 P = .069	Z = –1.260 P = .208	Z = –0.560 P = .575	T = –0.990 P = .355
	Z = –1.431 P = .156	Z = –1.985 P = .088	T = **–2.177** P = **.045**	T = –1.386 P = .208	Z = –1.828 P = .076

NOTE: Student's *T*-tests or Wilcoxon Signed Rank tests were used as means or maxima of the reduced series; they were not always disributed normally. Significant differences between control and social density feeding situation appear in boldface type. Degrees of Freedom equaled seven (df = 7) in all Student's *T*-tests.

mial test: $P = .024$; TABLE 3a) and maxima (significant repeats four-sample series: $N = 10$, three-sample series: $N = 6$; binomial test: $P < .001$; TABLE 3b). The comparison between the three- to two-sample series revealed a difference when comparing the maxima (significant repeats three-sample series: $N = 6$, two-sample series: $N = 2$; binomial test: $P = .012$; TABLE 3d), but not the means (significant repeats three-sample series: $N = 4$, two-sample series: $N = 0$; binomial test: $P = .242$; TABLE 3c).

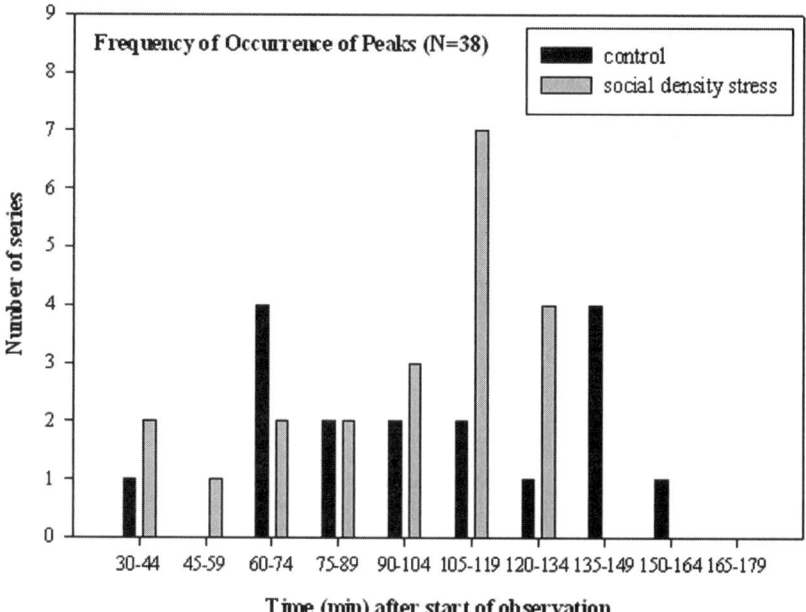

FIGURE 4. Histogram of the occurrences of maxima during control and social density stress feeding situations. Bins represent 14-min intervals, starting 30 min after the beginning of the feeding situation until more than 180 min after the beginning of the feeding situation have passed.

TABLE 3. 2 × 2 contingency tables of a number of significant and not significant test results either in the reduced 4- vs. 3-sample series (a and b) or 3- vs. 2-sample series (c and d)

	(a) 4- and 3-sample series			(b) 4- and 3-sample series		
Mean	Not significant	Significant	Maximum	Not significant	Significant
4 samples	2	8	4 samples	0	10
3 samples	6	4	3 samples	4	6
	(c) 3- and 2-sample series			(d) 3- and 2-sample series		
Mean	Not significant	Significant	Maximum	Not significant	Significant
3 samples	6	4	3 samples	4	6
2 samples	10	0	2 samples	8	2

NOTE: Means are displayed in panels (a) and (c); maxima are displayed in panels (b) and (d).

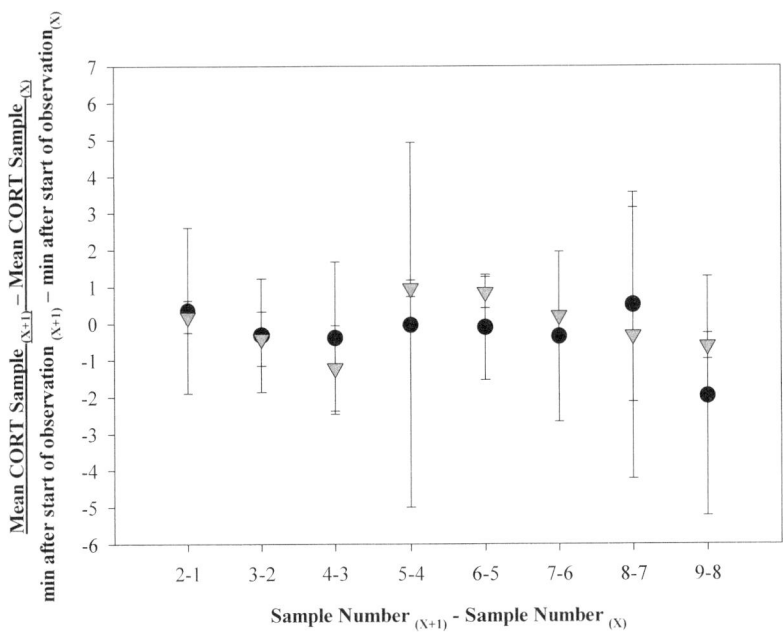

FIGURE 5. Concentration change in two consecutive CORT samples over time during control and social density stress. Data were calculated using the formula: Mean CORT Sample$_{(X+1)}$ − Mean CORT sample$_{(X)}$/min after start of observation$_{(X+1)}$ − min after start of observation$_{(X)}$.

DISCUSSION

We have demonstrated that, although CORT represents an integrated, proportional record of the corticosterone plasma level within a time frame of 2–4 h prior to defecation in geese,[5,7,8,26] it can be used to determine differences between individuals in response to an acute stressor if a sufficient number of samples are collected. We showed that means and ranges of the samples within a series as well as the maxima are suitable parameters for determining differences in control versus challenge situations. Similar results were found in black grouse, where Baltic et al.[30] found individual differences in means of excreted corticosterone immunoreactive metabolites despite large variation within the samples. In geese, the coefficient of variance, on the other hand, did not reveal a difference. This indicates that in geese, the variation around a mean in successive defecations was similar in control and social density stress feeding situations; therefore, a challenge does not change the scale of variance relative to the mean. In the case of geese, the aim to detect differences requires the collection of three samples and use of the maximum of these. One could also collect at least four samples and take the mean to determine individual differences. The more samples are used to calculate the mean, the more precise the results will be, and the greater the chance not to erroneously accept the null hypothesis (or commit a type II error). These findings apply to geese only, but might be different in other species.

Whether repeated fecal sampling is a useful alternative to collecting blood samples will depend not only on the study question and the feasibility of collection but also on several other important issues. For example, high defecation intervals in the relevant collection time are desirable. Also, the shorter the gut passage time of an animal, the more useful this method will be. It is particularly suitable for birds, in which gut passage times are shorter[5,8,12,13,26,31] than in mammals.[11,32] Consecutive urine[33] and fecal samples may show large variations in CORT, even in relatively short collection windows—for example, less than 1 h in chimpanzees and less than 20 min in geese. This may lead to over- or underestimation of differences both within individuals and between studies. Sufficient—and more standardized—sampling effort can level out this variation.

If fecal samples are used for measuring acute stress responses, several factors must be addressed before starting data collection. It should be determined how metabolite excretion profiles of the study species in question look. For example, not all avian species exhibit the excretion patterns found in geese. Chickens *(Gallus gallus domesticus)* show the expected, relatively smooth bimodal CORT excretion profile, with an early urine and later fecal peak measured over the course of the day.[34] The curve of the excretion profile will influence when and how many fecal samples should be collected. For example, domestic geese showed higher variation in consecutive fecal samples in the morning than in the afternoon (FIG. 2). This finding might indicate a biologically relevant difference in CORT excretion in the morning relative to the afternoon. It is also possible, however, that lower afternoon values result from geese habituating to the sampling procedure over the course of the day. In addition, greylag geese varied widely in their excretion of CORT: Whereas some series had wide variation between consecutive samples, others displayed only very little variation in CORT within a single series (e.g., FIG. 1, upper two panels). This was the case not only in control series but also in series collected during social density stress. Whether this lack of variability is biologically important, and how much variation constitutes biological relevance, needs to be determined. It should also be investigated whether this variability is a random effect between different series or if it is due to differences in individuals. If the latter were true, we would expect consistency in several series of the same individual. However, at this point we have too few series per individual to answer this question. An estimate of the number of samples necessary and which parameter to use is of uppermost importance to answer questions concerning acute stress responses. This will vary from species to species, but can vary even within the same species, depending on the type of stressor. Finally, for extraction, in general samples are weighed in at 0.5 g. Whether this is the best amount possible to even out heterogeneities within the sample itself is not known. It might be worthwhile to compare different amounts of the same fecal sample (i.e., 1 g, 2 g, etc.; Möstl, personal communication) to reduce the effect of a possible CORT concentration gradient within the sample. Ultimately, an ample fecal sampling scheme will result in more accurate assessments of physiological processes and their relationship to behavior.

ACKNOWLEDGMENTS

We gratefully acknowledge financial support from the FWF Project 15766-B03, from the "Verein der Förderer," and from the "Herzog von Cumberland Stiftung."

We are grateful to M. Kalas, M. Kirnbauer, V. Pilorz, and T. Stern for assistance in collecting fecal samples. The corticosterone enzyme immunoassays were conducted in the laboratories of E. Möstl at the Department of Biochemistry of the Veterinary University of Vienna as well as the laboratory ofJ. Dittami at the Department of Ethology, University of Vienna, and by A. Schöbitz and A. Aschauer. C. Schlögl, C. Pribersky-Schwab, and J. Hemetsberger offered statistical advice. S. Rettenbacher and C. Yussif helped collecting the domestic goose data, and the Möstl and Palme laboratories performed the biochemical validations of the new assay. B. Weiss. D. Frigerio, and the participants of the ESF technical workshop provided discussion on the topic. T. Bugnyar, W. Goymann, S. Jenni-Eiermann, and E. Möstl provided constructive criticism of the manuscript.

REFERENCES

1. TOUMA, C., R. PALME & N. SACHSER. 2004. Analyzing corticosterone metabolites in fecal samples of mice: a noninvasive technique to monitor stress hormones. Horm. Behav. **45:** 10–22.
2. MONFORT, S.L., J.L. BROWN & D.E. WILDT. 1993. Episodic and seasonal rhythms of cortisol secretion in male Eld's deer (*Cervus eldi thamin*). J. Endocrinol. **138:** 41–49.
3. THUN, R. *et al.* 1981. Twenty-four hour secretory pattern of cortisol in the bull: evidence of episodic and circadian rhythm. Endocrinology **109:** 2208–2212.
4. FULKERSON, W.J. & B.Y. TANG. 1979. Ultradian and circadian rhythms in the plasma concentration of cortisol in sheep. J. Endocrinol. **81:** 135–141.
5. HIRSCHENHAUSER, K. *et al.* 2000. Endocrine and behavioral responses of male greylag geese (*Anser anser*) to pairbond challenges during the reproductive season. Ethology **106:** 63–77.
6. FRIGERIO, D., B. WEISS, J. DITTAMI & K. KOTRSCHAL. 2003. Social allies modulate corticosterone excretion and increase success in agonistic interactions in juvenile hand-raised greylag geese (*Anser anser*). Can. J. Zool. **81:** 1746–1754.
7. HIRSCHENHAUSER, K., E. MÖSTL & K. KOTRSCHAL. 1999. Seasonal patterns of sex steroids determined from feces in different social categories of greylag geese (*Anser anser*). Gen. Comp. Endocrinol. **114:** 67–79.
8. KOTRSCHAL, K., K. HIRSCHENHAUSER & E. MÖSTL. 1998. The relationship between social stress and dominance is seasonal in greylag geese. Anim. Behav. **55:** 171–176.
9. KOTRSCHAL, K. *et al.* 2000. Effects of physiological and social challenges in different seasons on fecal testosterone and corticosterone in male domestic geese (*Anser domesticus*). Acta Ethol. **2:** 115–122.
10. SGOIFO, A. *et al.* 2001. Social stress: acute and long-term effects on physiology and behavior. Physiol. Behav. **73:** 253–254.
11. GOYMANN, W. *et al.* 1999. Noninvasive fecal monitoring of glucocorticoids in spotted hyenas, *Crocuta crocuta*. Gen. Comp. Endocrinol. **114:** 340–348.
12. CARERE, C. *et al.* 2003. Fecal corticosteroids in a territorial bird selected for different personalities: daily rhythm and the response to social stress. Horm. Behav. **43:** 540–548.
13. HIEBERT, S.M. *et al.* 2000. Noninvasive methods for measuring and manipulating corticosterone in hummingbirds. Gen. Comp. Endocrinol. **120:** 235–247.
14. HIRSCHENHAUSER, K. *et al.* 2000. Seasonal relationships between plasma and fecal testosterone in response to GnRH in domestic ganders. Gen. Comp. Endocrinol. **118:** 262–272.

15. FRIGERIO, D., J. DITTAMI, E. MÖSTL & K. KOTRSCHAL. 2004. Excreted corticosterone metabolites co-vary with ambient temperature and air pressure in male greylag geese (*Anser anser*). Gen. Comp. Endocrinol. **137:** 29–36.
16. PIAZZA, P.V. *et al.* 1993. Corticosterone in the range of stress-induced levels possesses reinforcing properties: implications for sensation-seeking behaviors. Proc. Natl. Acad. Sci. USA **90:** 11738–11742.
17. SCHWABL, H. 1995. Individual variation of the acute adrenocortical resonse to stress in the white-throated sparrow. Zoology **99:** 113–120.
18. SCHEIBER, I.B. *et al.* 2005. Active and passive social support in families of greylag geese (*Anser anser*). Behaviour. In press.
19. FRIGERIO, D., E. MÖSTL & K. KOTRSCHAL. 2001. Excreted metabolites of gonadal steroid hormones and corticosterone in greylag geese (*Anser anser*) from hatching to fledging. Gen. Comp. Endocrinol. **124:** 246–255.
20. LORENZ, K. 1988. Hier bin ichwo bist Du? Ethologie der Graugans. Piper Verlag. München.
21. HEMETSBERGER, J. 2001. Die Entwicklung der Grünauer Graugansschar seit 1973. *In* Konrad Lorenz und seine verhaltensbiologischen Konzepte aus heutiger Sicht. K. Kotrschal, G. Müller & H. Winkler, Eds.: 249–260. Filander Verlag. Fürth, Germany.
22. KOTRSCHAL, K., J. HEMETSBERGER & J. DITTAMI. 1992. Vigilance in a flock of semi-tame greylag geese (*Anser anser*) in response to approaching eagles *Haliaeetus albicilla* and *Aquila chrysaetos*. Wildfowl **43:** 215–219.
23. HEMETSBERGER, J. 2002. Populationsbiologische Aspekte der Grünauer Graugansschar (*Anser anser*). Ph.D. thesis, University of Vienna, Vienna.
24. KOTRSCHAL, K., J. HEMETSBERGER & B.M. WEISS. 2005. Homosociality in greylag geese. *In* Homosexuality Behaviour in Animals: An Evolutionary Perspective. P. Vasey & V. Sommer, Eds. Cambridge University Press. Cambridge. In press.
25. KOTRSCHAL, K., J. HEMETSBERGER & J. DITTAMI. 1993. Food exploitation by a winter flock of greylag geese: behavioral dynamics, competition and social status. Behav. Ecol. Sociobiol. **33:** 289–295.
26. KRAWANY, M. 1996. Die Entwicklung einer nichtinvasiven Methode zum Nachweis von Steroidhormonmetaboliten im Kot von Gänsen. Ph.D. thesis, Institute of Biochemistry, Veterinary University of Vienna, Vienna, Austria.
27. SCHÜTZ, K., B. WALLNER & K. KOTRSCHAL. 1997. Diurnal patters of steroid hormones from feces in greylag goslings (*Anser anser*). Adv. Ethol. **32:** 66.
28. MÖSTL, E. *et al.* 1987. Oestrogen determination in feces of mares by enzyme immunoassay on microtitre plates. Proc. Symp Analysis Steroids: 219–224. Sopron, Hungary.
29. SPSS. 2001. SPSS for Windows, version 11.0.1.
30. BALTIC, M. *et al.* 2005. A noninvasive technique to evaluate human-generated stress in black grouse. Ann. N.Y. Acad. Sci. **1046:** 81–95.
31. GOYMANN, W., E. MÖSTL & E. GWINNER. 2002. Corticosterone metabolites can be measured noninvasively in excreta of European Stonechats (*Saxicola torquata rubicola*). AUK **119:** 1167–1173.
32. WASSER, S., L. RISLER & R.A. STEINER. 1988. Excreted steroids in primate feces over the menstrual cycle and pregnancy. Biol. Reprod. **39:** 862–872.
33. ANESTIS, S.F. & R.G. BRIBIESCAS. 2004. Rapid changes in chimpanzee (*Pan troglodytes*) urinary cortisol excretion. Horm. Behav. **45:** 209–213.
34. RETTENBACHER, S. *et al.* 2004. Measurement of corticosterone metabolites in chicken droppings. Br. Poult. Sci. **45:** 704–711.
35. KLASING, K.C. 2005. Potential impact of nutritional strategy on noninvasive measurements of hormones in birds. Ann. N.Y. Acad. Sci. **1046:** 5–16.

Investigating Maternal Hormones in Avian Eggs: Measurement, Manipulation, and Interpretation

TON G. G. GROOTHUIS AND NIKOLAUS VON ENGELHARDT

Department of Behavioural Biology, University of Groningen, Haren, The Netherlands

ABSTRACT: The last decade has witnessed a surge in studies on steroid hormones of maternal origin present in avian eggs and affecting offspring development. The value of such studies for the understanding of maternal effects and individual differentiation is endorsed and a series of methodological and conceptual issues in the current approaches is discussed. First to be addressed is the topic of correct sampling of eggs or yolk for hormone analyses. Changes in yolk hormone levels during the incubation period and the uneven distribution of hormones within the egg are discussed. Different ways of calculating hormone levels and the importance of collecting data for specific *a priori* hypotheses are explained. Next to be discussed are the pros and cons of different techniques for manipulating yolk hormone levels and their proper scaling to naturally occurring levels. Third, several issues hampering the interpretation of results from descriptive and experimental studies are addressed. These concern biased embryonic mortality, clutch size, and egg quality that may confound the interpretation of the effect of egg position in the laying order, and the possibility of sex-specific effects and long-term effects. Also discussed are the probability of context-dependent results (due to, e.g., other egg components affecting egg quality, parental quality, and environmental factors), the difficulty in demonstrating adaptive effects due to individual optimization, and the lack of insight in the underlying physiological processes. Finally, it is concluded that this field has shown much progress but that it would profit from a more careful consideration of methodology and from a better integration of behavioral ecology and endocrinology.

KEYWORDS: maternal effects; maternal hormones; steroids; androgens; development; birds; egg quality, nongenomic inheritance

INTRODUCTION

The past decade has witnessed a surge in studies on patterns of maternal steroid hormone deposition and especially their effects on the offspring in avian species. Since the milestone paper by Hubert Schwabl,[1] which showed that substantial amounts of androgens are present in eggs of the canary and the zebra finch, androgens, estrogen, and corticosterone have been found in all avian eggs that have been analyzed for these hormones, from a number of both precocial and altricial species.[2]

These hormones are of maternal origin, since they are present in freshly laid eggs, long before the embryo would be able to produce hormones itself, and even in unfertilized eggs. Several studies have shown that the concentrations of these hormones, especially androgens, vary systematically with the position of the egg in the clutch and with several environmental factors such as social competition, food availability, mate attractiveness, laying date or photoperiod, and maternal parasite load (reviewed in Ref. 2). Moreover, a rapidly increasing number of studies demonstrated the wide array of the effects of these maternal hormones on the offspring, such as on hatching time, early muscular growth, early postnatal growth of body mass and structural size, early begging and competitive behavior, and prefledging survival. In addition, evidence is beginning to emerge that prenatal exposure to maternal androgens exerts long-lasting effects on morphology and behavior.[2]

These findings are of general importance. They demonstrate that parents can influence the development of their offspring not only by genetic inheritance but also by nongenetic inheritance. The advantage of the latter is that it can provide the parent with a relatively flexible tool to adjust offspring development to prevailing environmental conditions. Hormones are powerful tools for such a parental—or maternal— effect. Early exposure to steroid hormones can have a wide array of important organizing[3] and activating effects,[4] influencing not only sexual differentiation[3] but also the differentiation within the same sex,[5] perhaps leading to different personalities.[6] Furthermore, the production of steroid hormones such as androgens and glucocorticoids is strongly influenced by environmental factors,[7,8] and therefore offers an excellent pathway for the translation of these environmental factors, as perceived by the mother, to her offspring. Prenatal exposure to maternal hormones, such as androgens and glucocorticoids, has now been documented in several taxa of vertebrates other than birds (e.g., fish,[9] lizards,[10] turtles,[11] and mammals,[8,13] including humans[13]). Thus, the study of maternal hormones and their effects provides an excellent possibility to study general principles of mechanism, function, and evolution of nongenomic maternal effects.

Birds are especially adequate for such studies. The avian embryo develops in a sealed environment, the egg, outside the mother's body. Furthermore, avian eggs are relatively large, facilitating the sampling and manipulation of egg hormones without the interference of the mother. In addition, most avian species are studied under field conditions, facilitating the analysis of the function of maternal hormone deposition under natural circumstances. This has most likely stimulated the recent surge in studies of avian egg hormones. The time has now come to critically evaluate these studies, and this is the aim of two complementary contributions to this issue: one focusing on the methodology of the measurement of egg hormones,[14] and this paper, which focuses on conceptual issues and problems with the biological interpretation of the data.

COLLECTING DATA: WHERE, WHEN, AND WHAT?

Where and When

By far, most studies have measured egg steroid hormones in the yolk of the egg. These hormones are not evenly distributed within the yolk of freshly laid eggs but differ in their concentration among the concentric layers of the yolk that reflect the circadian pattern of yolk deposition.[15–17] This may have consequences for the interpretation of hormone levels in yolk samples.[14] However, most studies use samples of homogenized whole yolks. Because it seems that after a few days of incubation the structure of yolk layers disappears (own observation), and injection of a lipophilic substance revealed substantial diffusion of this substance over the whole yolk after a few days,[18] the functional significance of the difference in hormone levels among yolk layers may be absent. However, this distribution should be investigated in more detail by injecting labeled hormones into the egg and determining after a few days the recovery from several locations.

Such experiments should also check the possibility that hormones, deposited in the yolk during egg formation, diffuse to other parts of the egg in the course of incubation. If so, the analysis of hormone concentrations of the whole egg may be a more accurate measurement of embryonic exposure to these hormones than the analysis of the yolk only. Such diffusion may explain why yolk concentrations of steroid hormones substantially decrease after 1 day of incubation.[14,19,20] This decrease may lead to substantial bias in reported yolk hormone levels when timing of incubation is not carefully controlled for.

In addition, in the course of development embryos start to produce steroid hormones themselves in a sex-dependent manner. Substantial amounts are produced in, or after, the second week of incubation, and this may explain the increase in egg steroid hormone levels at the end of the incubation period.[19] In addition, lower levels may be present already in the first week because enzymes for the production of androgens and estrogens have been detected 2 days and 5 days, respectively, after the onset of incubation.[21] This may lead to an overestimation of levels of maternal hormones. However, data of embryonic steroid production in birds come from only a limited number of species, and to what extent and by what time in development the endogenously produced hormones reach the yolk is as yet unclear. Another possibility to consider is that the enzymes for hormone synthesis, being present early in development, may also convert the maternal hormones to other active or inactive metabolites. This may perhaps explain the function of the presence of large amounts of androstenedione in the egg, a precursor for other androgens that are more biologically active. Therefore, as long as we lack sufficient knowledge of the diffusion, production, and conversion of yolk steroid, eggs should be analyzed for the presence of maternal hormones as early in the incubation period as possible, ideally freshly laid. This would also avoid the potential problem that the mother deposit enzymes in the yolk that may metabolize the hormones in the period after the onset of incubation, a possibility so far not studied but that requires more attention.

What

Bird eggs contain several steroid hormones, such as testosterone, androstenedione, dihydrotestosterone, estradiol, and corticosterone.[2] They originate either from the mother or from the embryonic gonads. Although the effects of the hormones produced by the embryo itself on sexual differentiation and on the adult phenotype are well known,[22] information on the different effects of these hormones when of maternal origin is still very incomplete. There is a general tendency to focus on maternal testosterone. This is understandable since androstenedione is by itself biologically not very active, and testosterone can be converted to both 5-α-dihydrotestosterone and estradiol, all three being potent hormones. However, androstenedione is an important prohormone that can be converted to both testosterone and estradiol. Levels of maternal androstenedione are usually high compared with those of other hormones,[30] so androstenedione may constitute an important source for the embryo to produce testosterone and estradiol (T and E). Experiments manipulating specifically A4 levels and assessing yolk and embryonic plasma levels of T and E are urgently needed to test this hypothesis. Corticosterone and estradiol are present in the egg in much lower levels than androgens[2] and have received much less attention. The low levels of corticosterone and estradiol may be a protection against detrimental effects of these hormones on the vulnerable embryo, since persistent exposure to elevated levels of corticosterone is well known to be detrimental,[8] and estrogens affect sexual differentiation in birds.[22] Nevertheless, the small amounts of these hormones may still be biologically relevant for chick development. Recently, yolk corticosterone has received more attention, and the results suggest that it may have important effects on the offspring, too.[23]

In view of the presence of several steroid hormones in avian eggs, some studies have measured more than one hormone, especially all three androgens. However, without specific *a priori* hypotheses about the expected patterns of deposition or their effects on the chick, this has the serious risk of a positive bias in results, since multiple analyses increase the chance of yielding statistically significant findings, resulting in type II errors. Clearly, it is important to gain more knowledge about the specific effects of the different hormones by conducting a series of experiments to compare the effects of injections of the different hormones, separately and in combination, in eggs of the same species. On the basis of the question, these hormones should be injected in the similar concentrations (to compare the effectiveness of the hormones) or in the levels that occur in unmanipulated eggs (to study the effect of different maternal hormones as deposited by the mother).

Most studies present the hormone levels as the amount of hormone per milligram or milliliter of yolk. Other studies use the total amount of hormone in the whole yolk. This method requires the determination of total yolk weight for each egg. Because of the influx of water soon after incubation,[14] scaling to the start of incubation is important here. In blood plasma, the concentration of the hormone is generally the most appropriate parameter, but this is not necessarily the case for the study of yolk hormones. Yolk mass varies between and within clutches and may not always be associated with the body mass of the newly hatched chick. In such a case, the amount of hormone per unit of yolk may not be an adequate estimate of how much hormone reaches the embryo per unit body mass. Scaling the amount of hormone to total egg weight may not be a solution either, since weight differences between eggs may be due to the shell or to water content that do not contribute to the body mass of the chick.

In addition to yolk weight, other factors may influence hormone deposition, and their measurement and inclusion in the analysis can greatly enhance our understanding of the patterns in yolk hormone levels. First, the pattern of hormone deposition over the laying sequence as well as total levels of a certain hormone in a clutch can be substantially influenced by environmental or parental parameters (reviewed in Ref. 2). As a consequence, one study of a single population in a particular situation will provide a limited basis for the characterization of yolk hormone levels in that species if no additional contextual information is given. Furthermore, if the pattern across the laying sequence is indeed flexible, then one egg of a particular position in a clutch would not sufficiently characterize hormone deposition in that clutch. Second, such environmental or maternal parameters may mask interesting patterns. For example, we found in great tits (*Parus major*) that there was on average no pattern in hormone concentration over the laying sequence. However, when we analyzed the data in relation to the results of artificial selection on exploratory behavior, we found that one selection line showed a consistent increase, and the other a consistent decrease in androgen deposition over the laying sequence.[24]

Finally, evidence is beginning to emerge that levels of yolk hormone differ between eggs of male and female embryos,[25–28] which can become a confounder if avian mothers would be able to manipulate the primary sex ratio of their eggs. This problem can be solved by molecular sexing but would require a few days of incubation to allow sufficient embryonic growth for sampling DNA. Because incubation may affect yolk hormone levels (see earlier), it would be necessary to determine also the change of hormone levels across incubation time. Recent data suggest that after an initial decrease in hormone levels (24 h of incubation,[15,19] but see Ref. 20), these levels remain relatively stable for at least a few days. The importance of sex as a confounding variable is as yet unclear, since at the moment the evidence for substantial maternal control of the primary sex ratio in birds is ambiguous.[29]

How

The measurement of hormones in the egg; their extraction and separation; and the accuracy, specificity, precision, and sensitivity of the assay sometimes need more validation than provided, and we discuss this in a separate article in this issue.[14]

MANIPULATION OF EGG HORMONE LEVELS

To study the effects of yolk hormones, it is essential that the levels of these hormones be manipulated independently of other aspects of the egg and maternal quality. This is because yolk hormone levels may be associated with other aspects of egg quality, such as egg mass,[30] other yolk substances, such as carotenoids,[31] and aspects of maternal quality.[32] Thus, associations between unmanipulated levels of egg hormones with offspring performance can reveal only inconclusive results with respect to the effects of yolk hormones. Manipulating yolk hormone levels by manipulation of the mother (by modification of environmental conditions or by pharmacological tools) is inadequate, since it is likely to affect other aspects of egg quality and/or maternal quality as well, again confounding the results.

There are several techniques for treating avian eggs with hormones. An easy technique is dipping eggs in a solution containing the hormone, which then diffuses through the eggshell into the egg.[33,34] This has, however, the disadvantage that the effective dose that reaches the embryo, or at least the yolk and albumen, is unknown unless the effectiveness of the treatment is determined by assaying eggs before and after treatment or by studying the fate of a labeled hormone dissolved in the solution. Therefore, injection of known amounts into the egg appears to be more appropriate. This can be done in the air chamber, in the albumen, or in the yolk itself. In all cases, a known amount of hormone is dissolved in a solvent such as sesame oil. The first option may have the advantage that there is a relatively small chance that infections will affect the embryo in an early and vulnerable stage, before the bill penetrates the membrane of the air chamber. However, we (T.G.) found a relatively low hatching success with this technique, which may have been due to the oil that attaches to the membrane separating the air chamber from the rest of the egg, hampering air exchange. Furthermore, also in this case the effective dose is unknown. Therefore, the best option is injection of known amounts of hormone into the yolk, which is actually the location where the mother deposits the hormones. The distribution of the hormone after injection needs to be established and may depend on the solvent used. If the hormone does not dissolve in the yolk and is taken up in a high dose in a brief period of time, the injection of a natural amount of hormone may result in a pharmacological dosage.

Three issues need to be considered for all these techniques. First, it is essential that the dose used be scaled to the endogenous levels normally occurring in the species studied. In doing so, one must realize that the injected dose is added to the endogenous level already present. Unfortunately, such scaling is not always carried out. To complicate matters, it may be that levels of yolk hormones sampled in the year or population to which the injection is scaled are different from those in the year or population in which the experiment is actually carried out. A combination of biopsies and injection of the same egg would be ideal in this case.

Second, because injected levels are added to those already present, the resulting yolk level of the hormone will almost inevitably be at the upper level of the natural range. However, the hormone may have dose-dependent effects that are as of now hardly studied. The only study we are aware of that applied different doses of androgens in the same population of eggs[18] found dose-dependent effects on growth and immune function of the chick, suggesting that this issue deserves much more attention.

Third, the injection technique lowers hatchability of the egg, although this can in our experience improve considerably with practice. Although no difference in hatching success has yet been reported between eggs treated with the hormone in solvent and eggs injected with solvent only, a low hatching success may still be cause for some concern, not only for ethical reasons and statistical power but also for another reason: Using only well-hatched chicks may select for chicks of high quality. If the effect of the hormone depends on the quality of the chicks, or when there is a ceiling effect in measuring aspects of chick quality, such selection may bias the results to smaller or larger differences between experimental and control chicks.

In behavioral endocrinology, a standard approach to test the effect of a hormone is not only to enhance but also to lower exposure to the hormone. This result is usually achieved by using a receptor agonist or antagonist or by blocking the action of a specific enzyme needed for the production of the hormone. Unfortunately, we have as yet no tool to lower embryonic exposure to maternal hormones specifically. Lowering hormone deposition by affecting the egg-laying female can easily lead to confounding variables (see earlier). Injection of flutamide, a blocker of androgen receptors, in the egg has been applied,[35,36] but the half-life of this blocker in the egg and the timing of its uptake by the embryo is unknown. Therefore, it may block not only the effect of maternal androgens but also the effect of androgens produced by the embryo itself. To complicate this matter even further, it is conceivable that maternal androgens exert their effects very early in development, before the classic androgen receptors are present, by nongenomic mechanisms. In that case a classic receptor blocker would be of no use. Experiments in which the development of receptors is determined and in which androgen injections are conducted at different stages of development should provide the necessary information on this matter.

PROBLEMS OF INTERPRETATION

Apart from the problems of biased survival and uncontrolled covariates that have been mentioned previously, there are several other problems and pitfalls with the interpretation of results from descriptive and experimental studies. First, recent evidence indicates that yolk hormones may exert sex-specific effects on the offspring,[2,37] which requires sex-specific analysis of the data. For example, in a study on American kestrels it was found that experimentally elevated androgen levels in the first egg exerted detrimental effects on the chicks, whereas normally all eggs in clutches of this species have such elevated levels except for the first egg.[38] The European kestrel is a classicl example of a species showing primary sex ratio adjustment,[39] and it may be that the sex that is most vulnerable for the effect of elevated androgen levels was overproduced in the first egg.

Second, the effects of a specific hormone may depend on synergistic effects with other hormones or on other substances present in the egg. For example, maternal androgens may lower immune function of the chick,[18,28,40] but this effect may be counteracted by elevated levels of carotenoids enhancing immune function.[41] Therefore, the effect of yolk hormone manipulation may differ between species, populations, clutches, or even within clutches of the same female in case eggs differ in composition of other substances. For example, to study the effect of elevated androgen levels in last-laid eggs of black-headed gulls (*Larus ridibundus*), we used first-laid eggs of a clutch, which contain relatively low levels of androgens, and elevated their levels to those of last-laid eggs.[42] For our protocol, last-laid eggs could not be used because they contain high levels of androgens, and an additional elevation might have resulted in a supraphysiological level. We found results in the expected direction, but the effect of the induced androgen level in first-laid eggs may be different from similar levels in last-laid eggs due to the influence of other egg components that differ in concentration between these eggs.[28] This possibility may also explain the paradoxical effect found in the study on American kestrels, discussed previously.

This point relates to the important possibility of individual optimization in maternal hormone deposition that may severely hamper the interpretation of experimental results. Yolk hormones such as androgens can have not only beneficial but also detrimental effects on the chick (such as immune suppression[18,28,40]), and the production of the hormone may lead to elevated and detrimental levels in the mother, too (lowering, e.g., clutch size, reviewed in Ref. 2). Therefore, androgen deposition may be adjusted to the deposition of other egg substances and to maternal condition. Furthermore, in trading off costs and benefits of hormone deposition, avian mothers may optimize the latter in relation to environmental conditions, too. For example, avian mothers may adjust the transfer of androgens in relation to food availability (affecting the need for sibling competition, which is enhanced by maternal androgens,[42] reviewed in Ref. 2) and the risk of infectious diseases (androgens decrease

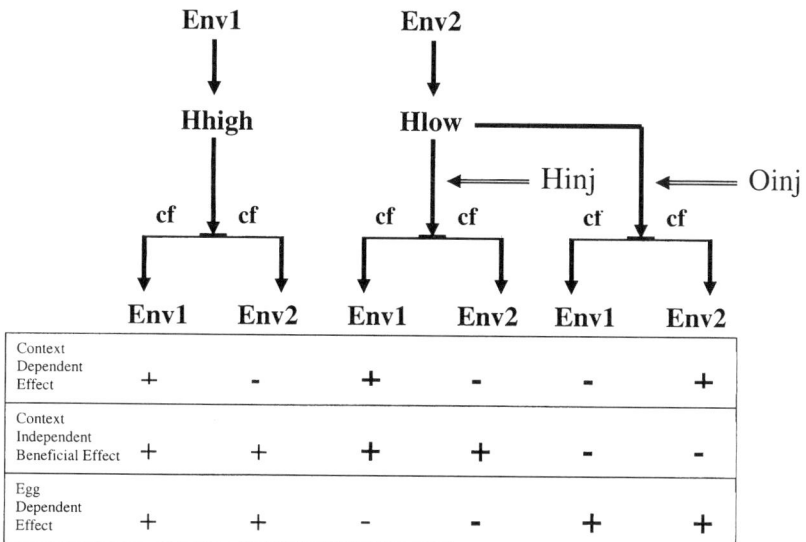

FIGURE 1. Experimental design to test context-dependent effects of elevated levels of maternal hormones assuming, that adjustment of hormone levels to the environment are beneficial for the chick. ABBREVIATIONS: Env: environment; Hhigh: high level of the yolk hormone; Hlow: low level of the yolk hormone; Hinj: hormone injection; Oinj: oil injection; cf: cross-fostering; + indicates beneficial effects on the offspring, − detrimental effects; bold signs indicate the most important part of the design.

immune function; see earlier). In theory, when all eggs contain the optimum level of androgens, experimental elevation of these hormones in the egg will always lead to detrimental effects on the chick, which would easily lead to the wrong conclusion that transfer of maternal androgens is not adaptive (FIG. 1, last row of the table). In the case where mothers adjust hormone deposition to their own (rearing) condition, one could avoid such paradoxical effects by cross-fostering the eggs to a random selection of other mothers. To investigate the adjustment of hormone deposition to environmental parameters, one could first test the effect of a certain environmental factor on the level of hormones in the yolk and then inject the hormone in those eggs that are produced under the condition that induced low hormone deposition. The effect of this treatment can then be analyzed in a design in which the eggs are cross-fostered to both environmental conditions (FIG. 1). Results of such experiments have not yet been published, but one such study is currently being conducted in our laboratory. It is also conceivable that the effect of yolk hormones depends on the genetic quality of the chick.[43] This could be tested by a combination of egg treatment and artificial insemination with semen of high- and low-quality males within the same design, as depicted in FIGURE 1.

Another issue that complicates the interpretation of the function of maternal hormones in the egg is the recent finding that they can have long-term consequences for the offspring, well after fledging and into adulthood.[15,37,44] If enough offspring survive to the reproductive stage, it would mean that monitoring short-term effects of the manipulation of yolk hormones is not sufficient for a conclusion about the functional significance of these hormones. For a reliable estimate of the fitness effects of exposure to maternal hormones in the egg, monitoring long-term survival and reproduction in the field is required, which is much more difficult than estimating survival until fledging.

A related issue is that maternal hormone deposition might be shaped by evolution to maximize fitness of the mother, not of her offspring. For example, in the case of deteriorating circumstances, mothers may benefit from reducing brood size at the cost of some individual chicks. In that case the number of chicks surviving until fledging in that particular reproductive bout may be the wrong estimate for maternal fitness, and her lifetime reproductive success should be estimated—again a much more challenging task than the estimation of current reproductive success.

Perhaps the most dominant functional hypothesis in the field is that maternal androgens adjust sibling competition, resulting from hatching asynchrony within the brood[1] (reviewed in Ref. 2). Female birds lay each egg with an interval of about 1 day or more, but incubation—and therefore embryonic development—generally starts before the last egg is laid, so that first-laid eggs hatch before late-laid eggs. This hatching asynchrony leads to age and size hierarchies among the nest members that compete for parental food provisioning. Therefore, many studies are interested in the patterns of maternal yolk hormones with respect to the position of the egg in the laying sequence within the clutch. Indeed, in most cases yolk hormone levels increase or decrease over the laying sequence.[2] A potential puzzle here is that clutches may substantially differ in number of eggs, and egg hormone levels may be related to clutch size. For example, if small clutches contain higher hormone levels than large clutches—even though laying order does not affect hormone levels at all—eggs of an early position appear to have on average higher levels than those of a later position when clutch size is not accounted for. On the other hand, if hormone levels increase with laying order, larger clutches appear to have on average higher levels than smaller clutches. Therefore, data should be plotted and analyzed, taking into account both laying position and clutch size. This is also important since in addition to the position of the egg in the laying sequence, clutch size itself may strongly affect hatching asynchrony (smaller in smaller clutches) and sibling competition (lower in smaller clutches).

In general, if one would like to test the hypothesis that the pattern of hormone deposition over the laying order functions as a tool to adjust sibling competition or the effect of hatching asynchrony (for a discussion of this possibility, see Ref. 2), then we need to know not only hormone levels in relation to the position of the egg in the laying sequence but also the actual degree of hatching asynchrony (that can be manipulated by the mother by varying her onset of incubation in the course of egg laying), as well as other factors that may affect sibling competition such as food availability. Such data are almost entirely lacking (but see Refs. 45 and 46). Moreover, the position of the egg in the laying sequence is often confounded with egg quality, since egg weight and composition usually change systematically with laying position.[28,31]

Finally, we invite researchers in this field to make a more cautious use of some terminology. Maternal hormone transfer into the egg is often referred to as an investment by the mother that reflects a certain strategy of allocating resources to different eggs. Our current knowledge does not justify this interpretation. There is currently no convincing evidence that hormones are costly resources or that their deposition into the egg is costly for the mother. The steroid hormones themselves are produced in very small amounts starting from cholesterol, which is abundantly present. It has also not been demonstrated that the transfer of hormones into the egg necessarily requires that the mother herself be exposed to these hormones too, which may inflict some cost on her. The mechanism of hormone deposition is as yet unclear, in particular whether hormone deposition into the egg can be regulated independently from the regulation of hormone levels in the circulation of the female. An independent regulation would suggest that avian hormone deposition is designed by evolution, indicating that it is an adaptation rather than an exaptation.[47] Although such an independent regulation is conceivable, evidence is ambiguous at the moment (for an extensive discussion of this issue, see Ref. 2). Some studies have found a positive relationship between maternal levels of androgens and the concentration of these hormones in the egg of these females,[15,48] but others have not.[49,50] More insight into the physiological mechanism of hormone deposition is urgently needed, which can be obtained both by sampling the plasma of females around the time of yolking the egg and sampling that egg directly after it is laid, as well as by experimental studies. The largest problem for the descriptive approach is the substantial temporal dynamics of plasma levels of hormones during egg laying[48] and of hormone deposition in the yolk,[17,51] although it may be of great help to use the different yolk layers to estimate the actual time of deposition.

CONCLUSION

Over the past decade it has been shown that avian species offer excellent opportunities to study hormone-mediated maternal effects, and the field has made great progress. Both descriptive and experimental studies suggest that the hormone deposition in avian eggs reflects a maternal tool to maximize her fitness. So far, androgens have received most attention, perhaps because they are present in the egg in large amounts. However, the time is ripe to study other hormones in more detail as well.

Also, there are as yet almost no studies regarding the physiological mechanisms and regulation of hormone deposition into the egg, and of the uptake and action on the developing embryo—such studies are now urgently needed. Optimally, the study of egg hormones should take place at the interface of the ultimate and proximate approach, integrating such diverse fields as behavioral ecology, animal physiology, and embryology. Finally, some aspects of the methods used, and some conceptual problems, especially with regard to the adaptive value of hormone deposition, need more careful consideration, and the field can benefit from new experimental approaches.

ACKNOWLEDGMENTS

We thank Wolfgang Goymann and Susi Jenni-Eiermann for organizing the inspiring workshop, "Analysis of Hormones in Droppings and Egg Yolk of Birds," and for editing this special issue. We thank all the members of the workshop for their open and constructive contributions to the discussions, and Claudio Carere for valuable suggestions that improved the manuscript.

REFERENCES

1. SCHWABL, H. 1993. Yolk is source of maternal testosterone for developing birds. Proc. Natl. Acad. Sci. USA **90:** 11446–11450.
2. GROOTHUIS, T.G.G. et al. 2005. Maternal hormones as a tool to adjust offspring phenotype in avian species. Neurosci. Biobehav. Rev. **29:** 329–352.
3. COOKE, B., C.D. HEGSTROM, L.S. VILLENEUVE & S.M. BREEDLOVE. 1998. Sexual differentiation of the vertebrate brain: principles and mechanisms. Front. Neuroendocrinol. **19:** 323–362.
4. ROS, A.F., S.J. DIELEMAN & T.G. GROOTHUIS. 2002. Social stimuli, testosterone, and aggression in gull chicks: support for the challenge hypothesis. Horm. Behav. **41:** 334–342.
5. RHEN, T. & D. CREWS. 2002. Variation in reproductive behaviour within a sex: neural systems and endocrine activation. J. Neuroendocrinol. **14:** 517–531.
6. GROOTHUIS, T.G. & C. CARERE. 2005. Avian personalities: characterization and epigenesis. Neurosci. Biobehav. Rev. **29:** 137–150.
7. WINGFIELD, J.C. et al. 1990. The challenge hypothesis—theoretical implications for patterns of testosterone secretion, mating systems, and breeding strategies. Am. Nat. **136:** 829–846.
8. WEINSTOCK, M. 1997. Does prenatal stress impair coping and regulation of hypothalamic-pituitary-adrenal axis? Neurosci. Biobehav. Rev. **21:** 1–10.
9. MCCORMICK, M.I. 1999. Experimental test of the effect of maternal hormones on larval quality of a coral reef fish. Oecologia **118:** 412–422.
10. LOVERN, M.B. & J. WADE. 2001. Maternal plasma and egg yolk testosterone concentrations during embryonic development in green anoles (*Anolis carolinensis*). Gen. Comp. Endocrinol. **124:** 226–235.
11. BOWDEN, R.M., M.A. EWERT & C.E. NELSON. 2000. Environmental sex determination in a reptile varies seasonally and with yolk hormones. Proc. R. Soc. Lond. B Biol. Sci. **267:** 1745–1749.
12. GLICKMAN, S.E. et al. 1987. Androstenedione may organize or activate sex-reversed traits in female spotted hyenas. Proc. Natl. Acad. Sci. USA. **84:** 3444–3447.
13. HINES, M. et al. 2002. Testosterone during pregnancy and gender role behavior of preschool children: a longitudinal, population study. Child Dev. **73:** 1678–1687.
14. VON ENGELHARDT, N. & T.G. GROOTHUIS. 2005. Measuring steroid hormones in avian eggs. Ann. N.Y. Acad. Sci. **1046:** 181–192.
15. EISING, C.M. 2003. Mother knows best? Costs and benefits of differential maternal hormone allocation in birds. Ph.D. thesis. University of Groningen. Groningen, the Netherlands.
16. HACKL, R. et al. 2003. Distribution and origin of steroid hormones in the yolk of Japanese quail eggs (*Coturnix coturnix japonica*). J. Comp. Physiol. [B] **173:** 327–331.
17. LIPAR, J.L., E.D. KETTERSON, V. NOLAN. JR. & M. CASTO. 1999. Egg Yolk layers vary in the concentration of steroid hormones in two avian species. Gen. Comp. Endocrinol. **115:** 220–227.
18. NAVARA, K.J., G.E. HILL & M.T. MENDONCA. 2005. Variable effects of yolk androgens on growth, survival, and immunity in eastern bluebird nestlings. Physiol. Biochem. Zool. In press.
19. ELF, P.K. & A.J. FIVIZZANI. 2002. Changes in sex steroid levels in yolks of the leghorn chicken, *Gallus domesticus*, during embryonic development. J. Exp. Zool. **293:** 594–600.
20. EISING, C.M., W. MÜLLER, C. DIJKSTRA & T.G. GROOTHUIS. 2003. Maternal androgens in egg yolks: relation with sex, incubation time and embryonic growth. Gen. Comp. Endocrinol. **132:** 241–247.

21. BRUGGEMAN, V., P. VAN AS & E. DECUYPERE. 2002. Developmental endocrinology of the reproductive axis in the chicken embryo. Comp. Biochem. Physiol. A **131:** 839–846.
22. BALTHAZART, J. & E. ADKINS-REGAN. 2002. Sexual differentiation of brain and behavior in birds. In Hormones, Brain and Behaviour. D. Pfaff et al., Eds.: 223–301. Academic Press. San Diego.
23. HAYWARD, L.S. & J.C. WINGFIELD. 2004. Maternal corticosterone is transferred to avian yolk and may alter offspring growth and adult phenotype. Gen. Comp. Endocrinol. **135:** 365–371.
24. CARERE, C. 2003. Personalities as epigenetic suites of traits. A study on a passerine bird. Ph.D. thesis. University of Groningen. Groningen, the Netherlands.
25. PETRIE, M., H. SCHWABL, N. BRANDE-LAVRIDSEN & T. BURKE. 2001. Sex differences in avian yolk hormone levels. Nature **412:** 498–499.
26. MÜLLER, W., C.M. EISING, C. DIJKSTRA & T.G. GROOTHUIS. 2002. Sex differences in yolk hormones depend on maternal social status in Leghorn chickens (Gallus gallus domesticus). Proc. R. Soc. Lond. B. Biol. Sci. **269:** 2249–2255.
27. RUTSTEIN, A.N. et al. 2005. Sex-specific patterns of yolk androgen allocation depend upon maternal diet in the zebra finch. Behav. Ecol. **16:** 62–69.
28. MÜLLER, W. 2004. Maternal phenotypic engineering: adaptation and constraint in prenatal maternal effects. Ph.D. thesis. University of Groningen. Groningen, the Netherlands.
29. EWEN, J.G., P. CASSEY & A.P. MØLLER. 2004. Facultative primary sex ratio variation: a lack of evidence in birds? Proc. R. Soc. Lond. B Biol. Sci. **271:** 1277–1282.
30. GROOTHUIS, T.G. & H. SCHWABL. 2002. Determinants of within- and among-clutch variation in levels of maternal hormones in Black-Headed Gull eggs. Funct. Ecol. **16:** 281–289.
31. ROYLE, N.J., P.F. SURAI & I.R. HARTLEY. 2001. Maternally derived androgens and antioxidants in bird eggs: complementary but opposing effects? Behav. Ecol. **12:** 381–385.
32. PILZ, K.M. et al. 2003. Interfemale variation in egg yolk androgen allocation in the European starling: do high-quality females invest more? Anim. Behav. **65:** 841–850.
33. PERRIN, F.M.R., S. STACEY, A.M.C. BURGESS & U. MITTWOCH. 1995. A quantitative investigation of gonadal feminization by diethylstilbestrol of genetically male embryos of the quail coturnix-coturnix-japonica. J. Reprod. Fertil. **103:** 223–226.
34. LAY, D.C. & M.E. WILSON. 2002. Development of the chicken as a model for prenatal stress. J. Anim. Sci. **80:** 1954–1961.
35. LIPAR, J.L. & E.D. KETTERSON. 2000. Maternally derived yolk testosterone enhances the development of the hatching muscle in the red-winged blackbird *Agelaius phoeniceus*. Proc. R. Soc. Lond. B Biol. Sci. **267:** 2005–2010.
36. MÜLLER, W. et al. 2005. An experimental study on the causes of sex-biased mortality in the black-headed gull—the possible role of testosterone. J. Anim. Ecol. In press.
37. VON ENGELHARDT, N. 2004. Proximate control of avian sex allocation. A study on zebra finches. Ph.D. thesis. University of Groningen. Groningen, the Netherlands.
38. SOCKMAN, K.W. & H. SCHWABL. 2000. Yolk androgens reduce offspring survival. Proc. R. Soc. Lond. B Biol. Sci. **267:** 1451–1456.
39. DIJKSTRA, C., S. DAAN & J.B. BUKER. 1990. Adaptive seasonal variation in the sex ratio of kestrel broods. Funct. Ecol. **4:** 143–147.
40. GROOTHUIS, T.G., C.M. EISING, C. DIJKSTRA & W. MÜLLER. 2005. Balancing between costs and benefits of maternal hormone deposition in avian eggs. Proc. R. Soc. Lond. B Biol. Lett. **1:** 78–81.
41. SAINO, N. et al. 2003. Experimental manipulation of egg carotenoids affects immunity of barn swallow nestlings. Proc. R. Soc. Lond. B Biol. Sci. **270:** 2485–2489.
42. EISING, C.M., C. EIKENAAR, H. SCHWABL & T.G. GROOTHUIS. 2001. Maternal androgens in black-headed gull (*Larus ridibundus*) eggs: consequences for chick development. Proc. R. Soc. Lond. B Biol. Sci. **268:** 839–846.
43. GIL, D., J. GRAVES, N. HAZON & A. WELLS. 1999. Male attractiveness and differential testosterone investment in zebra finch eggs. Science **286:** 126–128.

44. STRASSER, R. & H. SCHWABL. 2004. Yolk testosterone organizes behavior and male plumage coloration in house sparrows (*Passer domesticus*). Behav. Ecol. Sociobiol. **56:** 491–497.
45. ELLIS, L.A., D.W. BORST & C.F. THOMPSON. 2001. Hatching asynchrony and maternal androgens in egg yolks of house wrens. J. Avian Biol. **32:** 26–30.
46. MÜLLER, W. *et al.* 2004. Within-clutch patterns of yolk testosterone vary with the onset of incubation in black-headed gulls. Behav. Ecol. **15:** 893–897.
47. GOULD, S.J. & E.S. VRBA. 1982. Exaptation—a missing term in the science of form. Paleobiology **8:** 4–15.
48. SCHWABL, H. 1996. Environment modifies the testosterone levels of a female bird and its eggs. J. Exp. Zool. **276:** 157–163.
49. TANVEZ, A. *et al.* 2004. Sexually attractive phrases increase yolk androgen deposition in Canaries (*Serinus canaria*). Gen. Comp. Endocrinol. **138:** 113–120.
50. VERBOVEN, N., P. MONAGHAN, D.M. EVANS, *et al.* 2003. Maternal condition, yolk androgens and offspring performance: a supplemental feeding experiment in the lesser black-backed gull (*Larus fuscus*). Proc. R. Soc. Lond. B Biol. Sci. **270:** 2223–2232.
51. BAHR, J.M., S.C. WANG, M.Y. HUANG & F.O. CALVO. 1983. Steroid concentrations in isolated theca and granulosa layers of preovulatory follicles during the ovulatory cycle of the domestic hen. Biol. Reprod. **29:** 326–334.

Measuring Steroid Hormones in Avian Eggs

NIKOLAUS VON ENGELHARDT AND TON G. G. GROOTHUIS

Research Group Animal Behaviour, University of Groningen, 9750 AA Haren, the Netherlands

ABSTRACT: Avian eggs contain substantial levels of various hormones of maternal origin and have recently received a lot of interest, mainly from behavioral ecologists. These studies strongly depend on the measurement of egg hormone levels, but the method of measuring these levels has received little attention. This paper describes the sampling, extraction, and assay of hormones in avian eggs and related methodological problems. The method of sampling is important because hormones are not homogeneously distributed within the egg, and after onset of embryonic development their levels may decrease and increase due to changes in egg structure and secretion or uptake of hormones by the embryo. The extraction of hormones from the yolk and chromatographic separation of different hormones for immunoassays can strongly influence the results because such procedures remove interfering substances such as proteins, lipids, and other hormones and their metabolites, which can cross-react with the antiserum used. Finally, the assay itself needs more validation than many studies report, especially with respect to the accuracy and specificity of the hormone measurements. We conclude that the addressed issues need more attention for the correct interpretation of differences in hormone levels within and between studies.

KEYWORDS: maternal hormones; steroids; androgens; testosterone; validation; immunoassay; egg; yolk; albumen

INTRODUCTION

The discovery of hormones of maternal origin—such as androgens, estrogens, progesterone, corticosterone, and thyroid hormones—in avian eggs and their impact on offspring development[1-8] has stimulated many studies measuring and manipulating hormones in avian eggs (for recent reviews, see Refs. 9 and 10). Most studies focus on the intriguing question of whether deposition of maternal hormones reflects an adaptive maternal effect. For example, can a female bird influence offspring development by varying the deposition of hormones in her eggs, and in this way adjust offspring development to prevailing environmental conditions? Other questions include the assessment of maternal endocrine state using eggs instead of more invasive plasma samples,[11,12] the assessment of environmental contamination with steroids

Address for correspondence: N. von Engelhardt, Research Group Animal Behaviour, University of Groningen, P.O. Box 14, 9750 AA Haren, the Netherlands. Voice: +31-503637850; fax: +31-50-3632148.
nvengelhardt@gmx.de

or endocrine disruptors,[13] and the assessment of offspring endocrine state using hormones in allantoic fluid.[14,15]

This article deals with the methodology of hormone measurements in avian eggs. It is concerned primarily with (sex) steroid hormones of maternal origin in the yolk of eggs, but the general principles hold for the measurement of hormones in egg white and allantoic fluid of other origin. The paper complements others in this issue on conceptual questions in the study of maternal hormones,[16] the measurement of steroid hormones in feces,[17] the transfer of corticosteroids from the maternal circulation into the egg,[12] and steroid hormones in allantoic fluid.[15]

It has been known for a long time that avian egg yolk has hormone-like activity,[18] but only since Hubert Schwabl[1] demonstrated that steroid hormones are present in avian eggs, did they begin to attract a lot of attention, especially from behavioral ecologists interested in adaptive maternal effects. These studies use immunoassays to determine egg levels of hormones, both for describing patterns of maternal hormone deposition in relation to, for example, environmental factors, and for estimating dosages for experimental manipulation of egg hormone levels within the natural range. Clearly, such studies rely heavily on the proper use and validation of hormone quantification methods used, but these methods have not yet received the attention they deserve.

A comparison of nine different studies measuring androgens in egg yolks of zebra finches and black-headed gulls shows both the wide variation in methods used and the wide variation in the measured levels (TABLE 1; studies 6 and 9 show our unpublished data (N. von Engelhardt, and M. Lasthuizen, B. de Vries, and T. Groothuis, respectively), and part of the values of the other studies are calculated from the data in the original papers).[1,4,19-24] The differences in levels may partly be due to true differences in androgen content of eggs of different populations, years, or experimental conditions. This cannot be the case for three studies (3, 6, 9), in which differences of 150%–400% were found when assaying the same samples using simplified extraction with ether, additional purification with ethanol and hexane to remove proteins and lipids, or chromatographic purification and separation. The presence of proteins and lipids[25] or of steroids that cross-react with the antiserum used are the most important causes of the diverging results.

This example indicates the problems that can be encountered when measuring egg hormone levels. The qualitative and quantitative measurements of chemical analytes using indirect methods such as immunoassays are usually validated by determining their *accuracy* (closeness to "true" value), *specificity* (the degree to which the method also measures other compounds), *precision* (variability of repeated measures), and *sensitivity* (an indication of the smallest amount of analyte that can be detected with the method).

Unreliable estimates of hormone levels are a special concern for studies that use published data to compare absolute egg hormone levels or relative differences between species or populations. They also can be a problem for studies that intend to draw quantitative conclusions regarding causes or consequences of variation in egg hormones. In addition, they are problematic if one wants to accurately manipulate yolk hormone levels within the ranges that can be encountered under normal conditions in the species studied.

The goal of this paper is to give an overview of the currently used methods to sample, extract, and assay egg hormones; to point out which methodological issues

TABLE 1. Androgen/testosterone levels in yolks of zebra finch (1–6) and black-headed gull eggs (7–9)

Study/reference	Extraction[a]	Extraction efficiency (%)	Parallelism	Accuracy Known amounts (%)	Testosterone antibody	DHT	Specificity[b] A4	Mean ± SD (pg/mg)
1/Schwabl[1]	1,2,3,4	62	Yes	110	Wien	—	—	6 ± 2
2/Gil[19]	1,2	70	—	—	Biotrak (Amersham)	40–50	—	4.5 ± 3
3a/Gil[20]	1	—	—	—	DSL-4000	6	2.3	10.5 ± 4
3b/Gil[20]	1,2,3	—	—	—	DSL-4000	6	2.3	16.9 ± 7
4/Ward[21]	1	—	—	—	Coat-a-Count (Diagnostics)	0.04	0.01	6.2 ± 1
5/Rutstein[22]	1,2	76	—	—	Nash[53]	46	0.5	8 ± 4
6a/von Engelhardt	1,2	81	No	82	Esoterix T3-125	44	2.0	9.4 ± 3
6b/von Engelhardt	1,2,3	83	Yes	94	Esoterix T3-125	44	2.0	16.3 ± 6
6c/von Engelhardt	1,2,3,4	51	Yes	88	Esoterix T3-125	44	2.0	13.5 ± 6
7/Eising[4]	1,2,3,4	—	—	—	Esoterix T3-125	44	2.0	13 ± 4
8/Groothuis[23]	1,2,3,4	56	—	—	Wien	—	—	8.5 ± 7
9a/Müller[24]	1,2,4	49	—	—	Esoterix T3-125	44	2.0	17.1 ± 8.7
9b/Lasthuizen	1	85	—	—	Spectria (Orion)	4.5	—	64.0 ± 11
9c/Lasthuizen	1,2,3	74	—	—	Spectria (Orion)	4.5	—	79.3 ± 12
9d/Lasthuizen	1,2,3,4	63	—	—	Spectria (Orion)	4.5	—	17.8 ± 3

NOTE: In studies 3, 6, and 9, the same samples were measured using different extraction procedures. For the explanation of the table headers, see text.
[a]Extraction: 1 = ether; 2 = 90% ethanol; 3 = hexane; 4 = celite-chromatography.
[b]Specificity: Cross-reactivities with DHT and A4 at 50% binding of testosterone.

require particular attention and how problems can be tackled; and to demonstrate how important validation studies are.

SAMPLING

Inside the yolk of freshly laid eggs, hormones are not distributed homogeneously; rather, there is a radial gradient reflecting the concentric layering of yolk.[7,12,26–28] Hormones in the peripheral layers of the yolk may be taken up at different stages of development by the embryo than hormones in the center of the yolk, and therby serve different functions. Therefore, measurement of hormone levels in a random biopt or from a whole homogenized yolk would miss potentially valuable information. According to our own experience it may, however, in practice be very difficult—if not impossible—to sample yolk from specific locations of a live egg, because the yolk moves within the egg when touched by the needle used for taking the biopsy specimen.

The levels of different hormones in the yolk change rapidly after the onset of incubation, although the cause of this change is unknown.[28–30] Currently, there is little information concerning changes in yolk structure during embryonic development, the changes in the distribution of yolk hormones within the egg over time, and the stages at which different parts of the yolk are taken up by the embryo. There is an influx of water from the albumen in the first days after incubation, which dilutes and enlarges the yolk.[31,32] A decrease in yolk hormone concentrations is therefore expected, and hormones might also diffuse within the yolk and to albumen and allantois. Egg hormone levels may also decrease or increase during embryonic development due to the uptake, secretion, and metabolism of hormones by the developing embryo (see also Ref. 16). Because yolk hormone levels do not remain constant during development, one must be careful when drawing conclusions regarding maternal hormones when measuring hormone levels after the onset of embryonic development.

It is therefore advisable to assess hormone levels immediately after eggs are laid. In field research, this may not be possible, so at the very least, the developmental stage at which a sample has been taken should be assessed by, for example, measuring yolk and embryo size. The obvious problem here is that yolk hormones can accelerate or delay embryonic development,[4,5,33] so embryonic size may not indicate developmental stage independently of the hormone concentration in that egg.

Finally, a question that has hardly been addressed is the presence of hormones in the albumen—almost all studies focus on yolk hormones. This focus is based on the assumption that the yolk contains most of the hormone in the egg due to the lipophilic nature of steroid hormones and that the hormone is deposited during yolk formation. However, hormone levels in albumen can be similar to those in the yolk[11,32] and require further study because they can certainly also be relevant for offspring development. It is conceivable that hormone in the albumen is partly derived from the follicular wall (via diffusion from the yolk after ovulation) and partly deposited during albumen production in the magnum and/or when more water is added to the egg in the shell gland. Albumen is produced in about a day, whereas yolk deposition takes several days. Therefore, the levels of hormones in the albumen may better reflect short-term changes in the plasma levels of hormones than those in the yolk.

EXTRACTION

Hormones cannot be measured directly in the yolk, and extraction and purification is necessary to remove interfering substances. So-called matrix effects, caused by substances such as proteins and lipids that can bind hormones or interfere with the binding of the hormone to antibody and charcoal, can strongly influence the results of immunoassays.[25] The removal of lipids requires particular attention because many factors can influence both maternal hormones and the presence of lipids in avian eggs. Therefore, apparent differences in hormone levels might rather be due to differences in lipid content.

Most studies on avian eggs follow the extraction protocol published by Schwabl.[1] First, samples are extracted twice with 3 mL of petroleum ether/diethyl ether, 30:70 (vol:vol), a combination of solvents that extracts many different steroids from eggs. Samples are then dissolved in 90% ethanol and frozen overnight to precipitate proteins and neutral lipids. Finally, more lipids are removed by washing samples with hexane. Diethylether alone or 80% methanol has also been used for extraction of steroid hormones from the egg matrix.[6,7]

The physiochemical properties of different solvents determine the extent to which they extract different hormones and interfering substances (e.g., petroleum ether extracts nonpolar steroids such as progesterone, but not the polar corticosteroids)[34] and can be a cause of disagreement between studies in absolute levels of measured steroids. One should therefore choose an adequate extraction procedure according to the literature and test the extraction efficiency for the hormones of interest beforehand by adding radioactively labeled steroid hormones to yolk before extraction and assessing the amount of label recovered during each extraction step. This validation is also important because egg composition may differ between species, so different extraction methods may be optimal or required.

For quantitative measurements of absolute hormone levels, it is advisable to determine recovery for each sample, because the percentage of recovery can fluctuate substantially.

CHROMATOGRAPHY

Chromatography is used primarily to separate the various hormones present in the sample, but also leads to further sample purification. Chromatography is not necessary if no cross-reacting steroids are present in the sample and if samples are sufficiently clean that no other interfering substances are present.

When chromatographic separation is used, the presence of several hormones can be determined in a single sample, which is an advantage especially when samples (e.g., biopts) are too small and hormone levels too low to allow splitting of the sample for several immunoassays. Chromatography also allows the use of nonspecific antibodies, which can be used to measure more than one hormone. Testosterone and dihydrotestosterone are frequently assayed after chromatographic separation using an antiserum that cross-reacts with both hormones. The advantage of specific antisera is obviously that the relatively time-consuming chromatography may not be necessary, but these antisera are often more expensive and removal of interfering substances such as lipids is still necessary.

Avian endocrinologists use mostly celite column chromatography for the separation of steroid hormones.[1,34,35] Other methods using commercially available columns can be found in the literature but have not yet been used for hormones in eggs.[36-39]

The packing of the columns and the extraction, especially for the separation of several hormones, requires practice and careful validation following the published protocols.[1,34,35] The absolute volumes, mixtures, and succession of the eluting solvents (e.g., increasing polarities) influence the elution profile of the hormones. Also, the speed of elution and the temperature influence the quality of the separation and should remain constant.[34] It must first be established that the hormone of interest is eluted primarily in the fraction used for assaying this hormone; second, one must be certain that other hormones elute in this fraction only to a minor extent. This is important for two reasons. First, elution of hormones in other fractions will lead to an overestimation of recovery when using more than one labeled hormone at the same time, because the amount of radioactivity recovered in the eluate is used to calculate recovery and different hormones cannot be distinguished unless different labels (^{3}H, ^{14}C) are used. However, ^{14}C labels have a low specific activity (low radioactivity relative to the mass of the molecule), so the large amount of label required for recovery estimation can interfere with the assay. Second, elution of several hormones in the same fraction that cross-react with the antibody will lead to inflated hormone levels. The degree of potential interference can be estimated from the amount of hormone that is present in the sample, the percentage of cross-reactivity with the antibody used, and the proportion present in the eluate. A hormone that has a high cross-reactivity and elutes in the same fraction may not interfere with the assay if it is hardly present in the sample, whereas a hormone with low cross-reactivity present in much larger amounts than the hormone of interest may cause substantial.

Therefore, the quality of the separation has to be tested beforehand by adding radioactively labeled steroids—singly and in combination—to yolk samples and measuring recovery and separation in the different fractions. FIGURE 1 shows an example of our own validations for separating androstenedione, dihydrotestosterone, and testosterone on celite columns.

IMMUNOASSAY

Currently, steroid hormones in avian eggs are measured using immunoassays, mostly radioimmunoassays, which are readily available, cheap, fast, and very sensitive. They cannot, however, be used to biochemically identify a hormone because antisera cross-react to some extent with other hormones or metabolites. They are therefore an indirect method to measure amounts of a certain hormone, and for an accurate measurement they must be validated using an independent method such as a combination of gas chromatography and mass spectrometry (GC–MS), which can biochemically identify and quantify the hormone of interest.[40-43] Even these methods may have difficulty in distinguishing very similar analytes, such as isomeric

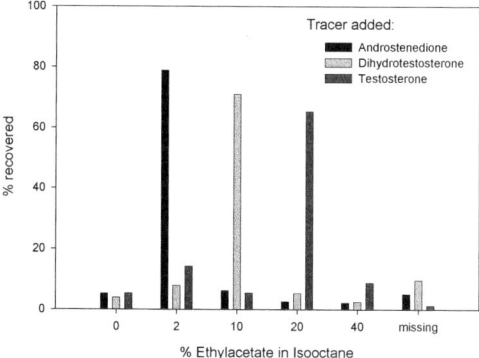

FIGURE 1. Recovery of radioactivity-labeled androstenedione (A4), dihydrotestosterone (DHT), and testosterone (T) successively eluted on celite columns with stepwise increasing percentages of ethylacetate in iso-octane (% EA/IO). For further details, see text.

metabolites.[44] Until now such an independent validation has been performed only for testosterone in canary eggs.[1]

The most commonly used parameters to describe the validity and reliability of a immunoassay are its accuracy, specificity, precision, and sensitivity, which can be evaluated statistically and should be part of every study measuring hormone levels on a new species or type of sample.[34,36,45–47] Accuracy (or bias) is the degree to which measured values correspond to the true values of the hormone of interest and can therefore be considered the most important measure of the validity of the assay.[48] Values may have an absolute bias, meaning that all values are larger or smaller than the true value by a constant amount. In that case, relative differences can still be used for quantitative analysis of treatment effects, but average levels between studies are not comparable. A relative bias exists when the degree to which the measured value deviates from the true value depends upon the amount of hormone in the sample. A relative bias is due to nonspecific interfering substances present in the sample that affect the measured values, depending upon the sample concentration. In such a case, hormone levels can still be ranked, but relative differences between treatments are not comparable because the values are not accurate on an interval scale (i.e., differences in hormone levels can be accurately measured, but absolute levels may still be biased because the "true" zero is not defined) or on a ratio scale (i.e., both relative differences and absolute levels can be accurately measured).

Two simple methods to assess accuracy are parallelism and recovery of known amounts. To assess parallelism, different amounts of a sample are extracted and assayed. If these differences in the amount of sample do not result in similar differences in the measured level of hormone—if the results of the sample dilutions do not run parallel to the results of the standard curve—there are interferences, which may be resolved by an improved extraction method. Parallelism therefore gives information about the degree of relative bias. It is frequently assessed by eye, but can be statistically evaluated.[45] Recovery of known amounts is assessed by adding known amounts of the hormone of interest to samples and measuring the amount of hormone present in the sample before and after addition of the hormone. If the difference is not equal to the amount of hormone added, then differences between samples do not accurately reflect true differences in endogenous hormone levels, most likely due to the presence of nonspecific interferences, such as proteins or lipids, which bind steroid hormones. This gives an indication of both absolute bias and relative bias, but an absolute bias of the hormone levels can ultimately be detected only by validation with an accepted reference method such as GC–MS.[40–43] Another relatively easy method to evaluate accuracy and specificity (see below) is to measure both the amount of radioactive label and the level of hormone in several fractions of a chromatographic separation.[49] A constant ratio of the amount of radioactivity and the amount of hormone measured in different fractions indicates that the assay measures the hormone of interest accurately and specifically.

Specificity indicates the extent to which the assay measures only one specific hormone or also other hormones or metabolites. For example, a nonspecific assay for testosterone may in fact measure several androgens, and conclusions are then limited to general conclusions regarding androgens.[50] Furthermore, no quantitative conclusions can be drawn because the different hormones bind with different affinities to the antibody. For example, a certain hormone level measured with an antiserum that binds testosterone twice as well as dihydrotestosterone can reflect that level of testosterone or twice as much dihydrotestosterone.

Increased specificity can be obtained by a chromatographic separation of the cross-reacting steroids or by finding an antiserum that has lower cross-reactivities with the interfering hormones. Even cross-reactivities below 1% may cause a problem for specific measurement if the cross-reacting steroid is present in much larger amounts than the hormone of interest. For example, yolk levels of androstenedione in gulls are 30-fold higher than testosterone,[23] and progesterone levels in starlings are 500 times higher than testosterone levels.[51] Unfortunately, studies with radiolabeled hormone injection in the mother, a standard validation technique in fecal hormone analyses,[17] are difficult for identification of reproductive hormones (and their metabolites) in the egg, because these steroids are produced in the follicles surrounding the egg. They therefore probably enter the yolk directly from the follicular wall,[10,16] and only a very small percentage of the injected hormone will reach the egg via the maternal circulation.[27] Radiometabolism studies will require *in vitro* experiments with follicles using labeled hormones and their precursors. Radiometabolism studies are much easier to perform and more valid for hormones, such as corticosteroids and thyroid hormones, produced in extragonadal glands.

Precision is the variability of the measures (either within or between assays) when a sample is assayed repeatedly. It is reported as the coefficient of variation [CV = (standard deviation/mean) × 100]. Precision is highest when samples are measured in a single assay, so many studies attempt to measure all samples of the study within a single assay. If samples must be measured in several assays, it is obviously very important that one does not measure samples of different experimental groups in different assays, because it would then be impossible to disentangle interassay effects from the effects of interest in the study.

Sensitivity or limit of detection gives an indication of the smallest amount of hormone that can be reliably detected. Determining sensitivity is methodologically and statistically nontrival, and there is no general agreement regarding its calculation.[46] Sensitivity is frequently defined as two or three standard deviations above the blank value—the apparent amount of hormone in a sample that is identical to the samples used in the assay but does not contain the analyte of interest. There is no optimal way of producing such a sample, because the method for removing the analyte may itself introduce error (e.g., removal of endogenous hormones with charcoal). Because most hormones are present in large amounts in avian eggs, samples can and should be measured at a concentration for which assay sensitivity is not an important issue, but some hormones, especially estradiol, are present in such low amounts that they are difficult to detect, and assay sensitivity must be assessed. Most current studies on yolk hormones report the precision and sensitivity of their method, but do not report or assess specificity and accuracy. Accuracy can be assessed to a certain extent relatively easily by demonstrating parallelism and recovery of a known amount of hormone added to samples. We recommend that this should become common practice for yolk hormone assays, as has been recommended for hormone assays in general.[48,52] If parallelism cannot be demonstrated and if known amounts are not recovered accurately, a given method is clearly not valid for quantitative comparisons of relative or absolute differences of hormone levels, but the measurements may still reflect the ranking and qualitative differences between samples.

CONCLUSION

Hormones in avian eggs vary in relation to various factors and can have strong effects on offspring development. However, contradictory results have been found,[10] and because these may partly be due to the methods used to measure yolk hormones it is important to validate and standardize the way yolk hormones are measured.

Accurate measurement of yolk hormones usually requires extraction and chromatographic purification and separation of samples because of the presence of substances that can interfere with the assay. Validation of the method is therefore necessary when setting up an assay for hormones in avian eggs in a new laboratory, for a different species, or for a different type of sample. Most endocrinologists agree that a minimal validation should include an assessment of the accuracy, specificity, precision, and sensitivity of the method.[34,46–48] The possibility of species-specific metabolites requires a validation for each new study species and potential adjustment of the method. In addition, a study evaluating the comparability of egg hormone measurements by analyzing samples from the same eggs at different laboratories would clearly be very useful.

If a quantitative assessment of specific yolk hormones at the level of an interval or ratio scale is not possible, more cautious conclusions with respect to ranks or qualitative differences can still be made. A study that is merely interested in the effect of a treatment on yolk androgen levels may not require separation of different steroid hormones by chromatography. The degree of validation needed therefore depends partly upon the questions that are being asked.

ACKNOWLEDGMENTS

We thank the organizers and participants of this inspiring workshop for many interesting discussions and useful comments. We greatly appreciate the collaboration in the laboratory and exchange of ideas with Corine Eising, Bonnie de Vries, Maarten Lasthuizen, and Wendt Müller.

REFERENCES

1. SCHWABL, H. 1993. Yolk is source of maternal testosterone for developing birds. Proc. Natl. Acad. Sci. USA **90:** 11446–11450.
2. SCHWABL, H. 1996. Maternal testosterone in the avian egg enhances postnatal growth. Comp. Biochem. Physiol. **114A:** 271–276.
3. MCNABB, F.M.A. & C.M. WILSON. 1997. Thyroid hormone deposition in avian eggs and effects on embryonic development. Am. Zool. **37:** 553–560.
4. EISING, C.M., C. Eikenaar, H. Schwabl & T.G. Groothuis. 2001. Maternal androgens in black-headed gull (*Larus ridibundus*) eggs: consequences for chick development. Proc. R. Soc. Lond. B Biol. Sci. **268:** 839–846.
5. SOCKMAN, K.W. & H. SCHWABL. 2000. Yolk androgens reduce offspring survival. Proc. R. Soc. Lond. B Biol. Sci. **267:** 1451–1456.
6. HAYWARD, L.S. & J.C. WINGFIELD. 2004. Maternal corticosterone is transferred to avian yolk and may alter offspring growth and adult phenotype. Gen. Comp. Endocrinol. **135:** 365–371.
7. MÖSTL, E., H. SPENDIER & K. KORTRSCHAL. 2001. Concentration of immunoreactive progesterone and androgens in the yolk of hen's eggs (*Gallus domesticus*). Wien. Tierarztl. Monschr. **88:** 62–65.
8. ADKINS-REGAN, E., M.A. OTTINGER & J. PARK. 1995. Maternal transfer of estradiol to egg yolks alters sexual differentiation of avian offspring. J. Exp. Zool. **271:** 466–470.
9. GIL, D. 2003. Golden eggs: maternal manipulation of offspring phenotype by egg androgen in birds. Ardeola **50:** 281–294.
10. GROOTHUIS, T.G.G. *et al.* 2005. Prenatally induced adaptive maternal effects: the case of maternal androgen deposition in avian eggs. Neurosci. Biobehav. Rev. **29:** 329–352.
11. DOWNING, J.A. & W.L. BRYDEN. 2002. A non-invasive test of stress in laying hens. RIRDC Publication No. 01/143.
12. RETTENBACHER, S. *et al.* 2005. Corticosterone in chickens' eggs. Ann. N.Y. Acad. Sci. **1046:** 193–203.
13. BRUNSTRÖM, B., J. AXELSSON & K. HALLDIN. 2003. Effects of endocrine modulators on sex differentiation in birds. Ecotoxicology **12:** 287–295.
14. WOODS, J.E., G.W. DE VRIES & R.C. THOMMES. 1971. Ontogeny of the pituitary-adrenal axis in the chick embryo. Gen. Comp. Endocrinol. **17:** 407–415.
15. BENOWITZ-FREDERICKS, M., A.S. KITAYSKY & J.C. WINGFIELD. 2005. Steroids in allantoic waste—an integrated measure of steroid exposure in ovo. Ann. N.Y. Acad. Sci. **1046:** 204–213.
16. GROOTHUIS, T.G.G. & N. VON ENGELHARDT. 2005. Investigating maternal hormones in avian eggs: measurement, manipulation, and interpretation. Ann. N.Y. Acad. Sci. **1046:** 168–180.
17. PALME, R. 2005. Measuring fecal steroids: guidelines for a practical application. Ann. N.Y. Acad. Sci. **1046:** 75–80.
18. ROMANOFF, A.L. 1960. The Avian Embryo. Macmillan. New York.
19. GIL, D., J. GRAVES, N. HAZON & A. WELLS. 1999. Male attractiveness and differential testosterone investment in zebra finch eggs. Science **286:** 126–128.
20. GIL, D. *et al.* 2004. Negative effects of early developmental stress on yolk testosterone levels in a passerine bird. J. Exp. Biol. **207:** 2215–2220.
21. WARD, B.C., E.J. NORDEEN & K.W. NORDEEN. 2001. Anatomical and ontogenetic factors producing variation in HVc neuron number in zebra finches Brain Res. **904:** 318–326.
22. RUTSTEIN, A.N. *et al.* 2005. Sex-specific patterns of yolk androgen allocation depend upon maternal diet in the zebra finch. Behav. Ecol. **16:** 62–69.
23. GROOTHUIS, T.G.G. & H. SCHWABL. 2002. Determinants of within- and among-clutch variation in levels of maternal hormones in black-headed gull eggs. Funct. Ecol. **16:** 281–289.

24. MÜLLER, W. et al. 2004. Within-clutch patterns of yolk testosterone vary with the onset of incubation in black-headed gulls. Behav. Ecol. **15**: 893–897.
25. RASH, J.M., I. JERKUNICA & D.S. SGOUTAS. 1980. Lipid interference in steroid radioimmunoassay. Clin. Chem. **26**: 84–88.
26. LIPAR, J.L., E.D. KETTERSON, V. NOLAN, JR. & J.M. CASTO. 1999. Egg yolk layers vary in the concentration of steroid hormones in two avian species. Gen. Comp. Endocrinol. **115**: 220–227.
27. HACKL, R. et al. 2003. Distribution and origin of steroid hormones in the yolk of Japanese quail eggs (*Coturnix coturnix japonica*). J. Comp. Physiol. B **173**: 327–331.
28. EISING, C.M. 2003. Mother knows best? Costs and benefits of differential maternal hormone allocation in birds. Ph.D. thesis, University of Groningen. Groningen, the Netherlands.
29. ELF, P.K. & A.J. FIVIZZANI. 2002. Changes in sex steroid levels in yolks of the leghorn chicken, *Gallus domesticus*, during embryonic development. J. Exp. Zool. **293**: 594–600.
30. EISING, C.M., W. MULLER, C. DIJKSTRA & T.G.G. GROOTHUIS. 2003. Maternal androgens in egg yolks: relation with sex, incubation time and embryonic growth. Gen. Comp. Endocrinol. **132**: 241–247.
31. ROMANOFF, A.L. 1967. Biochemistry of the Avian Embryo—A Quantitative Analysis of Prenatal Development. Wiley. New York.
32. WILSON, C.M. & F.M.A. MCNABB. 1997. Maternal thyroid hormones in Japanese quail eggs and their influence on embryonic development. Gen. Comp. Endocrinol. **107**: 153–165.
33. LIPAR, J.L. & E.D. KETTERSON. 2000. Maternally derived yolk testosterone enhances the development of the hatching muscle in the red-winged blackbird *Agelaius phoeniceus*. Proc. R. Soc. Lond. B Biol. Sci. **267**: 2005–2010.
34. ABRAHAM, G.E. 1974. Radioimmunoassay of steroids in biological materials. Acta Endocrinologica. **75**(Suppl. 183): 1–42.
35. WINGFIELD, J.C. & D.S. FARNER. 1975. The determination of five steroids in avian plasma by radioimmunoassay and competitive protein binding. Steroids **26**: 311–327.
36. JAFFE, B.M. & H.R. BEHRMAN. 1979. Methods of Hormone Radioimmunoassay. Academic Press. New York.
37. PAYNE, D.W., W.D. HOLTZCLAW & E.Y. ADASHI. 1989. A convenient, unified scheme for the differential extraction of conjugated and unconjugated serum C19 steroids on Sep-Pak cartridges. J. Steroid Biochem. **33**: 289–295.
38. VENKATESH, B., C.H. TAN & T.J. LAM. 1989. Blood steroid levels in the gold fish: measurement of six ovarian steroids in small volumes of serum by reversed-phase high-performance liquid chromatography and radioimmunoassay. Gen. Comp. Endocrinol. **76**: 398–407.
39. MORINEAU, G. et al. 1997. Convenient chromatographic prepurification step before measurement of urinary cortisol by radioimmunoassay. Clin. Chem. **43**: 786–793.
40. FITZGERALD, R.L. & D.A. HEROLD. 1996. Serum total testosterone: immunoassay compared with negative chemical ionization gas chromatography-mass spectrometry. Clin. Chem. **42**: 749–755.
41. DORGAN, J.F. et al. 2002. Measurement of steroid sex hormones in serum: a comparison of radioimmunoassay and mass spectrometry. Steroids **67**: 151–158.
42. TAIEB, J. et al. 2003. Testosterone measured by 10 immunoassays and by isotope-dilution gas chromatography-mass spectrometry in sera from 116 men, women, and children. Clin. Chem. **49**: 1381–1395.
43. WANG, C. et al. 2004. Measurement of total serum testosterone in adult men: comparison of current laboratory methods versus liquid chromatography-tandem mass spectrometry. J. Clin. Endocrinol. Metab. **89**: 534–543.
44. MAGNUSSON, M.O. & R. SANDSTRÖM. 2004. Quantitative analysis of eight testosterone metabolites using column switching and liquid chromatography/tandem mass spectrometry. Rapid Commun. Mass Spectrom. **18**: 1089–1094.
45. RODBARD, D. 1974. Statistical quality control and routine data processing for radioimmunoassays and immunoradiometric assays. Clin. Chem. **20**: 1255–1270.

46. CHARD, T. 1990. An Introduction to Radioimmunoassay and Related Techniques. Elsevier. Amsterdam.
47. COOK, B. & G.H. BEASTALL. 1987. Measurement of steroid hormone concentrations in blood, urine and tissues. *In* Steroid Hormones: a Practical Approach. B. Green & R.E. Leake, Eds.: 1–65. IRL Press. Oxford.
48. MATSUMOTO, A.M. & W.J. BREMNER. 2004. Serum testosterone assays—accuracy matters. J. Clin. Endocrinol. Metab. **89:** 520–524.
49. CEKAN, S.Z. 1979. On the assessment of validity of steroid radioimmunoassays. J. Steroid Biochem. **11:** 135–141.
50. TANVEZ, A. *et al.* 2004. Sexually attractive phrases increase yolk androgen deposition in Canaries (*Serinus canaria*). Gen. Comp. Endocrinol. **138:** 113–120.
51. LIPAR, J.L. 2001. Yolk steroids and the development of the hatching muscle in nestling European Starlings. J. Avian Biol. **32:** 231–238.
52. BUCHANAN, K.L. & A.R. GOLDSMITH. 2004. Noninvasive endocrine data for behavioural studies: the importance of validation. Anim. Behav. **67:** 183–185.
53. NASH, J. P. *et al.* 2000. An enzyme linked immunosorbant assay (ELISA) for testosterone, estradiol, and 17,20β-dihydroxy-4-pregnen-3-one using acetylcholinesterase as tracer: application to measurement of diel patterns in rainbow trout (*Oncorhynchus mykiss*). Fish Physiol. Biochem. **22:** 355–363.

Corticosterone in Chicken Eggs

S. RETTENBACHER,[a] E. MÖSTL,[a] R. HACKL,[b] AND R. PALME[a]

[a]Institute of Biochemistry, Department of Natural Sciences, University of Veterinary Medicine, A-1210 Vienna, Austria

[b]Clinic for Avian, Reptile, and Fish Medicine, Department of Farm Animals and Herd Management, University of Veterinary Medicine, A-1210 Vienna, Austria

ABSTRACT: Birds are discussed as models for prenatal stress. In this study, several experiments were conducted to gain basic knowledge of if, how, and when maternal adrenocortical activity is reflected by corticosterone concentrations in the egg. Radiolabeled corticosterone was administered to 10 laying hens to investigate the uptake into as well as the distribution within the eggs. The yolk was dissected in concentric layers and analyzed. Less than 1% of the administered radioactivity entered the egg but was, however, not evenly distributed. On the day after injection, highest radioactivity (Bq/g) was detected in the albumen and the outmost layer, whereas concentration peaked 4–7 days later in the inner layers. In two other experiments, increased plasma levels of corticosterone were induced by injection of adrenocorticotropic hormone (ACTH) or feeding of corticosterone. Again, yolk disks were cut in layers and analyzed with a corticosterone enzyme immunoassay. No effect of the ACTH administration was detected, whereas feeding of corticosterone resulted in increased immunoreactive corticosterone concentrations in the yolk. Straight-phase high-performance liquid chromatographic (HPLC) separations were also performed to characterize immunoreactive steroids in the yolk. Two close-eluting peaks at the approximate elution position of corticosterone could be observed after the feeding experiment, whereas in untreated control eggs they were absent. It was concluded that transfer from plasma to egg is low for corticosterone and that further investigations concerning the transport mechanisms and the exact nature of yolk steroids are necessary.

KEYWORDS: corticosterone; eggs; poultry; chicken

INTRODUCTION

Environmental perturbations elevate concentrations of glucocorticoids in the blood to maintain homeostasis and to trigger physiological and behavioral reaction patterns toward survival.[1] Plasma glucocorticoid concentrations are therefore widely used to monitor stress responses in various species.[2–5] If the adverse conditions causing higher hormone levels do not suppress reproduction, they may influence the phenotype of the offspring to maximize success under the constraints of the local en-

Address for correspondence: S. Rettenbacher, Institute of Biochemistry, Department of Natural Sciences, University of Veterinary Medicine, Veterinärplatz 1, A-1210 Vienna, Austria. Voice: +43-1-25077-4114; fax: +43-1-25077-4190.
sophie.rettenbacher@vu-wien.ac.at

vironment. In mammals, stress during gestation has profound deleterious effects on subsequent offspring, and the detrimental consequences of prenatal stress have been broadly shown.[6] In chicken, Lay and Wilson found that administration of corticosterone mimics some, but not all, of the effects of prenatal stress in mammals.[7]

In the blood, corticosterone is reversibly bound to proteins.[8] Basal plasma concentrations of corticosterone in chickens are around 1.3 ng/mL.[9] It is assumed that the amount of corticosterone in the yolk is influenced via passive diffusion by the levels of plasma corticosterone,[10] but it has never been verified that the corticosterone concentration in the yolk correlates with the circulating systemic corticosterone levels. Little is known about how corticosterone gets into the eggs, and basic knowledge about the distribution of stress hormones in the yolk and albumen is needed. Although the influence of maternal sex steroids on the developing embryo has been elucidated broadly,[11] it is not possible to draw analogous conclusions for corticosterone, because it is produced mainly in the adrenal glands, whereas sex steroids are synthesized by the ovary.

The aim of this study was to gain some basic information about the transport of corticosterone from the plasma into the egg and its distribution within the yolk. For this purpose, radiolabeled corticosterone was administered intravenously to mature laying hens. In addition, the concentration of corticosterone in the plasma was increased to elucidate consequences on the corticosterone levels in the eggs. A characterization of the yolk steroids was also performed.

MATERIALS AND METHODS

Animals

The experiment is described in detail by Rettenbacher *et al.*[12] The analyzed eggs came from laying hybrids, ISA brown, obtained from a commercial breeder (R. Schropper PLC, Gloggnitz, Austria). There were weekly intervals between experiments. Permission for performing the animal experiment was obtained from the Federal Ministry of Education, Science, and Culture (GZ 68.205/59-Pr/4/2002).

Radiometabolism Study

Ten birds were administered 1.7 MBq (i.e., 46 µCi) of ^{3}H-labeled corticosterone, dissolved in 1 mL of 0.9% NaCl solution containing 10% (vol/vol) ethanol into the vena cutanea ulnaris. The radiolabeled corticosterone (NET-399; [1,2,6,7-^{3}H(N)]corticosterone; 2830.5 GBq/mmol) was obtained from New England Nuclear (Perkin-Elmer, Boston, MA). During the next 12 days, eggs were collected and stored at −24°C. Previously laid eggs were used to determine background levels.

For analysis, the frozen eggs were incubated at room temperature for a few minutes, until the shells could be removed. During defrosting, the albumen changed its consistency to semifrozen and could be simply scraped off with a spatula. Amounts of albumen were quantified, and 0.5 mL was put directly into scintillation vials to measure radioactivity.

After weighing, the yolk was divided into five concentric layers. Therefore, a disk of approximately 3 mm was cut out, leaving two hemispheres. The radius of the disk was measured and divided by 5. With a pair of dividers, these five layers, approximately 2 to 3 mm thick, were separated from each other. In the following text, the first layer (layer 1) represents the outermost layer, and the subsequent numbers represent the following layers, with the fifth layer being the central layer. A total of 0.15 g of each layer was extracted with 4 mL of 80% (vol/vol) methanol by shaking for 30 min. After centrifugation, aliquots of the supernatant (0.5 mL in duplicates) were mixed with 6 mL of scintillation fluid (Quicksafe A, No. 100800; Zinsser Analytic, Maidenhead, UK) and measured in a liquid scintillation counter (Packard Tri-Carb 2100TR; Meriden, CT). Radioactivity levels were expressed as becquerels per gram of yolk.

In one animal, we calculated the mean radioactivity of the whole yolk as if it would have been homogenized prior to analysis. Therefore, the weight of each layer was calculated and multiplied with the measured layer concentration (Bq/g) to give the total radioactivity present. The sum of all five layers was divided by the weight of the whole yolk to calculate the mean concentration of the total yolk.

Injection of ACTH and Feeding of Corticosterone

To stimulate adrenocortical activity, we administered 2 mL (0.25 mg) of ACTH (Synacthen; Ciba-Geigy, Basel, Switzerland) to all 10 animals. Eggs were collected during the next four days and stored at –24°C. Previously laid eggs were used to determine background levels.

In a separate experiment, all hens were fed 0.1 g of corticosterone, mixed with a bit of moistened food. The ingestion was completed in about 30 min. Eggs were collected for 6 days.

The preparation of the yolk layers and the extraction were performed in the same way as described for the radiometabolism study. In the feeding experiment, after centrifugation, 10 μL aliquots of the supernatant, diluted with assay buffer (1:1.5), could be directly measured in a corticosterone enzyme immunoassay (EIA).[13] The sensitivity of the assay was 0.8 pg/well; the intra- and interassay coefficients of variation (CVs) were 10% and 13%, respectively. In the ACTH experiment, methanolic yolk extracts were purified with Sep-Pak C_{18} cartridges (1g; Waters, Milford, MA) as described by Möstl et al.[14] before analysis with the same EIA.

High-Performance Liquid Chromatography

One sample (extract of 0.15 g of yolk) from the feeding experiment, containing maximum corticosterone concentration, was used for high-performance liquid chromatography (HPLC). The supernatant of the extract was subjected to a cleanup procedure, as described by Rettenbacher et al.[12] Straight-phase HPLC separation was performed on a Lichrosorb Si 60 column (10 μm, 25 × 0.4 cm; Forschungszentrum Seibersdorf, Vienna, Austria), as described by Palme and Möstl.[13] Fractions were then analyzed by the corticosterone EIA.[13]

In another experiment, HPLC separations were performed after extraction of control eggs with diethyl ether. Therefore, 16 g of a homogenized yolk was divided into portions of 0.5 g, 0.5 mL of water was added, and the mixture was extracted twice with 5 mL of diethyl ether. The ether phase was transferred into a new vial and evaporated. A total of 2 mL of 100% methanol was added to resuspend the extracted substances. After addition of 400 μL of water and centrifugation for 10 min at 2500 × g, the supernatants were pooled. The cleanup procedure and the HPLC separation were performed as previously described.

RESULTS

Administration of Radioactivity

After injection of ^{3}H-labeled corticosterone, 108 eggs were collected and analyzed. Only 0.42 ± 0.07% (mean ± SD) of the total recovered radioactivity[12] was found in the yolk, and 0.25 ± 0.05% in the albumen. Peak concentrations of radioactivity in the eggs decreased during the sampling period. A stepwise transition of the

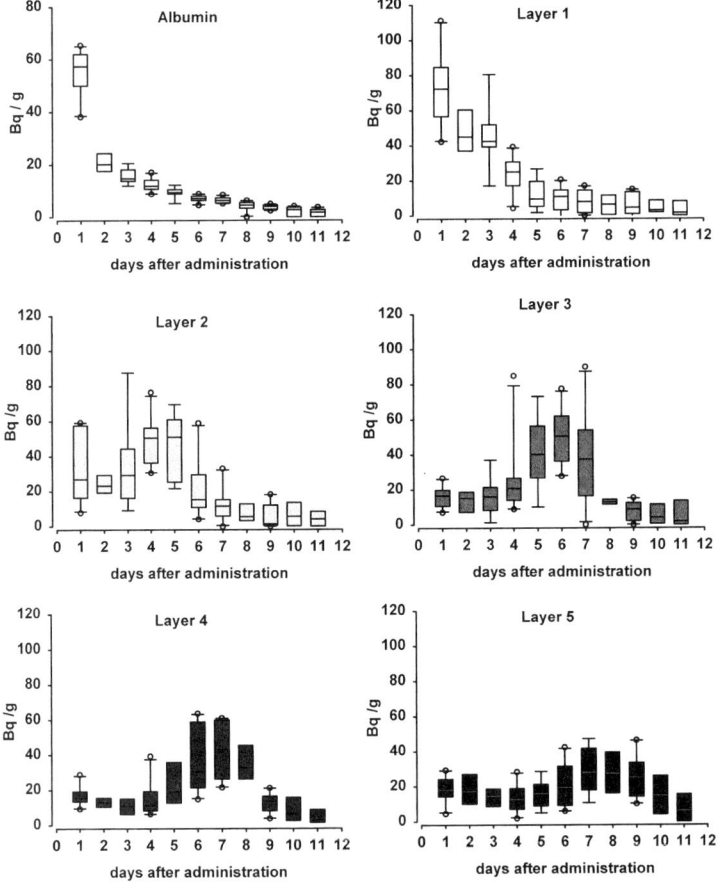

FIGURE 1. Radioactivity (Bq/g) in albumen and different yolk layers after administration of ^{3}H-labeled corticosterone to 10 laying hens. Data are given as box-plot diagrams showing medians (*lines in the boxes*), 25% and 75% quartiles (*boxes*), 10% and 90% ranges (*whiskers*), and outliers (*dots*).

daily maximal concentration of radioactivity from the surface layer of the first laid egg to the center of the tenth egg was observed (FIG. 1). Highest concentrations were found in layer 1 on day 1 after injection (116 Bq/g) and in the albumen (66 Bq/g). Peak concentrations in the second layer were reached on the third day. Between days 4 and 7, the concentration maxima could be found in layer 3. Layer 4 showed the highest concentrations of radiolabeled corticosterone on days 6 and 7, and layer 5 on days 7 and 8 (FIG. 1). Eleven days after administration, background levels were reached in all layers. Differences between layers as well as between days were statistically significant (one-way ANOVA on ranks, P < .001, h = 377). To isolate the groups that differ from the others, we applied a multiple-comparison procedure (Dunn's method). Due to the large number of groups, details are not given but can be obtained upon request.

The calculated results from the homogenized yolk showed a continuous decrease of radioactivity over the sampling period (FIG. 2).

Administration of ACTH

After administration of ACTH, no significant changes in corticosterone concentrations within the same layers could be monitored during the sampling period. In all the eggs analyzed, the highest concentrations were found in the outermost layers, the values decreased toward the center, and the lowest concentrations were found in the central layer of the analyzed yolks. This pattern remained unchanged throughout the sampling period (TABLE 1).

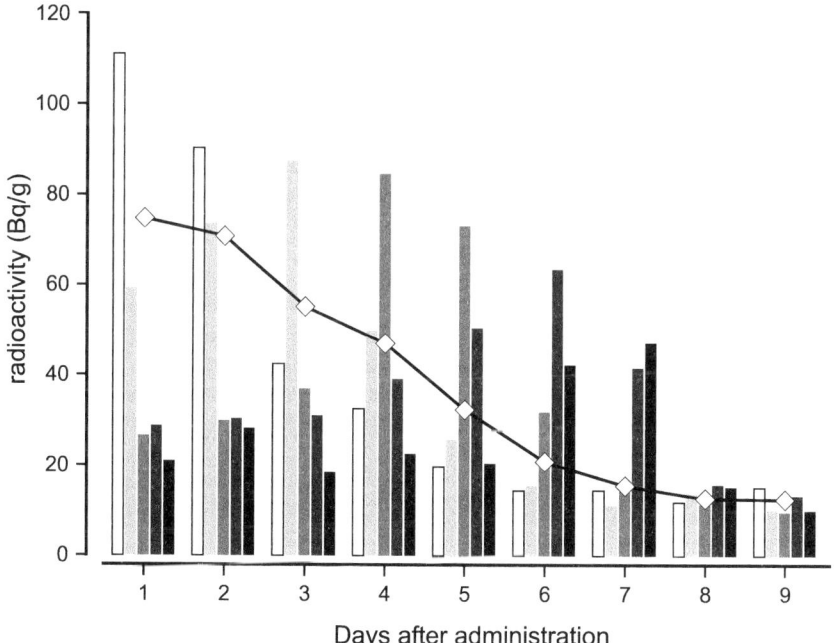

FIGURE 2. Radioactivity in the yolk of one animal. Data resulting from the consecutive layers are given as *bars;* data from the homogenized yolk (calculated) are given as line plot.

TABLE 1. Concentrations of immunoreactive corticosterone (ng/g; mean ± SD) in the different yolk layers after administration of ACTH

	Layer 1	Layer 2	Layer 3	Layer 4	Layer 5
Day 0	43.0 (±5.5)	46.7 (±9.6)	28.9 (±9.8)	16.2 (±2.6)	7.2 (±1.3)
Day 1	36.9 (±3.8)	35.5 (±10.4)	22.4 (±8.7)	16.1 (±10.4)	7.8 (±1.9)
Day 2	37.5 (±12.7)	35.2 (±2.9)	20.7 (±4.5)	9.8 (±4.1)	7.3 (±1.6)
Day 3	32.6 (±13.8)	30.1 (±7.2)	19.2 (±5.3)	9.0 (±3.1)	9.7 (±4.4)
Day 4	37.4 (±11.5)	35.3 (±5.9)	21.4 (±2.2)	11.2 (±4.4)	9.8 (±3.8)

TABLE 2. Concentrations of immunoreactive corticosterone (ng/g) in yolk layers after feeding corticosterone

	Layer 1	Layer 2	Layer 3	Layer 4	Layer 5
Day 0	43.1	43.3	28.3	16.4	7.9
	(37.7 – 49.6)	(37.7 – 61.1)	(18.6 – 44.9)	(11.9 – 18.7)	(5.5 – 15.5)
Day 1	96.1	31.0	12.0	9.5	9.0
	(58.3 – 182.1)	(11.7 – 70.9)	(5.3 – 14.4)	(4.8 – 14.6)	(6.7 – 11.7)
Day 3	55.2	65.9	33.5	15.5	9.8
	(17.9 – 91.8)	(59.8 – 119.8)	(20.7 – 52.9)	(7.8 – 18.3)	(5.1 – 14.8)
Day 4	62.8	88.8	104.0	45.3	16.1
	(56.4 – 178.1)	(55.7 – 132.5)	(20.2 – 122.1)	(14.0 – 49.4)	(15.7 – 19.9)
Day 5	58.3	100.3	93.2	71.6	28.4
	(49.4 – 122.7)	(55.1 – 169.4)	(54.7 – 407.1)	(15.5 – 131.1)	(18.8 – 50.1)
Day 6	72.7	107.6	120.1	88.2	39.7
	(59.2 – 86.1)	(93.5 – 121.7)	(81.1 – 159.1)	(36.6 – 139.8)	(22.3 – 57.2)

NOTE: The median and the range (min–max) is given.

Feeding Corticosterone

The distribution pattern of immunoreactive corticosterone was similar to that of radioactivity, but not so distinctive, because fewer eggs were obtained during the collection period due to a decreased laying performance. On the first day after feeding, the highest concentrations were measured in the outermost layers (58–182 ng/g; median = 96), and concentrations decreased toward the center of the yolk. Two days later, the highest concentration was detected in the second layer. The absolute highest concentration of corticosterone was found on day 5 after feeding in layer 3 (407 ng/g of yolk; TABLE 2).

High-Performance Liquid Chromatography

HPLC separations of one sample's extract of the feeding experiment containing peak concentrations measured by the corticosterone EIA revealed the presence of different immunoreactive substances (FIG. 3a). There were two main, close eluting peaks around the elution position of corticosterone and cortisone (fraction 42).

In the ether extracts of an untreated egg, the corticosterone EIA detected a sharp peak at fraction 6, resembling very apolar substances eluting closely to the solvent front (FIG. 3b), and some smaller apolar peaks between fractions 14 and 20. No peaks at the elution positions of corticosterone and cortisone were visible.

Levels of immunoreactive substances cannot be compared between both HPLC separations because there were different amounts of yolk processed. In the feeding experiment, only one sample (resulting from 0.15 g of yolk) was used, whereas the whole yolk of a normal egg was used for ether extraction, and methodological losses were not evaluated. Therefore, only levels within the HPLCs can be compared.

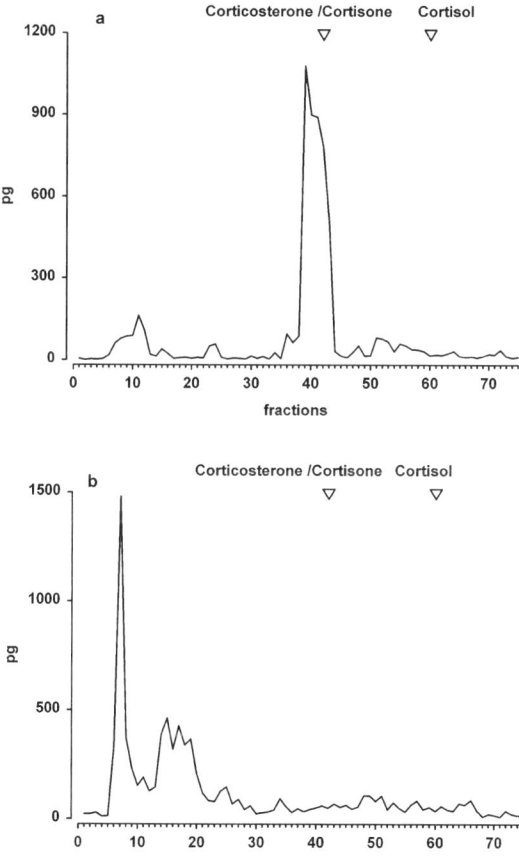

FIGURE 3. High-performance liquid chromatographic separations of yolk extracts obtained (**a**) from the feeding experiment and (**b**) from control eggs . Immunoreactivity was measured in a corticosterone EIA. *Open upside down triangles* mark the approximate elution positions of respective standards. Note that the absolute values cannot be compared between the two immunograms.

DISCUSSION

In this study, several investigations were conducted to gain basic knowledge about corticosterone transport into and distribution within the chicken's egg. We focused mainly on the situation in the yolk, and therefore albumen was not taken into particular consideration. Albumen was measured only in the radioactivity experiment, and the findings resembled the results from the outermost layer. We therefore conclude that for the other experiments, the situation in the albumen is partly reflected by the findings in the outermost layer of the yolk.

As shown by the radiometabolism study, the amount of recovered radioactivity in the eggs was less than 1% of the total recovery. This finding leads to the conclusion that the transfer of circulating corticosterone into the egg is rather low. The recovered radioactivity was distributed in a quite distinctive pattern, which can be explained by the fact that the injected radioactivity reaches follicles at different stages of their development. In follicles that are close to ovulation, radioactivity was found in the outermost layer and in the albumen, whereas follicles in an earlier stage of development accumulated radioactivity closer to the center. Hackl *et al.*[15] performed a similar experiment in quails. They found that 0.12% of radiolabeled testosterone was excreted via the yolk, whereas about 60% of the administered radioactivity was recovered in the feces. They also observed the same characteristic distribution pattern, with the radioactivity being deposited in layers, but the maximum concentrations were reached on the third day in the third layer. This difference could be explained by the slower resorption due to the intramuscular injection done by Hackl *et al.*,[15] whereas in this study, the intravenous administration resulted in an immediate increase of ^{3}H-labeled corticosterone in the blood. The fact that in the feeding experiment the maximum was also reached on day 3 supports this assumption.

Although administration of ACTH increased adrenocortical activity,[12] no changes of the corticosterone concentration could be monitored in the yolk. The corticosterone assay detected a characteristic pattern, with concentrations being higher in the outer layers and lowest in the central layers, but the absolute values remained more or less constant throughout the sampling period and did not differ from those of the control eggs. As Hackl *et al.*[15] found a similar distribution pattern for progesterone, our results could be explained by cross-reactions of the antibody used in the corticosterone EIA with other steroids, probably of gonadal origin. El-lethey *et al.*[16] showed that feeding of corticosterone increased plasma concentrations in chicken. In our study, after an orally administered dosage of 0.1 g of corticosterone, an increase of immunoreactive metabolites in the yolk could be detected by our corticosterone EIA. These results from the corticosterone feeding experiment indicate that high concentrations of circulating corticosterone result in a minor transfer of corticosterone into the yolk, which is detectable by the EIA. These findings are in accordance with the radiometabolism study, which showed that transfer from plasma to yolk is small. Although feeding corticosterone resulted in decreased laying performance, and therefore fewer eggs were available for analysis, a similar layer pattern as was found for radioactivity appeared.

In the HPLC separation of the feeding sample, the large peak detected by the corticosterone assay very likely reflects some of the administered corticosterone. Because we did not check for methodological losses, it was not possible to calculate the concentrations of the metabolites. Also, a comparison between the two HPLC separations concerning the amounts of the measured hormones cannot be performed. In the feeding experiment, one sample containing peak concentrations was used for the HPLC separations, whereas a whole yolk of an untreated egg was ether extracted for the other investigation. HPLC separations were done to characterize the hormones immunologically, not for their quantification. In the ether-extracted control eggs, no peak was visible at the elution position of corticosterone. This finding leads to the conclusion that in untreated eggs corticosterone is present only in trace amounts, if at all, and cannot be detected by our EIA.

From these findings, we conclude that unphysiologically high concentrations of plasma corticosterone, as obtained by feeding, are reflected in the yolk and that intravenously administered radioactivity can be recovered to some degree in the egg, whereas a short-term increase of adrenocortical activity, as provoked by ACTH administration, cannot be traced in the eggs.

Monitoring changes that are caused by increased adrenocortical activity is therefore not possible with the corticosterone EIA we used. In the literature, the reported concentrations are very much alike, although different species were investigated by different laboratory methods. Hayward et al.[10] found concentrations of 0.92 ng/g of corticosterone in yolk of quails and 2.06 ng/g after implantation of subcutaneous corticosterone deposits. Eriksen et al.[17] reported corticosterone concentrations between 1.17 and 1.55 ng/mL in unfertilized eggs. In the albumen, Downing and Bryden[8] measured concentrations between 1 and 1.5 ng/g of corticosterone. None of these authors characterized the measured immunoreactive steroids by means of HPLC immunograms. It is very likely that the reported corticosterone concentrations may, at least in part, be reflections of cross-reacting substances or blank values of the assays used.

Basic knowledge about the correlation between plasma and yolk corticosterone concentration is still lacking and difficult to establish. First, corticosterone concentrations in plasma can change quickly due to handling and manipulation stress and therefore rarely reflect basal values. As a better option, concentration of corticosterone metabolites in fecal samples, which reflect adrenocortical activity,[12] can be used as a reference in such studies where serial bleeding is necessary. Second, as this study points out, it is of major importance to take into account that the steroids are not evenly distributed within the yolk. Lipar et al.[18] pointed out that the exact definition of the sampling site is essential for interpreting results concerning gonadal steroids. In most studies, the yolk is homogenized or obtained via puncture, which can contribute to variations in the hormone concentrations. The occurrence of cross-reactions with sex steroids that are also not evenly distributed[15] must be kept in mind.

Another consideration might be that corticosterone is metabolized prior to the deposition in the egg and is therefore not detectable with corticosterone antibodies. Because of the low amount of radioactivity found in the eggs, HPLC separations to characterize these radioactive substances could not be performed.

The low transition of radioactivity from plasma to the egg, the unchanged distribution pattern after the ACTH administration, and the results of the HPLC, in which no corticosterone could be detected, are very strong indicators that a relationship between plasma and egg concerning corticosterone is difficult to establish. Our findings resemble those of Downing and Bryden,[8] who investigated plasma and albumen concentrations and did not find a correlation. There is some evidence that an influence of maternal stress on the offspring exists,[10] but this is probably more complicated to ascertain than simply measuring corticosterone concentrations in the eggs.

Therefore, further investigations concerning the yolk consumption of the embryo, the effects of steroids on the developing organism, and the transfer mechanism of adrenal steroids into the egg are necessary.

ACKNOWLEDGMENTS

We thank the staff of the Clinic for Avian, Reptile, and Fish Medicine for keeping the animals during the experiments, A. Kuchar-Schulz for performing the HPLC separations, and A. Zechner for excellent laboratory assistance.

REFERENCES

1. SILVERIN, B. 1998. Stress responses in birds. Avian Poult. Biol. Rev. **9:** 153–168.
2. BEUVING, G. 1983. Corticosteroids in welfare research of laying hens. *In* Indicators Relevant to Farm Animal Welfare. D. Smidt, Ed.: 47–53. Martinus Nijhoff Publishers. Boston.
3. MUNCK, A., P.M. GUYRE & N.J. HOLBROOK. 1984. Physiological functions of glucocorticoids in stress and their relation to pharmacological actions. Endocr. Rev. **5:** 25–44.
4. BROOM, D.M. & K.G. JOHNSON. 1993. Stress and Animal Welfare (Animal Behavior Series). Chapman & Hall. London.
5. VON HOLST, D. 1998. The concept of stress and its relevance for animal behavior. Adv. Study Behav. **27:** 1–131.
6. KAISER, S. & N. SACHSER. 2005. The effects of prenatal stress onbehaviour: mechanisms and function. Neurosci. Biobehav. Rev. **29:** 283–294.
7. LAY, D.C. & M.E. WILSON. 2002. Development of the chicken as a model for prenatal stress. J. Anim. Sci. **80:** 1954–1961.
8. DOWNING, J.A. & W.L. BRYDEN. 2002. A non-invasive test of stress in laying hens. RIRDC Publication No. 01/143. Available at: http://www.rirdc.gov.au/reports/EGGS/01-143.pdf (last accessed: December 9, 2004).
9. DEHNHARD, M., A. SCHREER, O. KRONE, *et al.* 2003. Measurement of plasma corticosterone and fecal glucocorticoid metabolites in the chicken *(Gallus domesticus)*, the great cormorant *(Phalacrocorax carbo)*, and the goshawk *(Accipiter gentilis)*. Gen. Comp. Endocrinol. **131:** 345–352.
10. HAYWARD, L.S. & J.C. WINGFIELD. 2004. Maternal corticosterone is transferred to avian yolk and may alter offspring growth and adult phenotype. Gen. Comp. Endocrinol. **135:** 365–371.
11. GIL, D. 2003. Golden eggs: maternal manipulation of offspring phenotype by egg androgen in birds. Ardeola **50:** 281–294.
12. RETTENBACHER, S., E. MÖSTL, R. HACKL, *et al.* 2004. Measurement of corticosterone metabolites in chicken droppings. Br. Poult. Sci. **45:** 704–711.
13. PALME, R. & E. MÖSTL. 1997. Measurement of cortisol metabolites in faeces of sheep as a parameter of cortisol concentration in blood. Int. J. Mammal. Biol. **62**(Suppl. II): 192–197.
14. MÖSTL, E., H. SPENDIER & K. KOTRSCHAL. 2001. Concentration of immunoreactive progesterone and androgens in the yolk of hen's eggs *(Gallus domesticus)*. Wien. Tieraerztl. Monschr. **88:** 62–65.
15. HACKL, R., V. BROMUNDT, J. DAISLEY, *et al.* 2003. Distribution and origin of steroid hormones in the yolk of Japanese quail eggs *(Coturnix coturnix japonica)*. J. Comp. Physiol. [B] **173:** 327–331.
16. EL-LETHEY, H., T.W. JUNGI & B. HUBER-EICHER. 2001. Effects of feeding corticosterone and housing conditions on feather pecking in laying hens *(Gallus gallus domesticus)*. Physiol. Behav. **73:** 243–251.

17. ERIKSEN, M.S., A. HAUG, P.A. TORJESEN & M. BAKKEN. 2003. Prenatal exposure to corticosterone impairs embryonic development and increases fluctuating asymmetry in chickens *(Gallus gallus domesticus)*. Br. Poult. Sci. **44:** 690–697.
18. LIPAR, J.L., E.D. KETTERSON, V. NOLAN, JR. & J.M. CASTO. 1999. Egg yolk layers vary in the concentration of steroid hormones in two avian species. Gen. Comp. Endocrinol. **115:** 220–227.

Steroids in Allantoic Waste

An Integrated Measure of Steroid Exposure *in Ovo*

Z MORGAN BENOWITZ-FREDERICKS,[a,b] ALEXANDER S. KITAYSKY,[b] AND JOHN C. WINGFIELD[a]

[a]*Department of Biology, University of Washington, Seattle 98195, Washington, USA*

[b]*Institute of Arctic Biology, University of Alaska, Fairbanks 99775, Alaska, USA*

ABSTRACT: Recent studies examining patterns and consequences of variation in maternally deposited steroids in avian egg yolk have demonstrated that these maternal hormones can have dramatic effects on chick phenotypes. However, maternal steroids are not the only source for avian embryos, which activate endocrine axes relatively early in development and are capable of producing substantial amounts of endogenous steroids. Although organizational effects of steroids have been demonstrated, the interactions between steroids from yolk and endogenous production have not been addressed. Steroids in the yolk are likely to alter development of the embryo's endocrine axes. The ability to assess total steroid exposure *in ovo* in a nonlethal fashion would improve our understanding of these interactions and help elucidate the mechanisms by which maternal steroids alter chick phenotype. Steroid levels in allantoic waste provide a cumulative measure of steroids excreted *in ovo* and may prove to be a useful tool. We present data from semiprecocial seabirds, common murres, demonstrating the presence of detectable steroids in allantoic waste and suggesting that some reflect differences in timing of hatching and may provide information about aspects of chick phenotype.

KEYWORDS: allantoic waste; allantois; avian embryo; chick development; common murre; corticosterone; egg waste; endogenous hormone production; maternal effects; organizational effects; testosterone; yolk steroids

INTRODUCTION

A whole suite of phenotypic traits, ranging from sexual behavior and morphology of the brain[1] and reproductive organs[1,2] to sexual behavior,[3,4] growth,[5] and cognition,[6] can be altered by the organizational or "programming" effects of exposure to steroid hormones during development. In birds, recent attention has been given to levels of maternally derived steroid in egg yolk and their effects on chick pheno-

Address for correspondence: Z. Morgan Benowitz-Fredericks, Department of Biology and Wildlife, Irving I, Rm. 211., University of Alaska, Fairbanks 99775, AK. Voice: 907-474-1983; fax: 907-474-6967.

ffzmb@uaf.edu

Ann. N.Y. Acad. Sci. 1046: 204–213 (2005). © 2005 New York Academy of Sciences.
doi: 10.1196/annals.1343.017

type.[1,7–10] Because the egg is essentially a closed system, embryos are exposed to these yolk hormones throughout development. Thus, the variation in maternally deposited steroids may have more drastic consequences for avian than for mammalian embryos, which may be less sensitive because exposure fluctuates substantially in tandem with maternal physiological state.[2]

There has been a great deal of recent interest and many advances in our understanding of the role of avian maternal yolk steroids—and many studies quantifying steroid levels in egg yolk. However, measuring these hormones in yolk does not provide a good measure of the embryonic steroid milieu. This is because avian embryos can produce substantial quantities of endogenous steroids.[11] Although maternal steroids are likely to influence levels of endogenous production, we do not understand the relationships between levels of steroids from these two sources in birds. Thus, our current understanding of embryonic steroid exposure and its consequences remains incomplete.

There is a substantial body of information about the capacity of avian embryos to produce steroids during development.[2] There are many studies showing temporal dynamics of embryonic plasma steroid concentrations or assaying steroid levels in whole embryonic endocrine glands.[11–13] In addition, steroid production is assessed by monitoring enzyme expression[14–16] and activity.[17] These and other studies have shown that the embryonic brain, adrenals, and gonads all produce steroids; such studies focus on differences between genders.[12,17] The caveats for these techniques are that they are lethal and provide measurements at only a single time point. Currently, there is very little information regarding integrated measures of steroid exposure *in ovo* for individuals. We discuss the use of "allantoic waste," first proposed by Bercovitz *et al.*,[18] as a noninvasive way to quantify the embryonic steroid milieu in birds. This measurement may serve as a valuable tool, allowing us to link yolk steroid levels with endogenous embryonic steroid production. Quantification of *in ovo* steroid levels may also help predict aspects of chick phenotype because many critical structures and pathways are programmed by early exposure.

What Is Allantoic Waste?

Because the egg is a mostly closed system that exchanges only gasses and water, all metabolic by-products are retained in the egg until the chick hatches. Avian embryos store waste products (including urates, amino metabolites, and metabolized steroids) in the allantois,[19] an embryonic membrane that buds off from the hindgut. A newly hatched chick is disengaged from the allantois and leaves it behind as a white-to-green sticky, pasty substance in the eggshell. This is "allantoic waste" and includes several components: allantois, extraembryonic membranes, blood vessels, and sometimes a "true" fecal/urine sample excreted by the embryo during hatching. Once membranes and vessels are removed, what remains is essentially an embryonic fecal and urine sample. However, rather than representing a short time course from the recent past (hours to days) like regular fecal samples, the steroid metabolites in the allantoic waste sample are a single, direct measure of all the steroids processed by the embryo, of both maternal and endogenous origin, during the weeks to month of development *in ovo*.

Case Study: Allantoic Waste Steroids from Common Murre (Uria aalge) Chicks

Common murres are seabirds that nest in large colonies and obtain protection from predation from the presence of neighbors. Murre females lay only a single egg, and thus are a good system in which to study maternal effects because measures of maternal effects and growth parameters can be interpreted free from the potential confounders of laying order, clutch size, and sibling competition. We previously found temporal differences in deposition of maternal steroids in yolk with later-laid eggs containing more androgens (unpublished data). We also found temporal variation in growth performance of captive chicks: later-hatched chicks grew more quickly than their earlier-hatching neighbors.[20] We collected allantoic waste from these chicks to validate the method.

METHODS

We collected and artificially incubated 24 common murre eggs from a colony in Cook Inlet, Alaska. We collected and froze allantoic waste from the 22 chicks that hatched. Chicks were eventually euthanized and sexed by necropsy.

Steroid Hormone Assay

All allantoic waste samples were lyophilized and stored at $-20°C$. Entire samples were powdered and homogenized by repeated sifting through a wire mesh and then frozen. Steroids in excretia are often in conjugated forms (glucuronides or sulfates), therefore we used acid solvolysis for deconjugation. All samples were deconjugated using acid solvolysis. Acid solvolysis was conducted by adding 400 μL of saturated NaCl solution, 200 μL of 2.5 M H_2SO_4, and 5 mL of distilled ethyl acetate to each sample following extraction with methanol. This was incubated overnight in a 40°C water bath, and then 2.5 mL of distilled water was added and vortexed for 5 min. The supernatant (containing deconjugated steroids) was retained, placed in a water bath and dried down with nitrogen gas, reconstituted in 10% ethyl acetate in isooctane, and placed on columns packed with diatomaceous earth for separation of steroids. We used standard radioimmunoassay procedures to measure corticosterone, testosterone, and estradiol levels.[21] Antibodies for corticosterone were from Esoterix Endocrinology (catalog #B3-163; Calabasas Hills, CA), testosterone from Wien laboratories (catalog #T-3003; Succasanna, NJ), and estradiol from Arnel (catalog #1702; New York, NY).

RESULTS

Validation Example: Testosterone

Dilutions of standard in nonsolvolyzed samples exhibited substantial interference (FIG. 1a); even stripping with concentrated charcoal did not eliminate the interfering substance (FIG. 1b). This problem was likely due to the presence of conjugated steroid metabolites,[19] as it was eliminated when samples were treated with acid solvol-

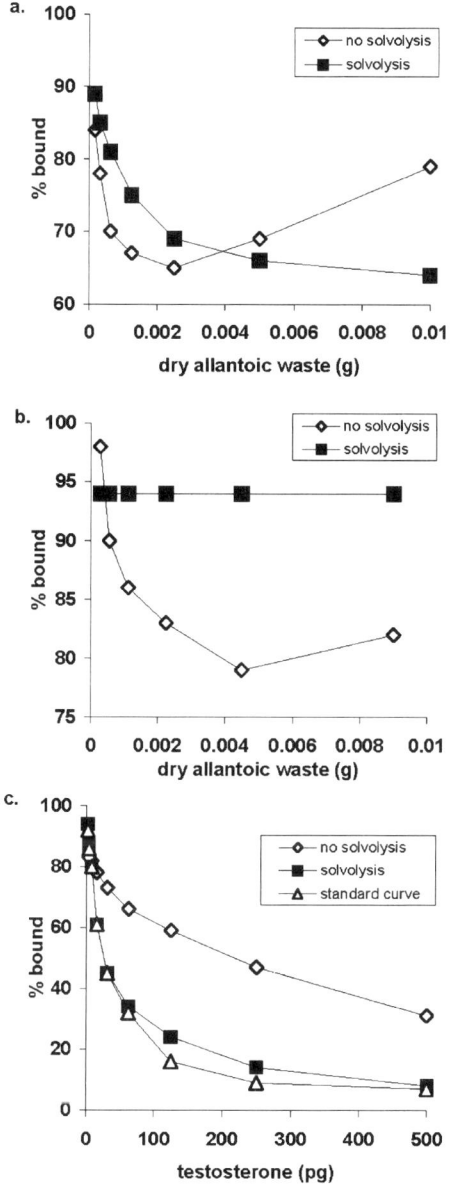

FIGURE 1. Validations of acid solvolysis for testosterone from allantoic waste. (**a**) Dilutions of solvolyzed and unsolvolyzed samples: percent bound in competitive binding assay vs. amount of dry allantoic waste in initial sample. (**b**) Dilutions of solvolyzed and unsolvolyzed sample that were stripped with concentrated charcoal: percent bound in competitive binding assay vs. amount of dry allantoic waste in initial sample. (**c**) Testosterone standard serially diluted in buffer (*standard curve*), in buffer containing solvolyzed sample, and in buffer containing unsolvolyzed sample: percent bound vs. amount of testosterone.

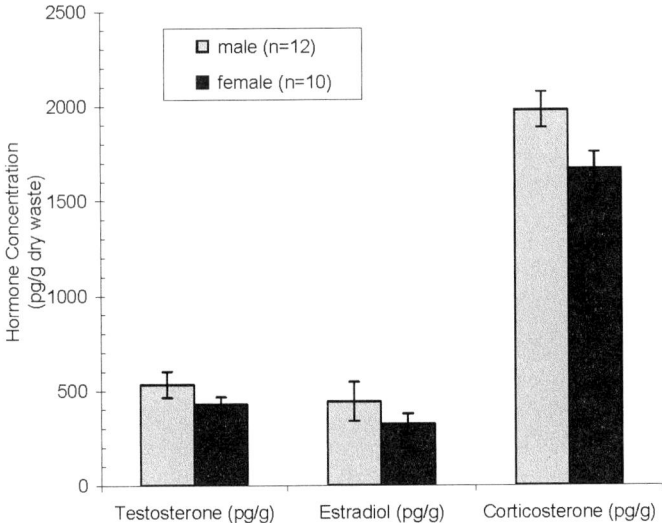

FIGURE 2. Concentrations of testosterone, estradiol, and corticosterone in allantoic waste from male and female common murre chicks. There are no differences between genders. Error bars indicate standard error.

ysis (FIG. 1b). Finally, when standard was diluted in solvolyzed sample it behaved identically to standard diluted in buffer, whereas binding was altered when it was diluted in nonsolvolyzed sample (FIG. 1c).

Gender and Hatch Date

Corticosterone, testosterone, and estradiol were all present in allantoic waste in detectable quantities that did not vary between genders (FIG. 2). We detected higher quantities of corticosterone than of testosterone or estradiol, which were present at similar levels. We found that levels of corticosterone were related to hatch date ($r = -0.56$, $P < .01$; FIG. 3a), whereas levels of testosterone were not (FIG. 3b).

DISCUSSION

Caveats and Further Validation of Allantoic Waste Hormones

It is critical to point out that by using column chromatography followed by an assay with antibodies and label developed for the parent hormone (rather than metabolites) to measure excreted steroids, we could detect only a fraction of the steroids present. Most studies have demonstrated that steroids are excreted almost entirely as metabolites, with little to none of the parent hormone ever evident in excreta.[28–31] At least one study of allantoic fluid from chicken embryos at embryonic day 13 found that conjugated steroids were present but free steroids were not.[19] Acid sol-

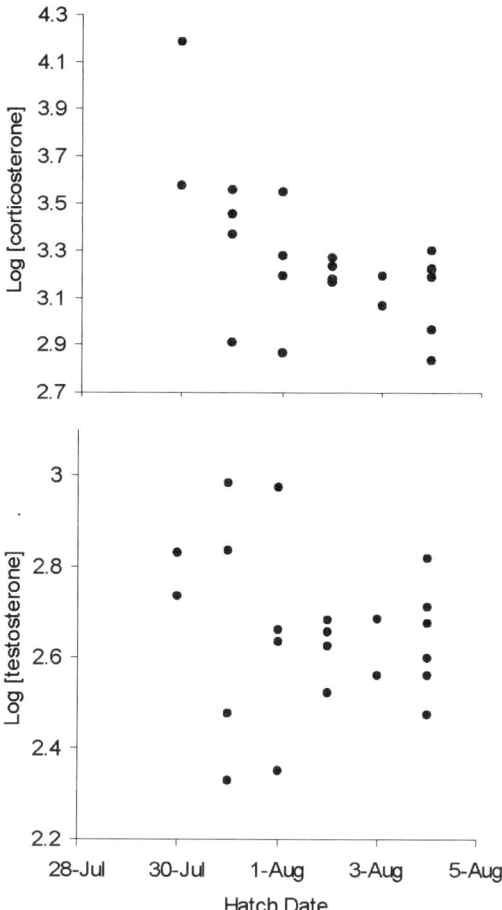

FIGURE 3. Relationships with hatch date of (a) corticosterone levels and (b) testosterone levels in allantoic waste from captive common murre chicks.

volysis probably returned most conjugated steroids to their parent form; however, because the specific metabolites have not yet been identified, it is possible that more distinct or even different patterns will emerge when an assay specific to the appropriate steroid metabolites is used. For example, sex differences may become apparent when different metabolites are examined. However, the strength of the correlations that we found using even this small portion of the total steroids excreted suggests that the relationships are robust (see below).

This method requires further validation. The next step is to determine which metabolites are present in allantoic waste by using radioinfusion techniques[28–31] and choose the antibodies most appropriate for the major metabolites being excreted.[28–31] Maximizing sensitivity will increase the chances that differences between individuals, sexes, or treatments can be detected.[29,31]

Finally, it may be feasible to collect allantoic waste only from species with relatively large eggs. Thus far, this method has been used to measure steroids in only four species that we are aware of, all with relatively large eggs: condors,[22] peregrines,[22] chickens,[18] and murres (this study). Whether or not samples from species with smaller eggs (such as most passerines) can be obtained and measured remains to be determined.

Potential Use of Allantoic Waste Hormones in Studies of Physiological Ecology of Birds

This pilot study identified a correlation between hormone levels in allantoic waste and an ecological parameter in common murres. Corticosterone levels were strongly related to hatch date, with later-hatching chicks excreting more corticosterone in their allantoic waste. There are several possible explanations for this pattern. The first is that it reflects maternal effects, either directly or indirectly. If it is a direct reflection of maternal steroid deposition, then it suggests that later-laying females deposited more corticosterone in their eggs. However, avian embryos are known to produce corticosterone in quantities sufficient for detection from a single plasma sample, particularly later in development.[23–25] In contrast, corticosterone is often at very low levels or even undetectable in yolk (Benowitz-Fredericks, unpublished data). This finding suggests that the total accumulated corticosterone in allantoic waste is likely to be weighted toward endogenously produced hormone. Therefore levels in allantoic waste are likely to reflect maternal steroids indirectly: Early exposure to maternal steroids affects the chick's hypothalamic–pituitary–adrenal (HPA) axis and endogenous production.[32] Which steroids and whether they exerted positive or negative feedback on endogenous production is not known. However, it has been clearly demonstrated that embryonic exposure to glucocorticoids in mammals alters the subsequent function of the HPA axis in juveniles and adults.[6,26,27] This kind of organizational effect seems probable if the patterns we observed reflect maternal effects. The second possibility is that corticosterone reflects the embryo's response to stressors. Because these eggs were collected partially incubated, the embryos were at different ages when they experienced the physical disturbance and uneven temperatures associated with egg collection and transport (although eggs never cooled to less than 35°C). Thus, it is possible that the older, more developed embryos had a higher capacity to respond to these disturbances by elevating corticosterone. If this is the case, it suggests that levels of corticosterone in allantoic waste could serve as a good indicator of exposure to stressors during incubation.

Testosterone levels showed no relationship to hatch date, suggesting that there were no temporal patterns of steroid deposition during laying, that existing patterns were obscured by feedback and endogenous production, or that there were no cues to alter production *in ovo*.

Measuring steroids in allantoic waste was first proposed as a noninvasive way to sex avian hatchlings.[18] However, we found no gender-related differences in testosterone, estradiol, or the ratio of testosterone to estradiol in allantoic waste from common murres. This result may be due to the nature of the metabolites in conjunction with the antibodies we used (see below), or it may be that different species exhibit different patterns of sex steroid production *in ovo*.

CONCLUSION

The use of allantoic waste as a noninvasive measure of embryonic endocrine regimes was first proposed in 1985 by Bercovitz *et al.*[18] They used ratios of estrogens to androgens to sex hatchling condors and falcons. We propose that steroid measurements from this substance may prove useful in the current effort to understand mechanisms of hormonal maternal effects in a variety of avian species. In addition, it may serve as a predictor of chick quality.

What Can We Learn from Allantoic Waste?

Using allantoic waste to measure total steroid exposure *in ovo* should allow us to assess the effects of maternal steroids and potentially other egg-related maternal effects (such as other egg contents or incubation behavior) on the physiological organization of the chick. As mentioned earlier, a likely consequence of the presence of maternal yolk steroids is alteration of the chick's developing endocrine axes.[32] Whether early exposure to steroids from yolk induces negative or positive feedback on the set points of various axes is unknown. A lack of information about dynamics of yolk uptake and the timing of embryonic exposure to steroids from yolk makes this even trickier to assess. By manipulating yolk steroid levels and assessing both levels in allantoic waste and steroid production by hatchlings, we can begin to understand some of the mechanisms by which hormonal maternal effects exert their influence on chick phenotype. In addition, if steroid levels in allantoic waste prove to be good predictors of chick growth or indicators of exposure to stressors during development, they may serve as a way to noninvasively predict chick quality. Finally, as demonstrated by Bercovitz and Sarver,[22] ratios of estrogens to androgens in allantoic waste can be a useful noninvasive sexing technique for some species, although this may not work for all species (e.g., in the common murre).

ACKNOWLEDGMENTS

This project was supported by a grant from the Exxon Valdez Oil Spill Trustee Council to A.S.K. Z.M.B.F. was supported by an NSF predoctoral fellowship and an Adkins ARCS fellowship.

We thank J. F. Piatt and M.T. Shultz for logistical support, and K. Hunt and T. Ziegler for assay advice and protocols. N. Benowitz provided helpful comments.

REFERENCES

1. ADKINS-REGAN, E., M.A. OTTINGER & J. PARK. 1995. Maternal transfer of estradiol to egg yolks alters sexual differentiation of avian offspring. J. Exp. Zool. **271:** 466–470.
2. BRUGGEMAN, V., P.V. AS & E. DECUYPERE. 2002. Developmental endocrinology of the reproductive axis in the chicken embryo. Comp. Biochem. Physiol. A **131:** 839–846.

3. ADKINS-REGAN, E., V. MANSUKHANI, R. THOMPSON & S. YANG. 1997. Organizational actions of sex hormones on sexual partner preference. Brain Res. Bull. **44:** 497–502.
4. CLARK, M.M. & B.G. GALEF, JR. 1995. Prenatal influences on reproductive life history strategies. Trends Ecol. Evol. **10:** 151–153.
5. GILL, J.W., B.J. HOSKING & A.R. EGAN. 1998. Prenatal programming of mammalian growth—a review of the role of steroids. Livest. Prod. Sci. **54:** 251–267.
6. WELBERG, L.A. & J.R. SECKL. 2001. Prenatal stress, glucocorticoids and the programming of the brain. J. Neuroendocrinol. **13:** 133–128.
7. SCHWABL, H. 1997. Maternal steroid hormones in the egg. *In* Perspectives in Avian Endocrinology. S. Harvey & R.J. Etches, Eds.: 3–13. Society for Endocrinology. Bristol, UK.
8. EISING, C.M., C. EIKENAAR, H. SCHWABL & G.G. GROOTHUIS. 2001. Maternal androgens in black-headed gull (*Larus ridibundus*) eggs: consequences for chick development. Proc. R. Soc. Lond. Ser. B **268:** 839–846.
9. SCHWABL, H. 1996. Maternal testosterone in the avian egg enhances postnatal growth. Comp. Biochem. Physiol. **114:** 271–276.
10. STRASSER, R. & H. SCHWABL. 2004. Yolk testosterone organizes behavior and male plumage coloration in house sparrows (*Passer domesticus*). Behav. Ecol. Sociobiol. **56:** 491–497.
11. TANABE, Y., T. NAKAMURA, K. FUJIOKA & O. DOI. 1979. Production and secretion of sex steroid hormones by the testes, the ovary and the adrenal glands of embryonic and young chickens (*Gallus domesticus*). Gen. Comp. Endocrinol. **39:** 26–33.
12. OTTINGER, M.A., S. PITTS & M.A. ABDELNABI. 2001. Steroid hormones during embryonic development in Japanese quail: plasma, gonadal, and adrenal levels. Poult. Sci. **80:** 795–799.
13. GUICHARD, A., L. CEDARD, T.M. MIGNOT, *et al.* 1979. Radioimmunoassay of steroids produced by chick embryo gonads cultured the presence of some exogenous steroid precursors. Gen. Comp. Endocrinol. **39:** 9–19.
14. NOMURA, O., O. NAKABAYASHI, K. NISHIMORI, *et al.* 1999. Expression of five steroidogenic genes including aromatase gene at early developmental stages of chicken male and female embryos. J. Steroid Biochem. Mol. Biol. **71:** 103–109.
15. IMATAKA, H., K. SUZUKI, H. INANO, *et al.* 1998. Sexual differences of steroidogenic enzymes in embryonic gonads of the chicken (*Gallus domesticus*). Gen. Comp. Endocrinol. **69:** 153–162.
16. FREKING, F., T. NAZAIRIANS & B.A. SCHLINGER. 2000. The expression of the sex-steroid synthesizing enzymes CYP11A1, 3B-HSD, CYP17 and CYP19 in gonads and adrenals of adult and developing Zebra finches. Gen. Comp. Endocrinol. **119:** 140–151.
17. IMATAKA, H., K. SUZUKI, H. INANO, *et al.* 1988. Developmental changes of steroidogenic enzyme activities in the embryonic gonads of the chicken: the sexual difference. Gen. Comp. Endocrinol. **71:** 413–418.
18. BERCOVITZ, A.B., A. MIRSKY & F. FRYE. 1985. Non-invasive assessment of endocrine differences in day-old chicks (*Gallus domesticus*) by analysis of the immunoreactive oestrogen excreted in the egg. J. Reprod. Fertil. **74:** 681–686.
19. EPPLE, A., B. GOWER, M. TEN BUSCH, *et al.* 1997. Stress responses in avian embryos. Am. Zool. **37:** 536–545.
20. BENOWITZ-FREDERICKS, Z.M. & A.S. KITAYSKY. 2004. Benefits and costs of rapid growth in common murre (*Uria aalge*) chicks. J. Avian Biol. In press.
21. WINGFIELD, J.C. & D.S. FARNER. 1975. The determination of five steroids in avian plasma by radioimmunoassay and competitive protein binding. Steroids **26:** 311–327.
22. BERCOVITZ, A.B. & P.L. SARVER. 1988. Comparative sex-related differences of excretory sex steroids from day-old Andean condors (*Vultur gryphus*) and Peregrine falcons (*Falco peregrinus*): non-invasive monitoring of neonatal endocrinology. Zoo Biol. **7:** 147–153.
23. KALLIECHARAN, R. & B.K. HALL. 1974. A developmental study of the levels of progesterone, corticosterone, cortisol, and cortisone circulating in plasma of chick embryos. Gen. Comp. Endocrinol. **24:** 364–372.
24. WISE, P.M. & B.E. FRY. 1973. Functional development of the hypothalamo–hypophyseal–adrenal cortex axis in the chick embryo, *Gallus domesticus*. J. Exp. Zool. **185:** 277–292.

25. TONA, K., R.D. BALHEIROS, F. BAMELIS, *et al.* 2003. Effects of storage time on incubating egg gas pressure, thyroid hormones, and corticosterone levels in embryos and on their hatching parameters. Poult. Sci. **82:** 840–845.
26. BERTRAM, C.E. & M.A. HANSON. 2002. Prenatal programming of postnatal endocrine responses by glucocorticoids. Reproduction **124:** 459–467.
27. MATTHEWS, S.G. 2002. Early programming of the hypothalamo–pituitary–adrenal axis. Trends Endocrinol. Metab. **13:** 373–380.
28. BALTIC, M., S. JENNI-EIERMANN, R. ARLETTAZ & R. PALME. 2005. A noninvasive technique to evaluate human-generated stress in the black grouse. Ann. N.Y. Acad. Sci. **1046:** 81–95.
29. GOYMANN, W. 2005. Noninvasive monitoring of hormones in bird droppings: physiological validation, sampling, extraction, sex differences, and the influence of diet on hormone metabolite levels. Ann. N.Y. Acad. Sci. **1046:** 35–53.
30. THIEL, D., S JENNI-EIERMANN & R. PALME. 2005. Measuring coticosterone metabolites in droppings of capercaillies (*Tetrao urogallus*). Ann. N.Y. Acad. Sci. **1046:** 96–108.
31. TOUMA, C. & R. PALME. 2005. Measuring fecal glucocorticoid metabolites in mammels and birds: the importance of validation. Ann. N.Y. Acad. Sci. **1046:** 54–74.
32. HAYWARD, L. & J.C. WINGFIELD. 2004. Maternal corticosterone is transferred to avian yolk and may alter offspring growth and adult phenotype. Gen. Comp. Endocrinol. **135:** 365–371.

Are There Specific Adaptations for Long-Distance Migration in Birds? The Search for Adaptive Syndromes

Outline of the European Science Foundation Workshop

ULF BAUCHINGER,[a] CHRISTIAAN BOTH,[b] AND THEUNIS PIERSMA[b,c]

[a]*Department of Biology II, University of Munich, D-82152 Planegg-Martinsried, Germany*

[b]*Animal Ecology Group, Centre for Ecological and Evolutionary Studies, University of Groningen, 9750 AA Haren, the Netherlands*

[c]*Department of Marine Ecology and Evolution, Royal Netherlands Institute for Sea Research (NIOZ), 1790 AB Den Burg, Texel, the Netherlands*

Twice a year, numerous species of almost all bird genera face the challenge of overcoming distances of up to several thousand kilometers between breeding and wintering grounds. Migrations enable birds to explore highly seasonal habitats and exploit resources during their peak abundance, primarily for reproduction. Time spent in migration can exceed that spent in other phases of the annual cycle. Metabolism is at its peak, due to high catabolism required during the energetically costly flight, while anabolism predominates during subsequent stopover periods. With respect to time and energy spent, and resources that are accumulated and depleted, migration is a dominating phase within the annual cycle of long-distance migratory birds. In the frame of the "Optimality in Bird Migration" program of the European Science Foundation, scientists were able to investigate, discuss, and exchange their thinking about the phenomenon of bird migration. The main issues of the "optimality approach" is that each individual is selected to maximize its individual fitness, and for the migration program it has been proposed to do so by either (1) minimizing time, (2) minimizing energy costs of transport, or (3) predator minimization.[1]

In recent years we have achieved a substantial increase in our knowledge about the ecophysiology of long-distance migration, but we are yet far from understanding the mechanisms that shape the phenomenon of long-distance migration. There is a long history of addressing the question whether different migratory traits should be seen as adaptations to specific needs, or as constraints set by the architecture of the birds and the laws of physics. It is clear, however, that most adaptations will not go without posing constraints on other potentially important traits, and that therefore we need to consider how different migratory traits are combined and traded off

Address for correspondence: Prof. Ulf Bauchinger Department of Biology II, University of Munich, Grosshaderner Str. 2, D-82152 Planegg-Martinsried, Germany.
bauchinger@zi.biologie.uni-muenchen.de

against each other to produce the optimal migratory phenotype. Investigating long-distance migration as an adaptive syndrome may give such a superordinate view and could simplify investigations of single aspects of the process.

Phenotypic changes of organs occurring during migration should thus be interpreted as adaptations that also may constrain flexibility in other traits. For example, the reduction in organ mass is a strategy by migrants to reduce flight and maintenance costs, but also reduces capacity for immediate refueling at new stopover sites. The optimal strategy for changes in organs during migration thus should depend on the species or specific possibilities of the individual to perform, for instance, short versus long flights, and therefore these selection pressures should not be ignored in explaining the adaptation. Compositional changes, such as increases in key catabolic enzymes in the flight muscles prior to the onset of migration, suggest an adaptive nature of these phenotypic changes. Phenotypic flexibility of organs during migration is likely to be a common adaptive syndrome, typical for vertebrates living in variable environments.[2]

In the workshop entitled, "Are There Specific Adaptations for Long-Distance Migration in Birds? The Search for Adaptive Syndromes," we aimed to discuss the indications for an adaptive syndrome from scientists offering different perspectives in the field of bird migration. The workshop was held from 6 to 8 January 2005 at the Max Planck Institute for Ornithology. Contributions from research ornithologists from a variety of backgrounds, including functional morphology, stopover ecology, behavioral ecology, evolutionary ecology, quantitative genetics, eco-physiology, dynamic modeling, and molecular biology were scheduled with the goal of investigating the existence of a syndrome for long-distance migration. We are convinced that the only way to understand adaptations for long-distance migration in the future is through combining these different perspectives, because specific adaptations will most likely reduce the degrees of freedom for other adaptations.

ACKNOWLEDGMENTS

We thank the European Science Foundation for financial support, the late Ebo Gwinner for encouragement and support, Franz Bairlein and the Bird Migration Program for giving us a reason to contemplate and organize the workshop, and the participants for making it such a stimulating event. The Max Planck Institute for Ornithology is gratefully acknowledged for its hospitality and logistic support, especially Lisa Trost, Andrea Wittenzellner, Helga Gwinner, Jasmin Keil, and Andi Gaiser, whose personal involvement created an intimate and productive atmosphere, which made this workshop lively and successful. Finally, we thank Lisa Trost and Nicole Hoiss for their brilliant organizational support before, during, and after the workshop.

REFERENCES

1. ALERSTAM, T. & Å. LINDSTRÖM. 1990. Optimal bird migration: the relative importance of time, energy, and safety. *In* Bird Migration: Physiology and Ecophysiology. E. Gwinner, Ed.: 331–351. Springer-Verlag. Berlin.
2. PIERSMA, T. & Å. LINDSTRÖM. 1997. Rapid reversible changes in organ size as component of adaptive behaviour. Trends Ecol. Evol. **12:** 134–138.

Flexible Seasonal Timing and Migratory Behavior

Results from Stonechat Breeding Programs

BARBARA HELM, EBERHARD GWINNER,[†] AND LISA TROST

Max Planck Institute for Ornithology, D-82346 Andechs, Germany

ABSTRACT: Rigid schedules of long-distance migrants could be among candidate traits for adaptive migratory syndromes. This prediction was tested on stonechats, passerines that differ widely in migratory behavior and seasonal schedules. Stonechats in Europe are short-distance migrants and multi-clutched, whereas African residents and Siberian long-distance migrants usually raise single broods. In captivity, all subspecies displayed endogenous cycles of reproductive development and molt. The subspecies differed in time afforded to life cycle stages. Under conducive aviary conditions, African stonechats were multibrooded, whereas Siberian stonechats did not add clutches. This difference in flexibility was exclusively related to the length of breeding windows. Stonechats also differed in premigratory preparations. Postjuvenile molt started early in Siberian stonechats, but in European and African stonechats, depended strongly on hatching date. In contrast, all subspecies shortened molt duration at the same rate when hatched from late broods. Plasticity of Zugunruhe timing was identical in Siberian and European subspecies and nearly compensated for hatching late. The stonechat data suggest a refined understanding of temporal plasticity in long-distance migrants. Overall, plasticity was not reduced, but was differently organized. Apparently rigid migrant schedules were related to short breeding cycles and inflexible molt onset. Short windows for breeding and juvenile development could provide safety measures for timely departure. Once molt was initiated, temporal plasticity of long-distance migrants matched that of less migratory conspecifics. In addition to adjusting endogenous programs, stonechats differed in implementing them in the field. Modifying the conditions under which programs are expressed may be an efficient way to enhance seasonal plasticity.

KEYWORDS: migration; breeding; flexible timing; photoperiod; plasticity; stonechat; molt

RIGID SCHEDULES OF LONG-DISTANCE MIGRANTS?

Long-distance migration has been associated with particularly precise seasonal timing. In the field and in captivity, long-distance migrants were observed to show

---

Address for correspondence: Barbara Helm, Ph.D., Max Planck Institute for Ornithology, von-der-Tann-Str. 7, D-82346 Andechs, Germany. Voice: 49-8152-373-114; fax: 49-8152-373-133.
helm@orn.mpg.de
[†]Deceased.

Ann. N.Y. Acad. Sci. 1046: 216–227 (2005). © 2005 New York Academy of Sciences.
doi: 10.1196/annals.1343.019

little variation in the timing of breeding, molt, and migratory activities. They typically responded less strongly to environmental perturbation and manipulation of temporal cues than did related, less migratory taxa.[1–7] Rigid schedules are therefore among candidate traits for "adaptive migratory syndromes." Avian seasonal timing, in turn, has been shown to involve endogenous clocks supplemented by additional information.[4–10] Hence, inflexible schedules of long-distance migrants could be based on specializations in the programs driving seasonal timing. Breeding, migration, and molt could be rigidly timed by endogenous calendars.[3,8,10,11] Temporal precision could also be achieved via a highly predictable timing cue: photoperiod, the light-fraction of the day.[4–7] In the first case, schedules of long-distance migrants would be largely unaffected by environmental cues; in the second, timing would be plastic to the predictable information conveyed by day length. In the following, we pursue these questions in model birds of seasonal timing, captive stonechats (*Saxicola torquata*). We test the general tenability of inflexible schedules of long-distance migrants and assess the importance of photoperiodic information in relation to migratory behavior.

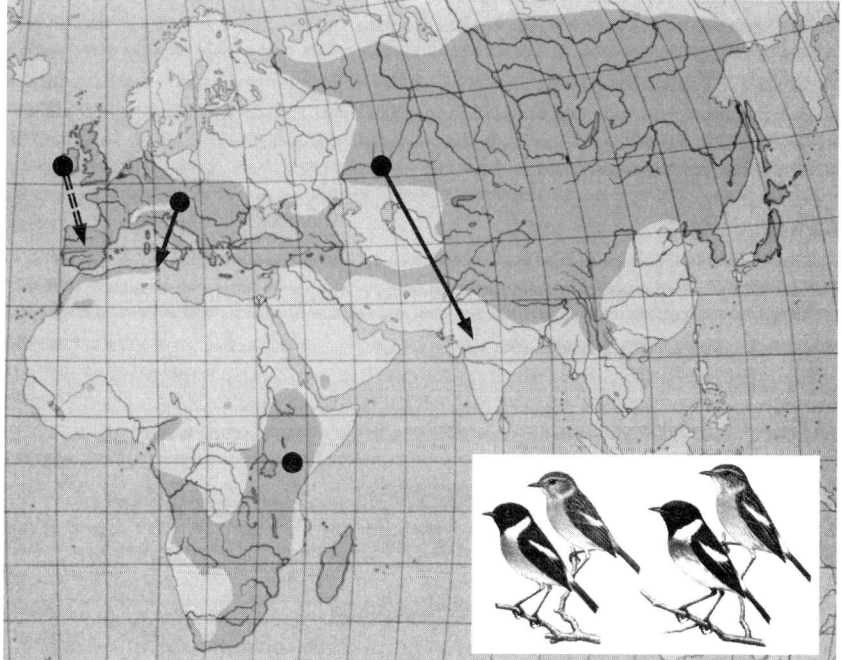

FIGURE 1. Breeding range of stonechats. Origins of birds studied in captivity are marked by *dots*; their migratory behavior is indicated by *arrows*. (For data of Irish partial migrants, see Ref. 16; modified after Ref. 36.)

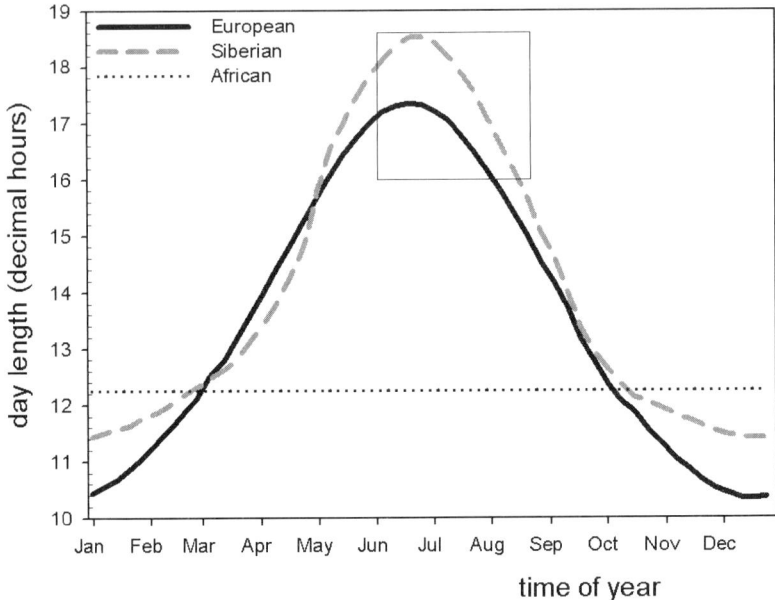

FIGURE 2. Seasonal course of photoperiod experienced by Siberian, European, and equatorial African stonechats. Equatorial stonechats experience nearly constant days year-round. For Siberian and European stonechats, day length during the breeding season simulates that of 52.5°N and 47.5°N, respectively; during migrations, the curves simulate day length en route; wintering day lengths are those of 25°N and 40°N, respectively. Hair-lined box indicates day-length range experienced by Siberian stonechats from hatching until molt onset. (Based on data from M. Raess, personal communication.)

STONECHATS AS A MODEL FOR STUDYING PLASTICITY OF SEASONAL TIMING

Stonechats, small turdid passerines, were selected as avian models to investigate seasonal timing by Eberhard Gwinner and colleagues in the early 1980s. A main reason for choosing this species was its extensive North–South breeding range, which is unmatched by other passerines (FIG. 1).[12,13] Stonechats breed from North Siberia to the Southern tip of Africa (~70°N to 30°S), including the immediate vicinity of the equator. Their distribution is partly disjunct, with large genetic distances between local populations.[12–14] Over vast parts of their breeding range, stonechats are at least partially migrant, with the peculiar habit of wintering in territorial pairs.[15] Migratory behavior tends to increase with latitude and continental influences on the climate.[13,16] Latitude also determines day length cues experienced by the birds (FIG. 2). Photoperiod fluctuates dramatically in higher-latitude ranges, but remains almost constant near the equator.[17] Since seasonal fluctuations of weather and resources are moderate, equatorial pairs maintain year-round territories. Migrant conspecifics renegotiate territories and pair bonds twice a year.[15,17]

Seasonal timing has been studied extensively in stonechats from different localities (FIG. 1; e.g., see Refs. 9 and 15–24). These include resident African, short-distance migrant European, and long-distance migrant Siberian stonechats. Until taxonomic relationships are resolved, we refer to handbook definitions of the investigated "subspecies" *Saxicola torquata axillaris* from East Africa, *S.t. rubicola* from Central Europe, and *S.t. maura* from North Asia.[25,26] Stonechats of all subspecies expressed clear, persistent circannual cycles of molt and gonadal development.[9,16,19,21] Comparisons of schedules under suites of experimental conditions have revealed influences of different endogenous programs and seasonal cues on the timing of life cycle stages. Detailed methods of assessing gonadal sizes, molt, and migratory restlessness (Zugunruhe) are provided elsewhere.[16,19,22] Here we review some results and present new data on migratory restlessness (Zugunruhe) and postjuvenile molt with a key interest in comparing seasonal flexibility in relation to migratory behavior.

MAKING TIME TO BREED

Avian breeding seasons tend to be longest at the equator and shorten with increasing latitude.[5–7,27] Stonechat subspecies, too, vary markedly in reproductive timing. In accordance with general patterns the northernmost taxon, Siberian stonechats, is single-brooded, whereas European conspecifics frequently lay three clutches.[13,28,29]

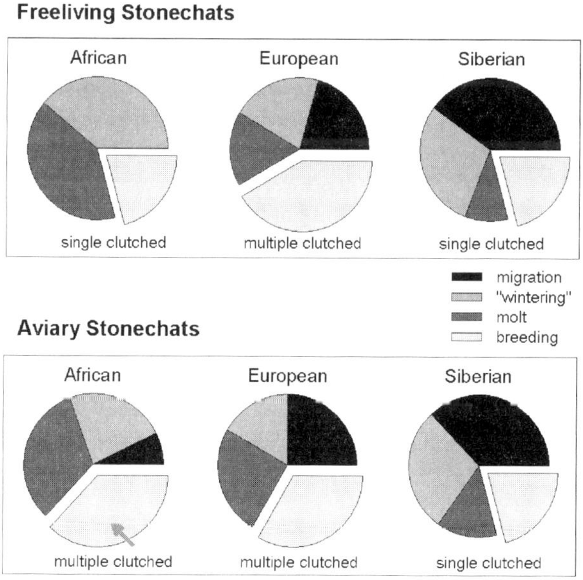

FIGURE 3. Seasonal activities of free-living (**upper panel**) and captive (**lower panel**) African, European, and Siberian stonechats; *pie slices* show the approximate proportion of the annual cycle spent on respective life-cycle stages. (Data are taken from Refs. 16, 17, 20, and 28; M. Raess, personal communication; and Helm and Gwinner, unpublished data.)

Stonechats from equatorial Kenya would be expected to breed over the longest time, but in the field are also single clutched (FIG. 3).[17] When stonechats were allowed to breed in captivity, the subspecies differed in their taking advantage of conducive aviary conditions. African stonechats doubled the number of clutches (FIG. 3).[20] In contrast, favorable breeding conditions did not induce additional broods in the other subspecies. As a consequence, aviary-breeding African stonechats matched their European conspecifics in broodedness. Siberian stonechats continued to produce only one clutch with occasional replacement after young were taken from the nest.[16]

The inability of Siberian stonechats to capitalize on breeding opportunities confirms the prediction of rigid timing in long-distance migrants. However, a closer look at underlying mechanisms refines this interpretation. Stonechats undergo circannual cycles of gonadal growth and regression that define the time of breeding ("reproductive windows").[18,19,21] African and European stonechats both have long breeding windows. In the field, only European stonechats use their entire reproductive life cycle stage for multiple clutches. Free-living African stonechats, in contrast, spend only part of the time set for reproduction on breeding. Aviary birds placed more clutches into their reproductive windows. Breeding stopped coterminously in European and African stonechats. Additional clutches of African stonechats were thus not achieved by prolonged gonadal activity or reinitiated gonadal growth, as described for some other passerine groups.[5,17,19,20] Siberian stonechats had short breeding windows typical of long-distance migrants. In aviaries as in the field they generally bred once and prepared for migration by an early molt (M. Raess, personal communication; Helm and Gwinner, unpublished data).[16] The shortness of their reproductive life cycle stage precluded additional clutches in aviaries. In all stonechats, therefore, breeding was limited by endogenous calendars. Reduced plasticity of Siberian migrants, due to short breeding windows, resulted in foregone breeding opportunities: Staying in time was costly.

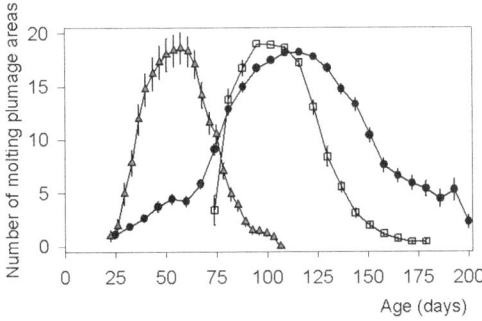

FIGURE 4. Time course of postjuvenile molt in African (*black dots*), European (*white squares*), and Siberian (*gray triangles*) stonechats; data show the median number of simultaneously molting plumage areas plotted against the age of young birds; *error bars* show standard errors of the median; African birds were kept under short days (LD12.25: 11.75 h), the other subspecies under changing light conditions of 47.5°N (see FIG. 2). (Data taken from Refs. 22 and 23.)

TIMING POSTJUVENILE MOLT: CUE USE DIFFERENTIATION

In many species, including stonechats, molt marks the termination of the breeding season and provides a seasonal link to migration. As a demanding regenerative stage of their annual cycle, passerines often molt precisely to avoid major overlap with migration.[4,11,30] Accordingly, birds from higher latitudes are generally expected to initiate molt earlier than those from more temperate or tropical ranges. In addition, juveniles hatching from late broods often initiate and finish molt at younger ages than offspring from early clutches. Speeding up molt late in the season, termed "calendar effect," is thought to allow late-hatched young to prepare for timely migration and wintering.[3,11,30] Calendar effects function via day length perceived by nestlings and can be simulated in captivity.[2,11] Birds breeding at higher latitudes, typically under predictable seasonality, are thought to rely particularly on day length cues for timing seasonal behavior.[4-7] Since long-distance migrants are thought to be under pressure to molt in time, calendar effects should increase with latitude and migratory behavior.

We thus expected stonechat molt to start earlier and be of shorter duration with increasing latitude and migratory inclination. We predicted calendar effects to be present in birds naturally exposed to seasonally changing day length, that is, Euro-

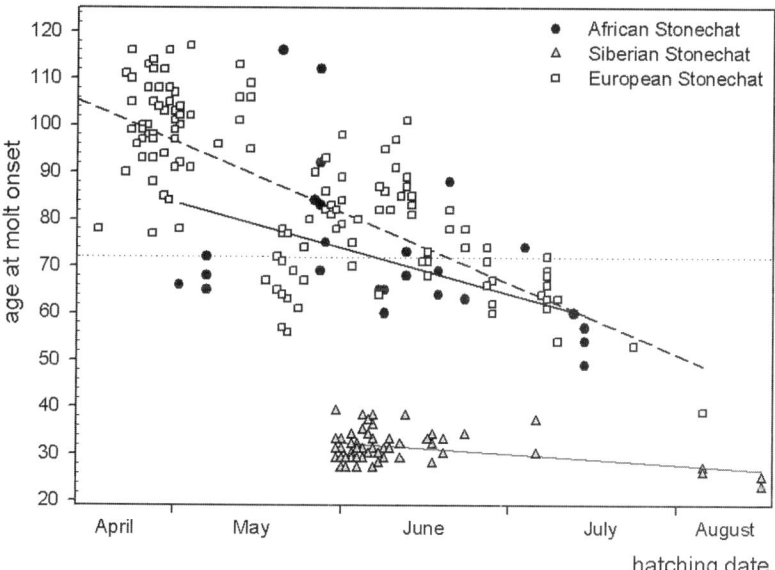

FIGURE 5. Effects of hatching date on the initiation of postjuvenile molt. The graph plots the age at which young African (*black dots*), European (*white squares*), and Siberian (*gray triangles*) stonechats molted at least five plumage areas against their date of hatching under changing light conditions of 47.5°N; all subspecies differed significantly in intercept and slope of calendar effect ($r^2 = 85.6$; $P < .001$; $n = 261$; mean overall slope = -0.40 ± 0.03); *dotted horizontal line* indicates median molt onset of African stonechats under equatorial day length. (Data taken from Refs. 22 and 23 and from Helm and Gwinner, unpublished.)

pean and Siberian stonechats, and to increase with latitude and migratory inclination. We tested these predictions on young birds, since adult molt is often affected by breeding behavior.[6] FIGURE 4 shows the time course of postjuvenile molt in immature African, European, and Siberian stonechats. Young were kept under simulations of their native photoperiods. In line with predicted patterns, Siberian stonechats commenced molt at the youngest ages, European stonechats at intermediate ages, and African stonechats at the oldest age. Molt duration was shortest in Siberian and intermediate in European stonechats. African stonechats took longest for molt. When raised under seasonally changing European day length, all stonechats modified molt timing depending on hatching date (FIG. 5). However, calendar effects on molt onset differed strikingly: Late-hatched European and African stonechats advanced molt onset markedly. Late-hatched Siberian stonechats, in contrast, initiated molt at only slightly but significantly younger ages (slope of subspecies: Siberian = -0.08 ± 0.11; $n = 70$; European = -0.49 ± 0.09; $n = 162$; African = -0.32 ± 0.08; $n = 30$). Only European stonechats therefore behaved in the predicted way. The unexpectedly strong calendar effect in African stonechats may be part of a suite of photoperiodic traits present in this population.[31,32] However, the photoperiodic response to shortening day length is apparently not implemented under native, constant short day length. In the field and in captivity, molt did not start earlier under equatorial days than under much longer but fluctuating European days (FIGS. 4, 5;[17] median molt onset $\pm$ SE: short day = 72 ± 2.03; $n = 73$;[22] European day = 68 ± 3.56; $n = 30$; Mann–Whitney U test: $U = 953.0$; $P = .306$). Conversely, the calendar effect would predict molt to start a month earlier than observed under equatorial day length. African stonechats thus distinguished between constant and seasonally changing photoperiod, advancing molt only when day length fluctuated and otherwise took time to change plumage.

Contrary to predictions, Siberian stonechats modified molt onset only little in response to hatching date (FIG. 5). We suggest two not necessarily mutually exclusive interpretations of the weak photoperiodic response. Early molt onset could provide a safety measure, securing timely preparations for long-distance migration. Low photoperiodic responsiveness could also be related to the day length range that Siberian hatchlings experience (FIG. 2).[23,33] Whereas young of multiclutched European stonechats grow up over a long season, most Siberian nestlings hatch around midsummer when day length is almost constant. Photoperiod may therefore be less informative for Siberian than European hatchlings. However, field data suggest that when young from replacement clutches are included (M. Raess, personal communication), the day length range experienced until molt onset is larger than previously assumed, exceeding 2.5 h (FIG. 2, hairline box). Therefore, a safety strategy to secure timely molt onset is a more likely explanation for weak calendar effects in Siberian stonechats.

Molt onset of stonechats seems to also confirm the hypothesis of more rigid seasonal timing in long-distance migrants. Despite their relatively high-latitude breeding areas, Siberian stonechats did not appear to rely much on photoperiodic cues for timing premigratory preparations. However, this conclusion is again refined by a closer look because photoperiodic responsiveness changed over the course of molt. Stonechats modified molt further by accelerating its rate when hatching late in the season. The overall magnitude of the calendar effect therefore increased over the course of molt from –0.40 days/day at onset to –0.51 days/day for molt peak to –0.61 days/day for molt completion. Consequently, molt duration was shortened by –0.21 days with each

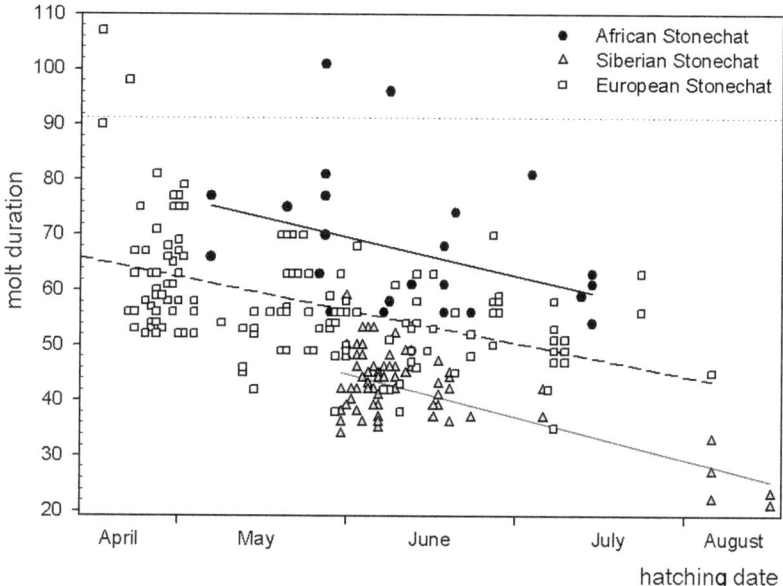

FIGURE 6. Effects of hatching date on duration of postjuvenile molt. The graph plots the number of days that young African (*black dots*), European (*white squares*), and Siberian (*gray triangles*) stonechats spent on molt against hatching date; all subspecies differed significantly in intercept, but not slope of calendar effect ($r^2 = 53.4$; $P < .001$; $n = 250$; mean overall slope = -0.21 ± 0.02); *dotted horizontal line* indicates median molt duration of African stonechats under equatorial day length. (Data as in FIG. 5.)

day a chick hatched later in the season (FIG. 6). In contrast to molt initiation, the subspecies did not differ in calendar effects on duration: Despite their fast molt, Siberian stonechats were able to shorten its duration at least as much as their conspecifics. Plasticity of molt duration was therefore not reduced in the long-distance migrants. African stonechats again appeared to distinguish between constant and fluctuating day length. Their molt under long European days was 1 month faster than under constant short days (FIG. 6, dotted line; median duration ± SE: short day = 91 ± 3.75; $n = 71$;[22] European day = 63 ± 3.27; $n = 25$; Mann–Whitney U test: $U = 258.5$; $P < .001$). Their relaxed molt under constant short day length provides further evidence for a photoperiodic specialization to native conditions.

COMMENCING MIGRATORY RESTLESSNESS

The onset of fall migration in individual passerines is generally thought to be related to molt completion.[30] However, young from early clutches have generally more time at the breeding grounds than late-hatched offspring. Only part of this excess time is spent on relaxed molt timing. Early hatched young can therefore afford to remain at the breeding grounds for some time after molt completion. In captive

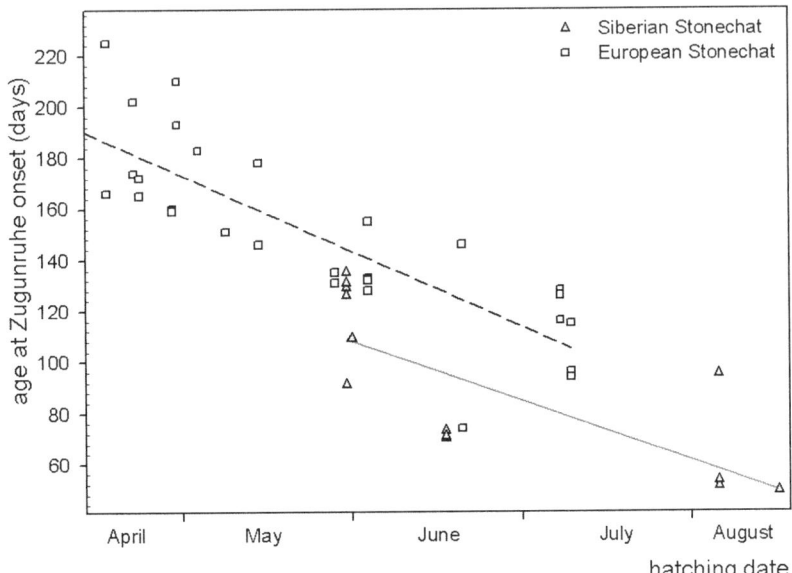

FIGURE 7. Effects of hatching date on the initiation of migratory restlessness. The graph plots the age at Zugunruhe onset in young European (*white squares*) and Siberian (*gray triangles*) stonechats; Siberian stonechats initiated Zugunruhe significantly earlier than European stonechats ($r^2 = 80.5$; mean overall slope = -0.88 ± 0.10; $P < .001$; $n = 44$; difference in intercept: -32.3 ± 7.44 days; $P < .001$). (Data from Ref. 16, and from Helm and Gwinner, unpublished.)

stonechats, Zugunruhe data are currently available for a subsample of the European and Siberian birds scored for molt. The amount of nocturnal restlessness was variable, but seasonal patterns were identified by time-series analyses in almost all individuals.[16] The interval between molt completion and onset of restlessness was longer in Siberian stonechats, but in both subspecies significantly decreased with later hatching dates ($r^2 = 23.3$; $P = .002$; $n = 44$; mean overall slope = -0.28 ± 0.10; difference in intercept of Siberian stonechats: 29.4 ± 7.84 days; $P = .001$). Zugunruhe of late-hatched young coincided with molt completion.

FIGURE 7 shows age at commencing nocturnal restlessness in relation to hatching date. The only difference between the subspecies was an overall earlier Zugunruhe onset in Siberian stonechats, typical of long-distance migrants.[1–3,10] Remarkably, calendar effects were identical among the subspecies, independently of their native grounds and migratory behavior. By the time Zugunruhe set in, European and Siberian stonechats had almost compensated for differences in hatching date. They advanced restlessness by -0.88 days for each day hatched later in the season. The Zugunruhe data allow once again the refinement of the concept of rigid schedules in long-distance migrants. Timing was plastic in response to hatching date, regardless of migratory habit. However, the exact way in which plasticity was achieved differed between subspecies by differentiated, graded timing responses throughout development from hatching to migration.

CONCLUSIONS

Comparative studies of stonechat subspecies revealed similar endogenous temporal organization, despite starkly different native environments, large genetic distances between taxa, and different migratory habits (estimated divergence time between African and northern Palearctic taxa, 1 million to 3 million years[13,14,31]). Siberian, European, and African stonechats displayed clear annual cycles of seasonal behavior. All accelerated postbreeding activities in response to shortening day length. However, the basic temporal organization was modified in at least three ways.

First, life cycle stages differed in length, with particularly long migration and short breeding seasons in Siberian stonechats typical of long-distance migrants. Reproduction was rigidly timed in all stonechats, but because time windows allotted to breeding differed, African residents but not Siberian long-distance migrants, were able to use conducive aviary conditions for additional clutches. Different breeding schedules were thus not based on greater flexibility of resident stonechats, but on their longer breeding windows. Observations from other taxa, including passerines, show that breeding seasons can be lengthened by different mechanisms, for instance, by extending or reinitiating gonadal activity under favorable conditions.[4] Stonechats may exemplify rigidly organized species with limited flexibility in transitions between life cycle stages.[6,7]

The second type of modification were differentiated, graded responses to photoperiodic cues. Siberian long-distance migrants differed from their conspecifics by reduced temporal plasticity of molt onset. Reduced plasticity was selective: As soon as molt started, Siberian stonechats developed strong responses to hatching date. Inflexible, early molt onset and subsequent increases in photoperiodic responsiveness, reported from the field and experimental studies, could indeed represent a typical strategy of long-distance migrants.[1,2,11,30] Initial reliance on endogenous clocks could provide a safety measure securing timely preparation for departure in birds from highly predictable but time-constrained habitats. Thus secured, long-distance migrants could time further premigratory preparations by the reliable cues provided by their habitats.[6,7]

The third type of modification was conditional implementation of endogenous programs. In resident African stonechats, photoperiodic responses typical of temperate species are apparently present as part of their endogenous programs.[21,31,32] However, the expression of these traits depended on conditions. African stonechats, which naturally breed under equatorial day length, showed pronounced calendar effects to shortening day length, but did not rush molt under constant short photoperiods. The data suggest that these birds show photoperiodic specializations, responding very differently to constant versus fluctuating day length conditions. In line with the idea of specialization to constant short days, Gwinner and Dittami[9,31] had noted particularly stable circannual rhythms in African stonechats, along with other modifications presumably related to tropical life.[21,24]

Evolutionary changes in the implementation of endogenous programs, for example, conditionality on specific environmental characteristics, can greatly affect life cycle timing. They imply no major changes of existing patterns and perhaps account for many of the observed differences between migrants and residents.[3] Clarifying the conditional expression of endogenous programs, for example, by comparing gene–

environment interactions[33–35] of taxa with different migratory habits, may therefore be a rewarding direction in future bird migration research.

ACKNOWLEDGMENTS

We gratefully acknowledge the great support of colleagues at our institute and for their patience in collecting the data presented here. Thanks to the European Science Foundation for financing the conference and to Ulf Bauchinger for organizing it.

REFERENCES

1. BERTHOLD, P., E. GWINNER & H. KLEIN. 1970. Vergleichende Untersuchung der Jugendentwicklung eines ausgeprägten Zugvogels (*Sylvia borin*) und eines weniger ausgeprägten Zugvogels (*S. atricapilla*). Vogelwarte **25:** 297–331.
2. GWINNER, E., P. BERTHOLD & H. KLEIN. 1971. Untersuchungen zur Jahresperiodik von Laubsängern. II. Einfluß der Tageslichtdauer auf die Entwicklung des Gefieders, des Gewichts und der Zugunruhe bei Fitis (*Phylloscopus trochilus*) und Zilpzalp (*Ph. collybita*). J. Orn. **112:** 253–265.
3. GWINNER, E. 1986. Circannual Rhythms. Endogenous Annual Clocks in the Organization of Seasonal Processes. Springer. Berlin.
4. HAHN, T. *et al.* 1992. Adjustments of the prebasic molt schedule in birds. Ornis Scand. **23:** 314–321.
5. HAHN, T. *et al.* 1997. Temporal flexibility in avian reproduction. Patterns and mechanisms. *In* Current Ornithology, Vol. 14. V. Nolan, E. Ketterson & C. Thompson, Eds.: 39–80. Plenum Press. New York.
6. WINGFIELD, J. *et al.* 1992. Environmental predictability and control of gonadal cycles in birds. J. Exp. Zool. **261:** 214–231.
7. WINGFIELD, J.C., T. HAHN & D. DOAK. 1993. Integration of environmental factors regulating transitions of physiological state, morphology and behavior. *In* Avian Endocrinology. P.J. Sharp, Ed.: 111–122. Society of Endocrinology, Journal of Endocrinology Ltd. Bristol, UK.
8. GWINNER, E. 1968. Circannuale Periodik als Grundlage des jahreszeitlichen Funktionswandels bei Zugvögeln. Untersuchungen am Fitis *(Phylloscopus trochilus)* und am Waldlaubsänger *(P. sibilatrix)*. J. Orn. **109:** 70–95.
9. GWINNER, E. & J. DITTAMI. 1990. Endogenous reproductive rhythms in a tropical bird. Science **249:** 906–908.
10. BERTHOLD, P. 2001. Bird Migration. A General Survey. Oxford University Press. Oxford.
11. NOSKOV, G.A., T.A. RYMKEVICH & N. IOVCHENKO. 1999. Intraspecific variation of molt: adaptive significance and ways of realisation. *In* Proceedings of the 22nd International Ornithological Congress Durban. N. Adams & R. Slotow, Eds.: 544–563. BirdLife South Africa. Johannesburg.
12. UNDERHILL, L.G. 1999. Avian demography: statistics and ornithology. *In* Proceedings of the 22nd International Ornithological Congress Durban. N. Adams & R. Slotow, Eds.: 61–70. BirdLife South Africa. Johannesburg.
13. URQUHART, E.D. 2002. Stonechats. Christopher Helm. London.
14. WINK, M., H. SAUER-GÜRTH & E. GWINNER. 2002. Evolutionary relationships of stonechats and related species inferred from mitochondrial-DNA sequences and genomic fingerprinting by ISSR-PCR. Br. Birds **95:** 349–355.
15. GWINNER, E., T. RÖDL & H. SCHWABL. 1994. Pair territoriality of wintering stonechats: behavior, function and hormones. Behav. Ecol. Sociobiol. **34:** 321–327.
16. HELM, B. 2003. Seasonal timing in different environments: comparative studies in stonechats. Ph.D. thesis. Ludwig-Maximilian University. Munich.

17. DITTAMI, J. & E. GWINNER. 1985. Annual cycles in the African stonechat *Saxicola torquata axillaris* and their relationship to environmental factors. J. Zool. A (Lond.) **207:** 357–370.
18. GWINNER, E. 1999. Rigid and flexible adjustments to a periodic environment: role of circadian and circannual programs. *In* Proceedings of the 22nd International Ornithological Congress Durban. N. Adams & R. Slotow, Eds.: 2366–2378. BirdLife South Africa. Johannesburg.
19. GWINNER, E., S. KÖNIG & M. ZEMAN. 1995. Endogenous gonadal, LH and molt rhythms in tropical stonechats: effect of pair bond on period, amplitude, and pattern of circannual cycles. J. Comp. Physiol. A **177:** 73–79.
20. KÖNIG, S. & E. GWINNER. 1995. Frequency and timing of successive broods in captive African and European stonechats *Saxicola torquata axillaris* and *S. t. rubicola*. J. Avian Biol. **26:** 247–254.
21. SCHEUERLEIN, A. & E. GWINNER. 1999. Proximate and ultimate aspects of photoperiodic sensitivity in equatorial stonechats: *Saxicola torquata axillaris*. *In* Proceedings of the 22nd International Ornithological Congress Durban. N. Adams & R. Slotow, Eds.: 1756–1766. BirdLife South Africa. Johannesburg.
22. HELM, B. & E. GWINNER. 1999. Timing of postjuvenile molt in African (*Saxicola torquata axillaris*) and European (*Saxicola torquata rubicola*) stonechats: effects of genetic and environmental factors. Auk **116:** 589–603.
23. HELM, B. & E. GWINNER. 2001. Nestling growth and post-juvenile molt under a tight seasonal schedule in stonechats *Saxicola torquata maura* from Kazakhstan. Avian Sci. **1:** 31–42.
24. WIKELSKI, M. *et al.* 2003. Slow pace of life in tropical sedentary birds: a common-garden experiment on four stonechat populations from different latitudes. Proc. R. Soc. Lond. B Biol. Sci. **270:** 2383–2388.
25. CRAMP, S. *et al.*, Eds. 1988. Handbook of the Birds of Europe, the Middle East and North Africa, Vol. 5. Oxford University Press. Oxford.
26. GLUTZ VON BLOTZHEIM, U.N. & K.M. BAUER, Eds. 1988. Handbuch der Vögel Mitteleuropas, Vol. 11. Aula. Wiesbaden, Germany.
27. BAKER, J.R. 1938. The evolution of breeding seasons. *In* Evolution: Essays on Aspects of Evolutionary Biology. G.R. De Beer, Ed.: 161–177. Oxford University Press. London.
28. DEMENTIEV, G. & N. GLADKOV. 1968. The Birds of the Soviet Union. [English transl.]. Israel Program for Scientific Translations. Jerusalem.
29. FLINKS, H. & F. PFEIFER. 1987. Brutzeit, Gelegegröße und Bruterfolg beim Schwarzkehlchen (*Saxicola torquata*). Charadrius **23:** 128–140.
30. JENNI, L. & R. WINKLER. 1994. Molt and Ageing of European Passerines. Academic Press. London.
31. GWINNER, E. & J. DITTAMI. 1985. Photoperiodic responses in temperate zone and equatorial stonechats: a contribution to the problem of photoperiodism in tropical organisms. *In* The Endocrine System and the Environment. B.K. Follett, S. Ishii & A. Chandola, Eds.: 279–294. Japan Scientific Societies Press. Tokyo/Springer-Verlag. Berlin.
32. STYRSKY, J.D., P. BERTHOLD & W.D. ROBINSON. 2004. Endogenous control of migration and calendar effects in an intratropical migrant, the yellow-green vireo. Anim. Behav. **67:** 1141–1149.
33. GWINNER, E. & B. HELM. 2003. Circannual and circadian contributions to the timing of avian migration. *In* Avian Migration. P. Berthold, E. Gwinner & E. Sonnenschein, Eds.: 81–95. Springer. Heidelberg.
34. VAN NOORDWIJK, A.J. 1989. Reaction norms in genetical ecology. BioScience **39:** 453–458.
35. SCHLICHTING, C. & M. PIGLIUCCI. 1998. Phenotypic evolution. A reaction norm perspective. Sinauer Associates. Sunderland, MA.
36. GWINNER, E. 2003. Circannual rhythms in birds. Curr. Opin. Neurobiol. **13:** 770–778.

Are Long-Distance Migrants Constrained in Their Evolutionary Response to Environmental Change?

Causes of Variation in the Timing of Autumn Migration in a Blackcap (*S. atricapilla*) and Two Garden Warbler (*Sylvia borin*) Populations

FRANCISCO PULIDO AND MICHAEL WIDMER[b]

Max Planck Institute for Ornithology, Vogelwarte Radolfzell, 78315 Radolfzell, Germany

ABSTRACT: Long-distance migratory birds often show little phenotypic variation in the timing of life-history events like breeding, molt, or migration. It has been hypothesized that this could result from low levels of heritable variation. If this were true, the adaptability of long-distance migratory birds would be limited, which would explain the vulnerability of this group of birds to environmental changes. The amount of phenotypic, environmental, and genetic variation in the onset of autumn migratory activity was assessed in two garden warbler (*Sylvia borin*) populations differing in breeding phenology and the length of the breeding season with the aim of investigating the effects of selection on the adaptability of long-distance migrants. High heritabilities and additive genetic variance components for the timing of autumn migration were found in both populations. Although genetic variation in the mountain population was lower than in the lowlands, this difference was not statistically significant. Moreover, no evidence was found for reduced levels of genetic variation in the garden warbler as compared to its sister species, the blackcap (*S. atricapilla*). Environmental variation, however, was markedly reduced in the garden warbler, suggesting that low levels of phenotypic variation typically found in long-distance migrants may be a consequence of environmental canalization of migratory traits. The buffering of environmental variation may be an adaptive response to strong stabilizing selection on the timing of migration. High environmental canalization of migration phenology in long-distance migrants could potentially explain low rates of immediate phenotypic change in response to environmental change.

KEYWORDS: heritability, zugunruhe, climate change, adaptation, canalization

Address for correspondence: Francisco Pulido, Netherlands Institute of Ecology, PO BOX 40, ZG 6666 Heteren, The Netherlands Voice: +31-26-4791247; fax: +31-26-4723227.
f.pulido@nioo.knaw.nl
[a]Current address: Netherlands Institute of Ecology, PO BOX 40, ZG 6666 Heteren, The Netherlands
[b]Current address: Orniplan AG, Wiedingstr. 78, CH-8045 Zurich, Switzerland

INTRODUCTION

We are currently witnessing rapid environmental changes that are causing changes in the migratory behavior of birds.[1] These changes are characterized by an increase in the proportion of resident individuals, a shortening of migration distances, and by earlier arrival on and later departure from the breeding grounds.[2–5] Alterations of avian migration patterns, particularly in phenology, repeatedly have been reported to be more pronounced in short-distance than in long-distance migrants.[6–10] At the same time, numbers of long-distance migrants have decreased, particularly among many species breeding in Europe[11,12] and North America.[13] Habitat loss, habitat fragmentation, and deterioration have been identified as the main causes for most of these population declines.[14] Increasing competition with short-distance migrants and resident birds on the breeding grounds, and low adaptability have been proposed as additional factors favoring the decline of long-distance migrants, particularly under the prospect of recent climatic change.[2–4,15] While ecological causes for population declines in long-distance migrants have deserved much attention, the potential role of reduced rates of adaptive evolution in this group of birds has only recently become the focus of research.[4,16,17] In a compilation of recent changes and presumptive instances of lack of change in avian migration patterns, all species with suboptimal migration routes were long-distance migrants.[18] This finding suggests that migration patterns of birds migrating long distances could be more resilient to environmental change than short-distance migrants, probably as a consequence of endogenous control, and genetically transmitted migration routes.

Constraints on the response to natural selection could be imposed by reduced levels of genetic variation in migratory traits due to strong stabilizing selection,[19] and by divergent selection on migration, on the breeding, and wintering grounds.[20–22] Webster *et al.*[23] recently predicted that in populations with high migratory connectivity—as found in many long-distance migrants—genetic variation may be limited and therefore "evolutionary response to large-scale climatic change could be severely hampered." In addition, unfavorable genetic correlations of migratory traits with other life-history traits,[24,25] and the disruption of the complex life cycles of long-distance migrants could limit their adaptive response to environmental changes.[26,27] However, it is unclear whether strong selection on migratory traits affects the evolutionary response of long-distance migrants. Although most theoretical models and empirical studies predict that stabilizing selection will erode phenotypic variance due to the loss of additive genetic variation,[28,29] theoretical studies have revealed that the conditions under which stabilizing selection may deplete genetic variation may be restrictive.[30] Stabilizing selection could also result in a reduction of environmental variance as a consequence of the fitness advantage of genotypes that are insensitive to environmental variation.[31–33]

In long-distance migrants, the timing of autumn migration is expected to be under strong stabilizing selection.[22,24,34] Compared to birds migrating short distances, long-distance migrants arrive at the breeding area late in the season, and are the first to leave. This group of birds is thus characterized by rapid juvenile development, which is an adaptation to the short breeding season.[26,35] Although rapid molt and growth is associated with high energetic and other fitness costs,[36–38] these costs are offset by the advantage of leaving the breeding area before conditions deteriorate. Birds migrating late in the season are at high risk of not finding enough food to build up energy reserves to overcome large barriers like seas or deserts. Encountering inclement weather before or on migration may be particularly detrimental for insectivorous migrants. Moreover, there are a number of species that establish territories in the nonbreeding season for which early arrival on the wintering ground may be highly advantageous [e.g., for the pied flycatcher (*Ficedula hypoleuca*)[39]].

Studies investigating genetic variation and evolutionary change in the timing of migration of long-distance migrants have yielded contradictory results. While there is evidence for a lack of additive genetic variation in the timing of spring arrival in pied flycatchers breeding in Spain,[40] high heritabilities for this trait have been found in barn swallows (*Hirundo rustica*) breeding in Denmark.[41] Moreover, cliff swallows (*Petrochelidon pyrrhonota*) in North America strongly responded to natural selection on spring arrival date.[42] These results and significant among-year repeatabilities for this trait in other species[17] are in line with a number of studies demonstrating changes in spring arrival in response to global warming.[43] In contrast to spring arrival, the timing of autumn migration has been little studied, probably because it is more difficult to measure in the field.[26] Despite methodological problems, there are a few studies reporting changes in the timing of autumn migration in a number of bird species, including some long-distance migrants.[10,44–46] However, it is unclear whether these changes are just plastic responses to environmental alterations (phenotypic plasticity) or if they involve changes in the genetic composition of populations (evolutionary change).

The presence of additive genetic variation in the timing of autumn migration has hitherto only been demonstrated in a short- to middle-distance migrant, the southern German blackcap (*Sylvia atricapilla*),[22] which predominantly winters in the western Mediterranean region. The sister species of the blackcap is the garden warbler (*S. borin*), a long-distance migrant, that breeds in the western Palearctic and winters south of the Sahara in subtropical and tropical Africa.[47] This species is an excellent candidate for studying the effects of stabilizing selection on the adaptability of long-distance migrants, as results can be compared with those in the well-studied blackcap.

Here, we estimate heritabilities and variance components for the timing of autumn migration in the garden warbler with the aim of evaluating whether adaptive evolution of migratory behavior in long-distance migrants may be constrained by the lack of additive genetic variation. Moreover, we compare two garden warbler populations differing in life-history and ecology with the blackcap to study the effects of stabilizing selection on phenotypic variance and its causal components, and to assess its consequences for adaptive evolution of migratory traits.

MATERIALS AND METHODS

In 1993 and 1994, we collected a total of 89 garden warblers at an age of 4–8 days from 30 nests in two central European populations. Fifty-one nestlings were sampled in the Urseren Valley (46°36′N, 8°31′E), a population dwelling in the central Swiss Alps at 1500 m above sea level. Thirty-eight individuals were collected in the Upper Rhine Valley (47°53′N, 7°34′E), in southwest Germany, at 200-m altitude. These garden warbler populations differ markedly in their breeding biology and phenology, due to differences in climatic conditions encountered in their respective breeding areas. The arrival of spring, for instance, is about 3 weeks later in the mountain than in the lowland population. As a consequence, the breeding season in the Rhine Valley is shorter by almost 4 weeks, which may cause particularly strong selection on the timing of autumn migration and on juvenile development in this population.[48,49] For a detailed description of the habitat and biology of the populations, see Widmer.[48,50,51]

All birds were hand-raised and kept in climate chambers under identical controlled conditions (see Berthold et al.[52] and Widmer[49] for details on rearing, feeding, and keeping conditions). To avoid common environment effects, nest mates were separated and randomly distributed to registration cages and rooms. Artificial light–dark cycles simulated the natural photoperiodic cycle experienced by these populations. Birds were exposed to photoperiodic conditions corresponding to 47°30′N until mid-August. Thereafter, and until the end of November, day-length changes simulated conditions on migration to the wintering areas in central Africa at about 5°N.[49]

Migratory activity was quantitatively recorded in registration cages during each bird's first autumn migratory period. The onset of migratory activity was defined as the first night on which the bird was active during at least five half-hour intervals. According to this criterion, activity onset could be determined in all but two birds. Because alternative definitions of the onset of migratory activity (e.g., the beginning of the first 10-day interval with mean activity exceeding one half-hour per night[49]) yielded very similar results, here we present only the results on the onset of migratory activity assessed by the same criterion as used in the heritability study of the blackcap.[22] It has been previously shown that the onset of this night activity in hand-raised birds is a reliable measure of the timing of autumn migration in the wild.[53]

We removed the effect of hatching date on the timing of autumn migration by regressing day at the onset of migratory activity on hatching date (mountain population: $b = 0.802$, $t_{49} = 3.41$, $P < .005$; lowland population: $b = 0.676$, $t_{36} = 2.81$, $P < .01$). Residuals obtained from these regressions were used in the subsequent statistical analyses. Residual onset of migratory activity did not deviate from normality (Shapiro-Wilks' Test; mountain population: $W = 0.9675$, $P > .5$; lowland population: $W = 0.9785$, $P > .5$). We found no effects of sex, year, or their interaction on the residual onset of migratory behavior neither in the mountain [analysis of variance (ANOVA): sex: $F_{1,45} = 0.12$, $P > .5$; year: $F_{1,49} = 1.31$, $P > .2$] nor in the lowland population (ANOVA: sex: $F_{1,32} = 0.17$, $P > .5$; year: $F_{1,36} = 1.86$, $P > .1$). Therefore, we pooled data for males and females, and for both years within each population.

We estimated causal variance components and heritabilities (h^2) by computing intraclass correlations among full sibs derived from one-way ANOVA.[54] Heritability estimates derived by this method have to be cautiously interpreted, as they may include large variance components attributable to common-environment and dominance effects.[54] However, we are confident that true narrow-sense heritabilities of the onset of migratory activity do not deviate significantly from estimates obtained in this study for two reasons: (1) The experimental setup was designed to minimize common environment effects. Therefore, we expect resemblances due to correlations between genotype and environment to be minimal and random. (2) In the blackcap, there is no indication for the presence of common environment (including maternal effects) or dominance effects either in the onset of autumn migration[22] or in the amount or intensity of migratory activity.[34,55] Considering the high similarity of phenotypic covariances among migratory traits in the blackcap and the garden warbler (unpublished results), we are confident that these conclusions also hold for the garden warbler.

For the estimation of heritabilities (h^2), phenotypic (V_P), additive genetic (V_A), and residual (V_R) variances, only families with at least two nestlings were considered. Residual variance (V_R) was the component of phenotypic variance not explained by genetic variation ($V_R = V_P - V_A$). Means, standard errors, and confidence intervals for all estimates obtained by the ANOVA approach were calculated using a bootstrapping procedure.[56] Empirical bootstrap distributions were generated by resampling the original number of families 100,000 times with replacement. These bootstrap distributions were used to estimate error probabilities for the hypothesis of equality of heritabilities and coefficients of additive genetic variance among populations. Bootstrap estimates were calculated with the software package H2BOOT.[57] In addition, we computed heritabilities and their standard errors from variance component estimates obtained using a restricted maximum likelihood (REML) approach as implemented in the VARCOMP routine in SAS.[58]

For comparison of heritability estimates and variance components in the two garden warbler populations with those in a species migrating shorter distances, we used the data on the onset of autumn migratory activity in a southern German blackcap population.[22] Blackcaps from this population predominantly winter in the western Mediterranean area, and can be considered as short- to middle-distance migrants.[55] Individuals were collected, maintained, and their activity measured by the same methods as used in the garden warbler. For this interspecific comparison of genetic variation we chose not to use standardized variance components, as is generally recomended,[59] as we measured the onset of migratory activity by the same methods and on the same scale in both species. Moreover, standardization of variance components by the mean biased our results (results not shown; see also Roff[28] for a critique of standardized variance components). We considered only heritability estimates and variance components derived from the resemblances among full sibs collected in the wild (see table 1 in Pulido et al.[22]). This sample was complemented with the data of 145 blackcaps collected in the same population in 2000 (F. Pulido & P. Berthold, unpublished). The total blackcap sample comprised the records of migratory activity of 676 individuals belonging to 178 families hatched in 12 different years. We tested the equality of heritabilities and variance components among species by using the distribution of cohort means in the blackcap and testing them against the estimates obtained in the garden warbler with a t-test. Equality of phenotypic variances was tested using the complete data set by Leven's test for equality of variances. Furthermore, we tested for among-species differences in heritabilities and variance components using weighted bootstrap distributions that were generated by resampling families within each cohort. The number of bootstraps was proportional to the number of individuals studied in each cohort.

RESULTS

Intraspecific Comparison

Garden warblers from the mountain population started migration at a mean age of 50.0 ± 7.3 days on August 10. Birds collected in the lowland population, initiated

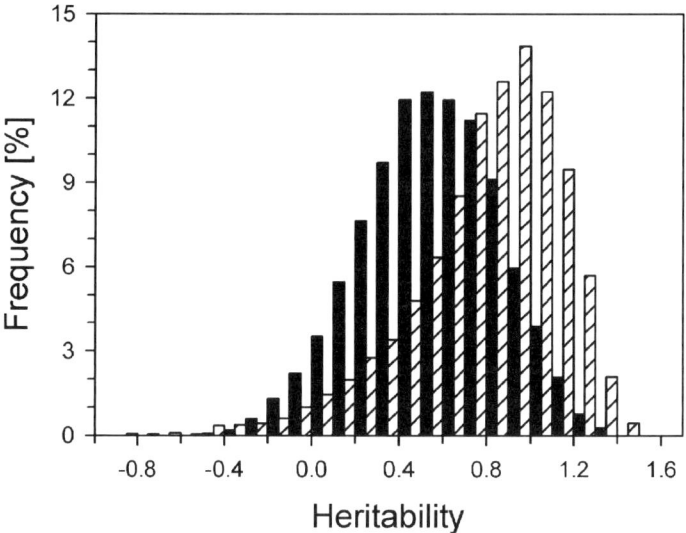

FIGURE 1. Bootstrap distribution of heritability estimates of the residual timing of autumn migratory activity in two garden warbler populations. Mean heritabilities for the mountain (*solid bars*) and the lowland population (*hatched bars*) are both significantly different from zero.

migratory activity at an age of 57.8 ± 9.6 days on August 8. While the difference in the mean age at onset of migratory activity was significant ($t_{87} = 4.3$, $P < .001$), populations did not differ in the mean date of onset of activity ($t_{87} = 1.17$, $P > .5$).

In both populations, heritability estimates obtained by bootstrapping over families were significantly different from zero (mountain: $h^2 = 0.552 \pm 0.308$, $P < .05$; lowland: $h^2 = 0.795 \pm 0.343$, $P < .05$), but not from one (FIG. 1). Estimates derived from variance components obtained by REML were in the same range, but means and standard deviations obtained by this method were a little higher than ANOVA results (mountain: $h^2 = 0.621 \pm 0.401$; lowland: $h^2 = 0.869 < .517$). Additive genetic variance was more than two times higher in garden warblers from the mountains than from the lowlands (TABLE 1). However, neither heritabilities nor additive genetic variance components differed among populations (bootstrap results: $P_{h^2} > .2$; $P_{VA} > .2$). Likewise, differences in phenotypic and residual variance (TABLE 1) were not statistically significant (V_P: $F_{1,85} = 1.64$, $P > .2$; V_R: bootstrap results: $P > .2$).

Interspecific Comparison of Variance Components

Mean phenotypic variance was significantly higher in the blackcap than in the garden warbler ($F_{1,771} = 4.96$, $P < .05$). This was due to low variance in the Urseren population, which significantly differed from the southern German blackcap population ($F_{1,735} = 5.29$, $P < .05$). Phenotypic variance in the Upper Rhine Valley population did not differ from the blackcap ($F_{1,720} = 0.64$, $P > .2$). Lower phenotypic variances in the garden warbler could, in principle, result from lower within-cohort

TABLE 1. Mean phenotypic (V_P), additive genetic (V_A), and residual (V_R) variance and standard errors (in parentheses) in two garden warbler populations and one blackcap population

Population	V_P	V_A	V_R
Mountain	51.7 (6.6)	28.9 (16.8)	22.9 (15.7)
Lowland	83.8 (21.9)	72.0 (41.2)	11.9 (23.1)
Garden warbler (mean)	67.8 (14.2)	50.4 (38.2)	17.4 (19.4)
Blackcap (mean)	119.1 (38.0)	54.6 (44.2)	64.5 (50.3)

NOTE: Means and standard errors for garden warblers were derived from bootstrap distributions. For the blackcap, weighted means and standard deviations were estimated from the distribution of the date at onset of migratory activity of 12 cohorts hatched in the wild from 1988 to 2000.

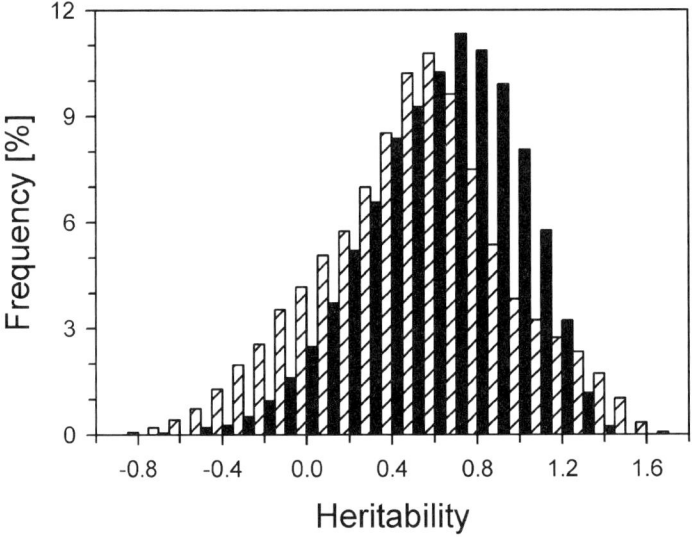

FIGURE 2. Pooled bootstrap distribution of heritability estimates of the residual timing of autumn migration in two garden warbler populations (*solid bars*), and combined bootstrap distribution of heritabilities in 12 blackcap cohorts from southern Germany (*hatched bars*). The total number of bootstrap samples used to generate the distribution of heritabilities was 65,000 for the blackcap and 20,000 for the garden warbler.

sample sizes as phenotypic variance increases with increasing sample sizes in the blackcap ($b = 0.75$, $t_{10} = 2.47$, $P < .05$), and mean within-cohort samples were smaller in the garden warbler than in the blackcap. However, if only cohorts with comparable sample sizes were considered ($N \leq 50$), mean phenotypic variance in the blackcap changed only little ($V_P = 106.6$), and was still significantly higher than in the garden warbler ($F_{1,531} = 5.72$, $P < .05$).

High phenotypic variance in the blackcap was due to high residual variance in this species (TABLE 1). Residual variance in southern German blackcaps was on average 3.6 times higher than in the garden warbler ($t_{10} = 2.86$; $P < .05$). As a consequence, mean heritability of the onset of autumn migratory activity in the blackcap was on average about 30% lower than in the garden warbler (FIG. 2). This difference is marginally significant if mean heritability in the garden warbler was tested against the among-cohort distribution of heritabilities in the blackcap ($t_{10} = 2.26$; $P < .05$). But if the blackcap mean or the pooled bootstrap distribution of heritabilities (see FIG. 2) was tested against the bootstrap distribution in the garden warbler, this result was not confirmed ($P > .2$). Moreover, we found no statistical evidence for interspecific differences in the amount of additive genetic variance, irrespective of the statistical method used (bootstrap test and t-test: $P > .5$).

DISCUSSION

Currently, climate is rapidly changing, and to cope with these changes organisms have to adapt.[1,60–62] It has repeatedly been reported that long-distance migrants may change migratory traits, particularly the timing of migration, at a slower rate than migratory birds wintering close to the breeding grounds.[6–10] In this study we investigate whether differences in the amount of genetic variation present in populations differing in migratory behavior could be the cause for differences in the rate of adaptive change. We predicted that bird populations subjected to strong stabilizing selection should maintain lower levels of genetic variation in migratory traits than populations in which selection on migration is weaker. Contrary to this prediction we found no evidence for a reduction of genetic variation in the timing of autumn migration in the species migrating long distances (the garden warbler) relative to its sister species migrating shorter distances (the blackcap). In both garden warbler populations, we demonstrated high, statistically significant heritabilities and genetic variance components for the onset of migratory activity. The amount of additive genetic variance expressed was practically identical in both species. Most remarkably, however, phenotypic variance in the garden warbler was lower than in the blackcap (on average by about 30%). This was primarily due to low amounts of residual variances found in both garden warbler populations.

In the intraspecific comparison, we expected the garden warbler population living in the mountain habitat to experience stronger stabilizing selection on the timing of autumn migration. This expectation was borne out of the fact that the breeding season of this population is almost four weeks shorter compared to the lowland population.[48,49] We predicted finding reduced levels of additive genetic variation in the mountains. As expected, heritability estimates and coefficients of additive genetic variation were both higher in the lowland population. The difference to the mountain population, however, was moderate and had a large variance; as a consequence, it was not statistically significant. Corresponding results were obtained for the timing of spring migration and the amount of migratory activity.[17,49] In this latter trait, genetic variation was larger in the mountains, though the difference was not statistically significant. This inconsistent pattern of among-population genetic variation in migratory traits could be a consequence of different traits being subjected to different selection intensities. However, large sampling variances suggest that inconsistent results are probably due to low, negligible differences in combination with small sample sizes and to considerable among-year fluctuations in the amount of additive genetic variation in these populations (see Pulido *et al.*[22] for results in the blackcap). If among-population differences in genetic variation exist, they are probably small. The intraspecific comparison suggests that differences in the intensity of stabilizing selection among populations did not lead to significant differences in genetic variation. Although this conclusion could be challenged by the fact that genetic drift and migration can potentially mask selection effects on genetic variation, we believe that our result cannot be explained by these processes. Because effective population size and rate of immigration are most probably larger in the Upper Rhine Valley than in the Urseren Valley population (M. Widmer, unpublished), we predict that these processes alone would cause a reduction of genetic variation in the mountain population. This would increase the expected effect of stabilizing selection on genetic variances rather than canceling it out. Since we find only moderate differences in genetic variation for the timing of autumn migration, we can exclude that stabilizing selection strongly reduces the amount of genetic variation maintained in the mountain population. Moreover, stronger isolation and lower effective population size could be the cause for the somewhat lower amounts of genetic variation found in the mountain population.

Environmental Canalization

There are several potential causes for low residual variation in the timing of migration found in garden warblers as compared to the blackcap. Because residual variance is composed of environmental and nonadditive (mostly dominance) variance,[54] both of these variance components could, in principle, be reduced in long-distance migrants. However, there are a number of arguments supporting the idea that it is the environmental variance component that is reduced in garden warblers:

(1) Theoretical considerations and empirical findings suggest that in traits closely correlated with fitness, nonadditive genetic variance components, and above all dominance variance, should increase.[29,59,63,64] As a consequence, we expect high residual variances in traits closely correlated with fitness and low residual variances in "nonfitness" traits. This is opposite to our finding of reduced residual variance in the garden warbler, that is, the species in which this trait is supposedly subjected to stronger stabilizing selection.

(2) Environmental and genetic canalization increase under stabilizing selection.[31–33,65] "Environmental canalization is the reduction in the phenotypic variation caused by an environmental perturbation."[66] Thus, we would expect a reduced response to environmental variation in the species under stronger selection.

(3) The finding of low phenotypic variance and environmental canalization in the timing of autumn migration in the garden warbler is in accord with field observations, and experimental results suggesting that long-distance migrants generally display less variation in migratory traits, particularly in the timing of spring migration, than short-distance migrants.[4,26]

In British and eastern European blackcaps, ranges and standard deviations of the distribution of among-year means of spring arrival date are about two times larger than in garden warblers.[8,67,68] Moreover, Berthold *et al.*[69] studied *zugunruhe* under controlled laboratory conditions and found higher coefficients of among-individual variation of the amount of migratory activity in the blackcap than in the garden warbler. They concluded that "this result could be interpreted by assuming that migratory activity is less rigid endogenously programmed in the blackcap than in the garden warbler," and that environmental factors have a stronger impact on migratory activity in the blackcap. Similar results were obtained in a comparative study of migratory activity in the chiffchaff (*Phylloscopus collybita*), a short-distance migrant, and the willow warbler (*Phylloscopus trochilus*), a long-distance migrant. Coefficients of variation for duration, amount, and maximum intensity of migratory activity were smaller in the short-distance migrant.[70] In a comparative study of blackcap and garden warbler migration based on ringing recoveries, Klein *et al.*[71] found stronger among-year and geographic variations (correlated with isotherms) of the timing of autumn migration in blackcaps than in garden warblers. They concluded that "there are numerous differences between the migration of *S. borin* and *S. atricapilla* that fit into the picture of a lower environment-dependent migration in *S. borin* as compared to *S. atricapilla*." Moreover, two other studies found that distributions of autumn passage dates of long-distance migrants are characterized by small variances and high skewness. These findings were interpreted as indicators for the predominantly endogenous control of the timing of autumn migration in this group of birds.[72,73]

In line with the hypothesis of environmental canalization of migratory traits in long-distance migrants, we found a reduction of phenotypic variance in the garden warbler for the timing of molt and the onset of vernal migration, but not for the termination, duration and amount of autumn migratory activity (F. Pulido and M. Widmer, unpublished). These latter traits are not expected to be subjected to stronger selection in the garden warbler than in the blackcap, as the large winter range of the garden warbler permits considerable variation of migration distances in this species.[71] The preliminary analysis of other migratory traits in the garden warbler and the blackcap thus seems to corroborate the conclusion derived from this study that strong stabilizing selection reduces phenotypic variation by increasing environmental canalization. As migratory traits, particularly those traits related to the timing of migration, are likely to be under strong selection in long-distance migrants, we expect to find reduced levels of phenotypic variation in this group of birds as a consequence of environmental canalization. We further predict that the degree of canalization in long-distance migrants will decrease if populations are subjected to strong directional selection, as currently caused by rapid environmental changes.

Consequences for Adaptive Evolution

Environmental canalization is equivalent to the lack of plasticity in response to environmental variation.[74] Phenotypic plasticity could in principle accelerate or slow down adaptive responses. Simulations using genetic algorithms, however, have shown that under a large spectrum of conditions phenotypic plasticity will slow down evolutionary change.[75] Although populations with high levels of phenotypic plasticity may show fast adaptive responses to environmental changes, these changes will be individual adjustments and will not involve genetic changes in the population. If directional selection persists and the optimal phenotype shifts beyond the range of environmentally inducible phenotypes, then evolutionary change is to be expected. Populations with a high degree of environmental canalization will show more immediate responses to selection. However, their degree of genetic adaptation will be lower, and the time needed to reach an adaptive peak will be considerably larger. For the adaptation of migratory traits in long-distance migrants, this means that compared to short-distance migrants we should expect lower rates of phenotypic change as immediate response to environmental change. Rapid immediate changes in short-distance migrants, however, would not involve changes in the genetic composition of the population, but would result from phenotypic plasticity. In the long term, populations of long-distance migrants could theoretically evolve at a faster rate than short-distance migrants, if they are large and survive the initial phase of selection.

CONCLUSIONS

There is currently no evidence that the amount of additive genetic variance present in populations of long-distance migrants is smaller than in short-distance migrants. Our results suggest, however, that differences in the speed of adaptive phenotypic change between long-distance and short-distance migrants could potentially be caused by differences in phenotypic plasticity. Unfortunately, strong evidence for this hypothesis is currently lacking. Although there are a number of studies suggesting low plasticity of migration timing in long-distance migrants and lack of response to climatic variation,[76] other studies indicate that long-distance migrants can adjust migration speed to weather conditions and plant phenology.[77] To date all evidence is at best circumstantial, as it is based on correlations and means of groups of migrating birds. To understand if and how short- and long-distance migrants differ in their mode to adapt to changing environmental condition, we urgently need longitudinal studies and quantitative genetic studies of migratory behavior in the wild.

ACKNOWLEDGMENTS

We are grateful to Peter Berthold for financial support and critical discussion. We thank Timothy Coppack, Barbara Helm, Arie van Noordwijk, Thomas Städler, and an anonymous reviewer for helpful comments on previous versions of the manuscript.

REFERENCES

1. WALTHER, G.-R. *et al.* 2002. Ecological responses to recent climate change. Nature **416:** 389–395.
2. BERTHOLD, P. 1991. Patterns of avian migration in light of current global "greenhouse" effects: a central European perspective. Acta XX Congr. Int. Ornithol.: 780–786.
3. BERTHOLD, P. 1998. Vogelwelt und Klima: gegenwärtige Veränderungen. Naturw. Rundsch. **51:** 337–346.
4. BERTHOLD, P. 2001. Bird Migration, 2nd ed. Oxford University Press. Oxford.
5. FIEDLER, W. 2003. Recent changes in migratory behaviour of birds: a compilation of field observations. *In* Avian Migration. P. Berthold, E. Gwinner & E. Sonnenschein, Eds.: 21–38. Springer-Verlag. Berlin.

6. GATTER, W. 1992. Zugzeiten und Zugmuster im Herbst: Einfluß des Treibhauseffekts auf den Vogelzug? J. Ornithol. **133:** 427–436.
7. JENKINS, D. & A. WATSON. 2000. Dates of first arrival and song of birds during 1974–99 in mid-Deeside, Scotland. Bird Study **47:** 249–251.
8. TRYJANOWSKI, P., S. KUZNIAK & T.H. SPARKS. 2002. Earlier arrival of some farmland migrants in western Poland. Ibis **144:** 62–68.
9. BUTLER, C.J. 2003. The disproportionate effect of global warming on the arrival dates of short-distance migratory birds in North America. Ibis **145:** 484–495.
10. MILLS, A.M. 2005. Changes in the timing of spring and autumn migration in North American migrant passerines during a period of global warming. Ibis **147:** 259–269.
11. BÖHNING-GAESE, K. & H.-G. BAUER. 1996. Changes in species abundance, distribution, and diversity in a central European bird community. Conserv. Biol. **10:** 175–187.
12. BERTHOLD, P. *et al.* 1998. 25-year study of the population development of Central European songbirds: a general decline, most evident in long-distance migrants. Naturwissenschaften **85:** 350–353.
13. BALLARD, G. *et al.* 2003. Long-term declines and decadal patterns in population trends of songbirds in western North America, 1979–1999. Condor **195:** 737–755.
14. BAUER, H.-G. & P. BERTHOLD. 1997. Die Brutvögel Mitteleuropas. Bestand und Gefährdung, 2nd ed. Aula Verlag. Wiesbaden.
15. LEMOINE, N. & K. BÖHNING-GAESE. 2003. Potential impact of global change on species richness of long-distance migrants. Conserv. Biol. **17:** 577–586.
16. COPPACK, T. & C. BOTH. 2002. Predicting life-cycle adaptation of migratory birds to global climate change. Ardea **90:** 369–378.
17. PULIDO, F. & P. BERTHOLD. 2003. Quantitative genetic analysis of migratory behaviour. *In* Avian Migration. P. Berthold, E. Gwinner & E. Sonnenschein, Eds.: 53–77. Springer-Verlag. Berlin.
18. SUTHERLAND, W.J. 1998. Evidence for flexibility and constraint in migration systems. J. Avian Biol. **29:** 441–446.
19. BERTHOLD, P. 1995. Microevolutionary aspects of bird migration based on experimental result. Isr. J. Zool. **41:** 377–385.
20. MYERS, J.P. & R.T. LESTER. 1992. Double jeopardy for migrating animals: multiple hits and resource asynchrony. *In* Global Warming and Biological Diversity. R.L. Peters & T.E. Lovejoy, Eds.: 193–200. Yale University Press. New Haven.
21. BOTH, C. & M.E. VISSER. 2001. Adjustment to climate change is constrained by arrival date in a long-distance migrant bird. Nature **411:** 296–298.
22. PULIDO, F. *et al.* 2001. Heritability of the timing of autumn migration in a natural bird population. Proc. R. Soc. Lond. B **268:** 953–959.
23. WEBSTER, M.S. *et al.* 2002. Links between worlds: unraveling migratory connectivity. Trends Ecol. Evol. **17:** 76–83.
24. COPPACK, T., F. PULIDO & P. BERTHOLD. 2001. Photoperiodic response to early hatching in a migratory bird species. Oecologia **128:** 181–186.
25. PULIDO, F. & T. COPPACK. 2004. Correlation between timing of juvenile moult and onset of migration in the blackcap (*Sylvia atricapilla*). Anim. Behav. **68:** 167–173.
26. BERTHOLD, P. 1996. Control of Bird Migration. Chapman & Hall. London.
27. MOSS, S. 1998. Predictions of the effects of global climate change on Britain's birds. Br. Birds **91:** 307–325.
28. ROFF, D.A. 1997. Evolutionary Quantitative Genetics. Chapman & Hall. New York.
29. MERILÄ, J. & B.C. SHELDON. 1999. Genetic architecture of fitness and nonfitness traits: empirical patterns and development of ideas. Heredity **83:** 103–109.
30. BURGER, R. & A. GIMELFARB. 1999. Genetic variation maintained in multilocus models of additive quantitative traits under stabilizing selection. Genetics **152:** 807–820.
31. STEARNS, S.C. & T.J. KAWECKI. 1994. Fitness sensitivity and the canalization of life-history traits. Evolution **48:** 1438–1450.
32. STEARNS, S.C., M. KAISER & T.J. KAWECKI. 1995. The differential genetic and environmental canalization of fitness components in *Drosophila melanogaster*. J. Evol. Biol. **8:** 539–557.
33. WAGNER, G.P., G. BOOTH & H. BAGHERI-CHAICHIAN. 1997. A population genetic theory of canalization. Evolution **51:** 329–347.

34. PULIDO, F. 2000. Evolutionary Quantitative Genetics of Migratory Restlessness in the Blackcap (*Sylvia atricapilla*). Tectum Verlag. Marburg.
35. GWINNER, E. 1986. Circannual Rhythms. Springer-Verlag. Berlin.
36. JENNI, L. & R. WINKLER. 1994. Moult and Aging in European Passerines. Academic Press. London.
37. HALL, K.S.S. & T. FRANSSON. 2000. Lesser whitethroats under time-constraint moult more rapidly and grow shorter wing feathers. J. Avian Biol. **31:** 583–587.
38. Dawson, A. *et al.* 2000. Rate of moult affects feather quality: a mechanism linking current reproductive effort to future survival. Proc. R. Soc. Lond. B **267:** 2093–2098.
39. SALEWSKI, V., F. BAIRLEIN & B. LEISLER. 2002. Different wintering strategies in Palearctic migrants in West Africa—a consequence of foraging strategies? Ibis **144:** 85–93.
40. POTTI, J. 1998. Arrival time from spring migration in male pied flycatchers: individual consistency and familial resemblance. Condor **100:** 702–708.
41. MØLLER, A.P. 2001. Heritability of arrival date in a migratory bird. Proc. R. Soc. Lond. B. **268:** 203–206.
42. BROWN, C.R. & M.B. BROWN. 2000. Weather-mediated natural selection on arrival time in cliff swallows (*Petrochelidon pyrrhonota*). Behav. Ecol. Sociobiol. **47:** 339–345.
43. LEHIKOINEN, E., SPARKS, T. H. & M. ZALAKEVICIUS. 2004. Arrival and departure dates. Adv. Ecol. Res. **35:** 1–31.
44. BEZZEL, E. & W. JETZ. 1995. Verschiebung der Wegzugperiode bei der Mönchsgrasmücke *(Sylvia atricapilla)* 1966–1993—Reaktion auf die Klimaerwärmung? J. Ornithol. **136:** 83–87.
45. JENNI, L. & M. KÉRY. 2003. Timing of autumn migration under climate change: advances in long-distance migrants, delays in short-distance migrants. Proc. R. Soc. Lond. B **270:** 1467–1471.
46. COTTON, P.A. 2003. Avian migration phenology and global climate change. Proc. Nat. Acad. Sci. USA **21:** 12219–12222.
47. SHIRIHAI, H., G. GARGALLO & A.J. HELBIG. 2001. Sylvia warblers. *In* Identification, Taxonomy and Phylogeny of the Genus Sylvia. Christopher Helm. London.
48. WIDMER, M. 1998. Altitudinal variation of phenology and reproductive traits of the garden warbler (*Sylvia borin*). Biol. Conserv. Fauna **102:** 128–134.
49. WIDMER, M. 1999. Altitudinal Variation of Migratory Traits in the Garden Warbler *Sylvia borin*. Ph.D. thesis, University of Zurich, Switzerland.
50. WIDMER, M. 1993. Brutbiologie einer Gebirgspopulation der Gartengrasmücke Sylvia borin. Ornithol. Beob. **90:** 85–113.
51. WIDMER, M. 1996. Phänologie, Siedlungsdichte und Populationsökologie der Gartengrasmücke *Sylvia borin* in einem subalpinen Habitat der Zentralalpen. J. Ornithol. **137:** 479–501.
52. BERTHOLD, P., E. GWINNER & H. KLEIN. 1970. Vergleichende Untersuchung der Jugendentwicklung eines ausgeprägten Zugvogels, *Sylvia borin*, und eines weniger ausgeprägten Zugvogels, *S. atricapilla*. Vogelwarte **25:** 297–331.
53. BERTHOLD, P. 1990. Wegzugbeginn und Einsetzen der Zugunruhe bei 19 Vogelpopulationen—eine vergleichende Untersuchung. Proc. Int. 100. DO-G Meeting, Current Topics Avian Biol., Bonn, 1988, pp. 217–222.
54. FALCONER, D.S. & T.F.C. MACKAY. 1996. Introduction to Quantitative Genetics, 4th ed. Longman. Harlow, UK.
55. BERTHOLD, P. & F. PULIDO. 1994. Heritability of migratory activity in a natural bird population. Proc. R. Soc. Lond. B **257:** 311–315.
56. EFRON, B. & R.J. TIBSHIRANI. 1993. An Introduction to the Bootstrap. Chapman & Hall. New York.
57. PHILLIPS, P.C. 1998. H2BOOT: Bootstrap Estimates and Tests of Quantitative Genetic Data. Software available at: http://darkwing.uoregon.edu/~pphil/software.html.
58. SAS INSTITUTE. 1990. SAS/STAT User's Guide, Version 6. SAS Institute. Cary, NC.
59. HOULE, D. 1992. Comparing evolvability and variability of quantitative traits. Genetics **130:** 195–204.
60. HOLT, R. D. 1990. The microevolutionary consequence of climate change. Trends Ecol. Evol. **5:** 311–315.

61. BERTEAUX, D. *et al.* 2004. Keeping pace with fast climate change: Can Arctic life count on evolution? Integr. Comp. Biol. **44:** 140–151.
62. PULIDO, F. & P. BERTHOLD. 2004. Microevolutionary response to climatic change. Adv. Ecol. Res. **35:** 151–183.
63. CRNOKRAK, P. & D.A. ROFF. 1995. Dominance variance: associations with selection and fitness. Heredity **75:** 530–540.
64. MERILÄ, J. & B.C. SHELDON. 2001. Avian quantitative genetics. Current Ornithol. **16:** 179–255.
65. GIBSON, G. & G. WAGNER. 2000. Canalizing in evolutionary genetics: a stabilizing theory? BioEssays **22:** 372–380.
66. WADDINGTON, C.H. 1942. Canalization of development and the inheritance of acquired characters. Nature **150:** 563–565.
67. MASON, C.F. 1995. Long-term trends in the arrival dates of spring migrants. Bird Study **42:** 182–189.
68. SOKOLOV, L.V. *et al.* 1998. Long-term trends in the timing of spring migration of passerines on the Courish Spit of the Baltic Sea. Avian Ecol. Behav. **1:** 1–21.
69. BERTHOLD, P. *et al.* 1972. Beziehungen zwischen Zugunruhe und Zugablauf bei Garten- und Mönchsgrasmücke (*Sylvia borin* und *S. atricapilla*). Z. Tierpsychol. **30:** 26–35.
70. GWINNER, E. 1968. Artspezifische Muster der Zugunruhe bei Laubsängern und ihre mögliche Bedeutung für die Beendigung des Zuges im Winterquartier. Z. Tierpsychol. **25:** 843–853.
71. KLEIN, H., P. BERTHOLD & E. GWINNER. 1973. Der Zug europäischer Garten- und Mönchsgrasmücken (*Sylvia borin* und *S. atricapilla*). Vogelwarte **27:** 73–134.
72. DORKA, V. 1966. Das jahres- und tageszeitliche Zugmuster von Kurz- und Langstreckenziehern nach Beobachtungen auf den Alpenpässen Cou/Bretolet. Ornithol. Beob. **63:** 165–223.
73. BERTHOLD, P. & V. DORKA. 1969. Vergleich und Deutung von jahreszeitlichen Wegzugs-Zugmustern ausgeprägter und weniger ausgeprägter Zugvögel. Vogelwarte **25:** 121–129.
74. DEBAT, V. & P. DAVID. 2001. Mapping phenotypes: canalization, plasticity and developmental stability. Trends Ecol. Evol. **16:** 555–561.
75. BEHERA, N. & V. NANJUNDIAH. 1995. An investigation of the role of phenotypic plasticity in evolution. J. Theor. Biol. **172:** 225–234.
76. HUBÁLEK, Z. 2003. Spring migration of birds in relation to North Atlantic Oscillation. Folia Zool. **52:** 287–298.
77. MARRA, P.P. *et al.* 2004. The influence of climate on the timing and rate of spring migration. Oecologia **142:** 307–315.

Spatial Behavior of Medium and Long-Distance Migrants at Stopovers Studied by Radio Tracking

NIKITA CHERNETSOV

Biological Station Rybachy, Rybachy 238535, Kaliningrad Region, Russia

ABSTRACT: Spatial behavior and range of movements at daytime stopovers of three species of passerine nocturnal migrants (European robins, sedge warbler, and pied flycatchers) were studied by radio tracking. Both in spring and in fall, 94% of European robins remained within 350–400 m of their landing location ($n = 51$ and 65, respectively). Movements of robins became more area-restricted with more time spent at stopover. Sedge warblers never moved more than 335 m ($n = 12$). A reason for this could be their relatively narrow habitat use. However, four of seven pied flycatchers tracked covered at least several kilometers during their daytime movements. Pied flycatchers that moved longer distances also showed higher linearity of movements. Spatial behavior of three species of migrants was probably a result of a complex interaction between their habitat use, foraging habits, and weather conditions. No evidence is available that long-distance versus short-distance migratory habits are involved in shaping the spatial behavior of passerine migrants at stopovers.

KEYWORDS: stopover; spatial behavior; movements; radio tracking

INTRODUCTION

Events that occur during stopovers shape the organization of migration of avian individuals and populations.[1] A very important aspect of migrant stopover ecology and behavior is spatial behavior. Some passerines occupy temporary territories at stopovers,[2–5] whereas others move broadly across their stopover areas.

In this study, the pattern of movements during migratory stopovers is compared among three avian species: one short- to medium-distance migrant, the European robin (*Erithacus rubecula*), and two long-distance migrants, the sedge warbler (*Acrocephalus schoenobaenus*) and the pied flycatcher (*Ficedula hypoleuca*). During fall migration, European robins are known to occupy small stopover areas that might be territories.[6,7] Pied flycatchers have been shown to occupy small defended territories at stopovers during fall migration.[3] Sedge warblers were reported to share their home ranges with conspecifics during migratory stopovers.[4,8] No data on the range of their movements at daytime stopovers were available.

Address for correspondence: Nikita Chernetsov, Biological Station Rybachy, Rybachy 238535, Kaliningrad Region, Russia. Voice: +7- 01150-41251; fax: +7-01150-41345.
nchernetsov@bioryb.koenig.ru

The aim of this project was to study the pattern of spatial movements in several species with different migratory habits (short- to medium-distance vs. long-distance) and various foraging techniques and diet. Pied flycatchers and sedge warblers are trans-Saharan migrants, whereas European robins spend their winter mainly within Europe, occasionally as far north as the study area (Chernetsov, unpublished data). The study area lies within the breeding range of all three species; therefore, a mixture of transient migrants and local breeders might be present at the study site. However, we made every effort to make sure that birds tagged departed by nocturnal flight—that is, that they were stopover migrants. In all three species, the distance to go to the breeding locations of individual birds could vary between several dozens of kilometers (one nocturnal flight) and more than 1000 km.

Earlier analyses of recaptures within a stopover area allowed us to suggest that the main factor governing the spatial behavior of migrants at stopovers is the distribution of their food.[8–10] The species whose prey is more or less evenly distributed tend to occupy defined stopover areas and sometimes to defend territories, whereas species that use food distributed unpredictably in space and time move broadly in search for prey aggregations. To test this hypothesis, it was necessary to increase the variety of species studied and to include a short-distance migrant (e.g., the European robin). Pied flycatchers feed extensively on aerial insects ("flycatching"), European robins forage mainly on the ground for a large variety of different invertebrates,[11,12] and sedge warblers in spring feed mainly on insects that emerge after hibernation in reed stems and insects that develop in aquatic/moist habitats.[13]

MATERIAL AND METHODS

The study was conducted during migratory seasons 2002–2003 on Cape Rossitten on the Courish Spit on the southeastern Baltic coast (55°09 N, 20°51 E). This site is the study area of the Biological Station Rybachy, where a long-term banding project aimed at capturing small passerines is being carried out. Movements of the birds were studied by radio tracking. This technique makes it possible to control the position of a bird within a certain area regardless of its local movements, and thus of its probability of recapture.

In spring, European robins were tagged between April 1 and 30, 2002 (the last tagged bird departed in the night of May 3/4, 2002) and between April 13 and 26, 2003 (the last bird departed in the night of May 6/7, 2003). In fall, European Robins were tagged between September 2 and October 17, 2002 (tracking of the last bird stopped on October 29, 2002 due to transmitter failure) and between September 6 and October 20, 2003 (tracking of the last bird stopped on November 8, 2003). All birds tagged in fall were hatching-year individuals. Pied flycatchers were radio tagged in spring 2003, between April 28 and May 12. The last tagged bird departed in the night of May 12/13. Sedge warblers were tagged between May 21 and 30, 2003; the last birds departed in the night of May 30/31. From European robins and sedge warblers, one location per hour was taken between the onset of daytime activity at dawn and evening civil twilight. Pied flycatchers were tracked continuously, with occasional gaps. The locations were plotted on a digitized map of the study area. At night, all birds were surveyed continuously, so that their nocturnal departures were directly observed when they occurred. With the exception of some pied flycatchers that moved long distance from the release site, I did not have to assume that birds departed by nocturnal flight when they were not relocated in the morning: they were actually observed departing.

The daily dynamics of captures of small passerines on the Courish Spit shows a clear wavelike pattern, like at many other coastal sites.[14,15] Seniority analysis showed that most European robins initially captured during the first day of a migratory wave had indeed just arrived.[14,15] In 2002, some European robins were radio tagged on the first day of a wave of arrivals; others were tagged on recapture on the second or third day. All tagged robins were initially captured and ringed in the first days of the arrival of a wave and had most probably arrived on that day. In 2003, all birds were tagged in the first days of a migratory wave. The same was true for sedge warblers. All European robins and sedge warblers included in this study departed by nocturnal flights that were observed directly—that is, they were indeed migrants at stopovers and not local breeders/winterers. For pied flycatchers, we could not be sure that the birds were tagged just after arrival. Furthermore, some individuals of this species were lost during their long-range diurnal movements; therefore, their nocturnal departure was not directly observed. In pied flycatchers, we know only the minimum stopover duration.

The tags were fitted as backpacks with a Rappole harness; the weight of a tag with harness was 0.61 g. Body mass of the tagged robins varied between 14.8 and 19.2 g, of the pied flycatchers, between 12.3 and 14.6 g, and of the sedge warblers, between 11.5 and 14.7 g. Mass added in just one case slightly exceeded 5% of the birds' body mass (5.2%), which is believed to be the upper permissible limit.[16,17]

In most cases, the position of the birds was identified to the nearest 5 m; therefore, a standard deviation of 10 m was assumed. The number of locations per individual was too low in most pied flycatchers and sedge warblers to estimate their home range area. Therefore in this study the analysis was restricted to movements range—that is, the maximum distance between locations of an individual. To estimate the aggregation of locations yielded from birds that were followed during short periods, I used the linearity index, which is a measure of area-restricted movement. The linearity index is the linear distance between the first and the last location points divided by the total distance moved (the sum of all distances between the subsequent location points). I use this measure rather than meander ratio,[18] which is reciprocal

TABLE 1. Numbers of European robins radio-tagged in different seasons

Season	Stopover migrants tagged	Followed from the first day of stopover	Followed from the first to the last day of stopover
Spring			
2002	21	12	10
2003	30	30	29
Total spring	51	42	39
Fall			
2002	29	25	24
2003	36	36	35
Total fall	65	61	59

to linearity index, because the linearity index by definition varies between 0 (area restricted) and 1 (linear movement).

The number of European robins tagged in each year and season, and those followed from the first until the last day of stopover, is given in TABLE 1. All pied flycatchers tagged were males. It was not possible to identify sex in either European robins or sedge warblers.

RESULTS

Robins: Spring

Thirty-nine European robins were tracked since the day of their presumed arrival until departure, so that their stopover length measurements were unbiased. Their mean stopover duration was 2.4 days (SE = 0.31, median 2 days, FIG. 1). Thirty-three robins were tracked from the first until the last day of stopover, with locations plotted on the map. The linearity index varied between 0.008 (very aggregated locations) and 0.65 (nearly straight-line movement). No significant interannual difference was recorded ($t = 0.57$, $P = .57$); therefore, results from both years were pooled for analysis. Linearity index was negatively related to both number of locations taken from a bird (Spearman's rank correlation: $r_S = .69$, $P < .0001$) and stopover duration expressed in days ($r_S = -.58$, $P = .0004$). The longer a robin remained at stopover, the more aggregated its locations were.

I also calculated the linearity index for each stopover day. During the first day of stopover, the robins moved widely, and from the second day on they remained in a more restricted area. The difference in linearity index between the first and the second day was significant ($t = 3.00$, $P = .004$); between other days, it was not (one-way analysis of variance [ANOVA]: $F = 1.76$, $P = .09$). This finding implies that during

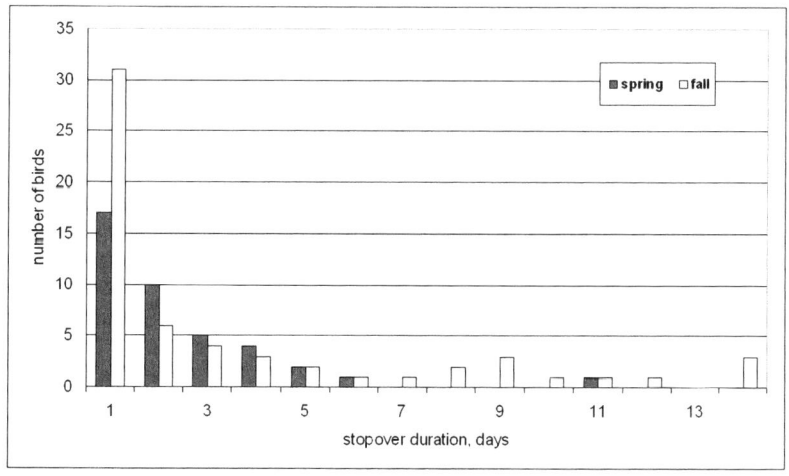

FIGURE 1. Frequency distribution of stopover durations of European robins as assessed by radio tracking in spring and in fall.

the first day of stopover, movements were less area-restricted than during subsequent days.

I calculated linearity index for each stopover day for European robins for which the respective day was the last day of stopover and for which it was not (FIG. 2A). In all cases the difference between these two categories was not significant (*t*-test, $P <$ .20). The difference in the mean linearity index between the first and the second day was significant in the birds for which departed after the respective day [$t = 2.61$, degrees of freedom (d.f.) = 21, $P = .016$], but not in the birds for which it was not the last stopover day ($t = 1.76$, d.f. = 23, $P = .09$). Birds that stay for just one day tend to show less area-restricted movements during stopover than those that remain longer.

The range of robins' movements during spring stopovers did not exceed 350 m in 94% of birds ($n = 51$). The remaining three birds covered 1.3, 1.7, and 1.8 km.

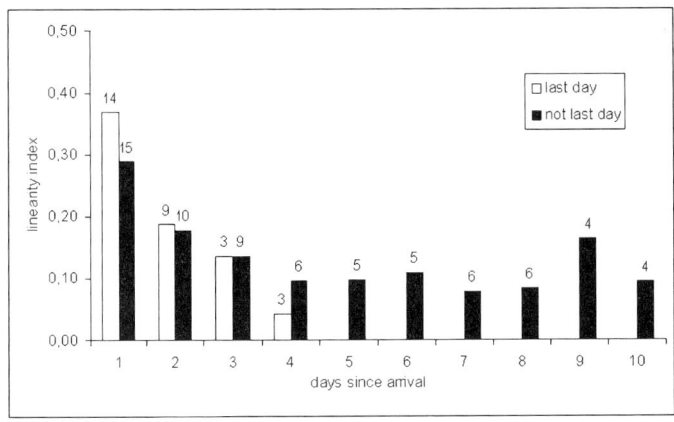

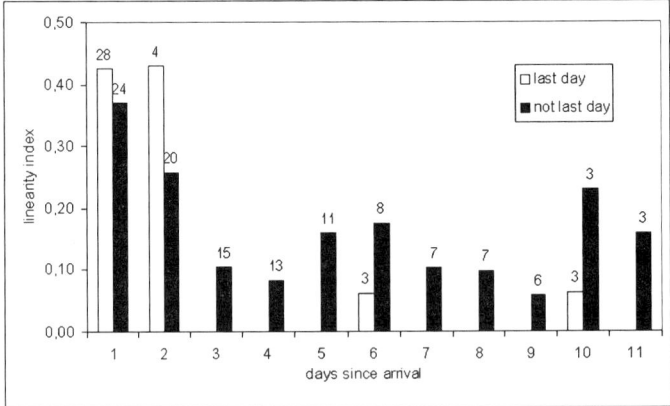

FIGURE 2. Daily linearity index values of European robins at stopover. *x*-Axis, day since arrival; *y*-axis, mean linearity index. Data above the histogram bars show sample size.

Robins: Fall

Fifty-nine European robins were tracked since arrival until departure; their mean stopover duration was 3.4 days (SE = 0.50, median 1 day, FIG. 1). Forty-two robins were tracked with their locations plotted on the map from the first until the last day of stopover. Linearity index of their movements varied between 0.003 and 0.93. No significant interannual difference was recorded ($t = 0.04$, $P = .97$); therefore, results from both years were pooled for analysis. Linearity index was negatively related to both number of locations taken from each bird (Spearman's rank correlation, $r_S = -.55$, $P < .0002$) and stopover duration in days ($r_S = .56$, $P < .0001$). Robins that stopped over for longer periods showed more area-restricted movements.

In fall, the linearity index did not differ significantly between the first and the second day of stopover ($t = 1.80$, $P = .079$), but both the first and the second day were significantly different from the third one ($t = 3.61$, $P < .001$; $t = 2.14$, $P = .039$, respectively). Since the third day of stopover, no significant between-day variation existed (one-way ANOVA: $F_{7,59} = 1.00$, $P = .44$). Thus, in fall, robins' movements became significantly more area-restricted since the third day after arrival. When the linearity index is analyzed for the birds for which the respective day was the last day of stopover and for which it was not (FIG. 2B), the difference between these two categories was not significant in any case (t-test, $P < .16$).

The bulk of birds (94%, $n = 65$) did not cover more than 350 m during their daytime movements; the remaining four individuals covered 1.0, 1.1, 3.4, and 3.7 km.

Sedge Warblers and Pied Flycatchers: Spring

Stopover duration of radio-tagged sedge warblers varied between 1 and 3 days, on average 1.6 days (SE = 0.23, median 1 day, $n = 12$). The maximum distance between the locations of the same individual varied between 44 and 335 m, on average 97 m (SE = 22.9, median 75 m, $n = 12$). Home ranges of sedge warblers were clearly not defended territories, as they overlapped broadly.

Of the seven pied flycatchers whose spatial behavior was tracked by radio telemetry, exact stopover duration has been measured for four. Three of them stopped for 1 day, and one, for 2 days. Three pied flycatchers were lost during their long-range daytime movements; therefore, only their minimum stopover duration could be estimated: 1, 3, and 4 days, respectively. If minimum estimates are accepted, the mean stopover length was 1.9 days (SE = 0.46, median 1 day, $n = 7$).

The range of pied flycatchers' movements varied from 0.27 to 4.0 km, on average 2.0 km (SE = 0.63, median 2.3 km, $n = 7$). The difference between pied flycatchers and sedge warblers in the range of their stopover movements was highly significant ($t = 3.73$, $P < .002$); the difference in variance was also significant ($F_{6,11} = 492$, $P < .001$). It should be stressed that in three cases the range of pied flycatchers' movements was underestimated.

The pied flycatcher with the tag 037 moved for at most 260 m during the first 3 days after tagging; its home range area estimated by 95% kernel[19] comprised 13,650 m^2. In the morning of the fourth day (at 8:26 local time, i.e., 2.5 h after local sunrise) the bird started moving toward the north very rapidly. It was lost 2.3 km from the site where it had spent 3 previous days. All pied flycatchers that made longer movements (covering at least 2.3, 3.1, 3.8, and 4.0 km) did so under relatively cold weather con-

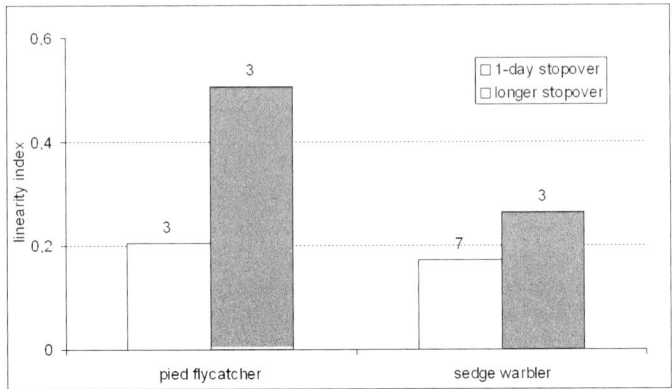

FIGURE 3. Linearity index values during the first day at stopover in pied flycatchers and sedge warblers. *Open bars*, birds that stopped for just 1 day; *hatched bars*, birds that stopped for more than 1 day. Data above the histogram bars show sample size.

ditions when air temperature at 12 P.M. local time was below 10°C. Pied flycatcher 037 started its movements on May 5, when the temperature fell below 10°C after being higher on previous days. Conversely, pied flycatchers that stayed near the tagging site did so during sunshine, with air temperature being above 10°C at noon.

The linearity index of sedge warblers' movements varied between 0.010 and 0.315, on average 0.147 (SE = 0.026, median 0.142, $n = 12$). In pied flycatchers this index varied more broadly, from 0.061 to 0.910, on average 0.421 (SE = 0.130, median 0.474, $n = 7$). Interspecific difference between the mean values was significant ($t = 2.65$, $P = .017$), as well as in difference in variance ($F_{6,11} = 14.2$, $P < .001$).

In either species, no significant difference was observed in linearity index during the first stopover day between the birds stopping for just one day and those remaining longer (*t*-test, $P > .15$ in both bases; FIG. 3). In the pied flycatcher, there was a tendency for birds stopping for just 1 day to show more area-restricted movements, but it was not significant, perhaps due to the small sample size (FIG. 3).

DISCUSSION

European robins showed similar patterns of movements, as indicated by linearity index values, in both years of study (which made it possible to pool the data for analysis) and in both seasons. The reason for this was in fact rather variable spatial behavior, with linearity indices ranging from 0.008 to 0.65 in spring and 0.003 to 0.93 in fall. This was the reason for the lack of significant interannual or seasonal difference. European robins, and especially sedge warblers, remained within very limited home ranges throughout their stopovers. Sedge warblers never moved for more than 335 m. The reason for this could be a comparatively narrow range of habitats used by this species during migratory stopovers.[20–22] On Cape Rossitten, reedbeds are rather small and restricted; therefore, longer movements were likely to bring sedge warblers to suboptimal habitats. Because stopover duration of sedge warblers was low (maximum 3 days, median 1 day, $n = 12$), it is difficult to make interspecific comparisons.

Despite much intraspecific variation, some patterns still were discernible in the movements of European robins. They were less area-restricted during the first stopover day (in spring) or first two stopovers days (in fall) than in subsequent days, as shown by the linearity index. Furthermore, in spring, birds that stopped just for 1 day moved more broadly (were less area-restricted) during their only stopover day than birds that remained for a longer period. This pattern [broad movements during the first day(s) and defined home range in subsequent days] has been suggested on the basis of analysis of recaptures[6,7] and is consistent with the idea of exploration just after landing and subsequent settlement.[23] These movements probably reflect search/settling costs, which are known to be an important parameter in optimal migration modeling.[24–26] It has been shown that European robins do not start to show positive fuel deposition rates until they settle in a small defined stopover area.[7] However, it remains unclear whether European robins that moved broadly and failed to gain a defined stopover area decided to emigrate (to depart) during the nearest night or if birds that were going to stop over for just one day anyway did not bother to gain a restricted home range.

At least some European robins covered up to 3.7 km in fall and at least 1.8 km in spring during daytime movements. However, the bulk of the birds (94% in both seasons) remained within 350 m during their stopovers. The latter figure might be an underestimate, because tracking often did not cover the first hours after landfall when longer movements might occur. However, most European robins, at least after a short initial search/settling period, clearly showed rather sedentary habits. Direct visual observations of radio-tagged European robins suggested that their home ranges were not defended territories, either in spring or in fall. I frequently observed "intruders" in the core part of an occupied home range, quite near the owner, causing no aggression. In this study, home ranges of neighbors often overlapped to various degrees, which has also been reported in another study.[27]

Stopover duration of the few pied flycatchers included in this study was low, and in this species I could not rule out its underestimation. Together with the small sample size, this fact precludes making interspecific comparisons of stopover duration. The range of movements by some pied flycatchers was by an order of magnitude larger than that of sedge warblers and most European robins. One could argue that pied flycatchers that made long-range movements during daytime stopovers were not transients but rather local breeders searching for their exact breeding location. Unlike sedge warblers and European robins, which breed on the Courish Spit in small numbers, pied flycatchers in large numbers breed in nest-boxes available along the whole spit. Adult male pied flycatchers are known to exhibit a high degree of breeding philopatry.[28] Yearling males breed on the Courish Spit within 6 km of their natal site with a greater than chance probability.[29] Thus, pied flycatchers might be searching for their local breeding area after the final nocturnal flight.

The local population of pied flycatchers in the southern (Russian) part of the Courish Spit is controlled in the framework of the project to study their breeding biology and dispersal.[29,30] No radio-tagged pied flycatchers were found breeding in the area. One pied flycatcher, which moved for 3.1 km, disappeared during the night. The bird was located at 9:30 P.M. local time (i.e., 17 min after local sunset), but not relocated during the checkup at 1:20 A.M. This bird is most likely to have departed by migratory flight. Thus, long-range movements by pied flycatchers probably refer to their stopover behavior. The pattern of movements of pied flycatchers observed in this study contrasted sharply with the pattern reported from fall stopovers in Portugal, where pied flycatchers occupied small defended territories.[3] In this study, some pied flycatchers were making long-range movements and none was doing anything like defending territory. It has been suggested that the main factor governing the spatial behavior of avian migrants at stopovers is the distribution pattern of their food.[8–10] All pied flycatchers that made long-range movements recorded in this study did so under adverse weather, which probably decreased prey availability: low air temperature, sometimes rain and strong wind. The birds were probably trying to locate patches with higher density of prey, for example, in areas with locally weaker wind and higher temperature. Pied flycatchers should have moved more under poor weather conditions because of their aerial foraging style and utilization of flying insects. This species is believed to be the least "flycatching" one within its guild,[31] but our data from the Courish Spit suggest that pied flycatchers do take a substantial proportion of their prey from the air during spring passage (Chernetsov, unpublished). European robins and sedge warblers showed no weather dependence in spatial behavior, probably due to their foraging styles, which are less dependent on high air temperatures.[11–13]

It seemed that all three species studied shaped their behavior mainly to maximize their fueling rate. Theoretically, another important aim should be predation risk minimization.[32,33] However, I found no evidence of the influence of this factor on the behavior of European robins, sedge warblers, or pied flycatchers during their vernal migratory stopovers on the Courish Spit. Certainly this does not rule out that under different circumstances [e.g., higher density of (migrating) sparrowhawks (*Accipiter nisus*)], the situation may be different.

ACKNOWLEDGMENTS

This study was supported by the Russian Foundation for Basic Research (grant no. 02-04-48608).

I am most grateful to Andrey Mukhin and Pavel Ktitorov for their help in the field and many stimulating discussions. Dmitry Leoke did much of the routine telemetry work. Comments by Franz Bairlein, Theunis Piersma, Ulf Bauchinger, and two anonymous referees helped a lot to improve an earlier draft of the manuscript.

REFERENCES

1. LINDSTRÖM, Å. 2003. Fuel deposition rates in migrating birds: causes, constraints and consequences. *In* Avian Migration. P. Berthold, E. Gwinner & E. Sonnenschein, Eds.: 307–320. Springer-Verlag. Berlin.
2. KODRIC-BROWN, A. & J.H. BROWN. 1978. Influence of economics, interspecific competition, and sexual dimorphism on territoriality in migrant hummingbirds. Ecology **49:** 285–296.
3. BIBBY, C.J. & R.E. GREEN. 1980. Foraging behaviour of migrant pied flycatchers, *Ficedula hypoleuca*, on temporary territories. J. Anim. Ecol. **49:** 507–521.

4. BIBBY, C.J. & R.E. GREEN. 1981. Autumn migration strategies of reed and sedge warblers. Ornis. Scand. **12:** 1–12.
5. JOHNSTONE, I. 1998. Territory structure of the robin *Erithacus rubecula* outside the breeding season. Ibis **140:** 244–251.
6. TITOV, N. 1999. Individual home ranges of robins *Erithacus rubecula* at stopovers during autumn migration. Vogelwelt **120:** 237–242.
7. TITOV, N. 1999. Home ranges in two passerine nocturnal migrants at a stopover site in autumn. Avian Ecol. Behav. **3:** 69–78.
8. CHERNETSOV, N. & N. TITOV. 2001. Movement patterns of European reed warblers *Acrocephalus scirpaceus* and sedge warblers *A. schoenobaenus* before and during autumn migration. Ardea **89:** 509–515.
9. CHERNETSOV, N. 2002. Spatial behaviour of first-year blackcaps (*Sylvia atricapilla*) during the pre-migratory period and during autumn migratory stopovers. J. Ornithol. **143:** 424–429.
10. CHERNETSOV, N. & C.V. BOLSHAKOV. Spatial behavior of some nocturnal passerine migrants during stopovers. Acta Zool. Sin. In press.
11. TITOV, N. 2000. Interaction between foraging strategy and autumn migratory strategy in the robin *Erithacus rubecula*. Avian Ecol. Behav. **5:** 35–44.
12. CHERNETSOV, N.S. & N.V. TITOV. 2003. Foraging and spring migratory strategy of the robin *Erithacus rubecula* (Aves, Turdidae) in the southeastern Baltic Sea region. Zool. Zhurn. **82:** 1525–1529.
13. CHERNETSOV N. & A. MANUKYAN. 2000. Foraging strategy of the sedge warbler (*Acrocephalus schoenobaenus*) on migration. Vogelwarte **40:** 189–197.
14. TITOV, N.V. & N.S. CHERNETSOV. 1999. Stochastic models as a new method for estimating length of migratory stopovers in birds (in Russian). Usp. Sovrem. Biol. **119:** 396–403.
15. CHERNETSOV, N. & N. TITOV. 2000. Design of a trapping station for studying migratory stopovers by capture-mark-recapture analysis. Avian Ecol. Behav. **5:** 27–33.
16. CACCAMISE, D.F. & R.F. HEDIN. 1985. An aerodynamic basis for selecting transmitter loads in birds. Wilson Bull. **97:** 306–318.
17. NAEF-DAENZER, B. 1993. A new transmitter for small animals and enhanced methods of home range analysis. J. Wildl. Manag. **57:** 680–689.
18. WILLIAMSON, P. & L. GRAY. 1975. Foraging behavior of the starling (*Sturnus vulgaris*) in Maryland. Condor **77:** 84–89.
19. HOOGE, P.N. & B. EICHENLAUB. 2000. Animal movement extension to Arcview. Version 2.0. Alaska Science Center—Biological Science Office, U.S. Geological Survey. Anchorage, AK.
20. ORMEROD, S.J. 1990. Time of passage, habitat use and mass change of *Acrocephalus* warblers in a South Wales reedswamp. Ringing Migr. **11:** 1–11.
21. CHERNETSOV, N. 1998. Post-breeding and post-fledging movements in the reed warbler (*Acrocephalus scirpaceus*) and sedge warbler (*A. schoenobaenus*) depend on food abundance. Ornis Svecica **8:** 77–82.
22. CHERNETSOV, N. 1999. Migration strategies of *Acrocephalus* warblers within Europe (in Russian). Ph.D. thesis. Zoological Institute. St. Petersburg.
23. MOORE, F.R. & D.A. ABORN. 2000. Mechanisms of *en route* habitat selection: how do migrants make habitat decisions during stopover? Stud. Avian Biol. **20:** 34–42.
24. WEBER, T.P. & A.I. HOUSTON. 1997. Flight costs, flight range and stopover ecology of migrating birds. J. Anim. Ecol. **66:** 297–306.
25. HOUSTON, A.I. 1998. Models of optimal avian migration: state, time and predation. J. Avian Biol. **29:** 395–404.
26. CHERNETSOV, N. *et al.* 2004. Optimal stopover decisions of migrating birds under variable stopover quality: model predictions and the field data. Zh. Obsch. Biol. **65:** 211–217.
27. LAJDA, M. 2001. Telemetrische Untersuchung zum Rastverhalten des Rotkehlchens (*Erithacus rubecula*) in Südwestdeutschland während des Herbstzuges. Diploma thesis. University of Zürich. Zürich.
28. LUNDBERG, A. & R. ALATALO. 1992. The Pied Flycatcher. T & AD Poyser. London.

29. SOKOLOV, L. *et al.* 2004. Spatial distribution of breeding pied flycatchers *Ficedula hypoleuca* in respect to their natal sites. Anim. Biodiv. Conserv. **27.1:** 355–356.
30. SOKOLOV, L.V. 2000. Spring ambient temperature as an important factor controlling timing of arrival, breeding, post-fledging dispersal and breeding success of pied flycatchers *Ficedula hypoleuca* in Eastern Baltic. Avian Ecol. Behav. **5:** 79–104.
31. ALATALO, R. & R. ALATALO. 1979. Resource partitioning among a flycatcher guild in Finland. Oikos **33:** 46–54.
32. DIERSCHKE, V. 2003. Predation hazard during migratory stopover: are light or heavy birds under risk? J. Avian Biol. **34:** 24–29.
33. LANK, D. & R. YDENBERG. 2003. Death and danger at migratory stopovers: problems with "predation risk". J. Avian Biol. **34:** 225–228.

Ecomorphology of the External Flight Apparatus of Blackcaps (*Sylvia atricapilla*) with Different Migration Behavior

WOLFGANG FIEDLER

Max Planck Institute for Ornithology, Vogelwarte Radolfzell, Germany

ABSTRACT: An analysis of the external flight apparatus of 700 blackcaps from eight different populations (sedentary to long-distance migrators) is presented. With increasing migration distances of populations, (1) wing length, aspect ratio, and wing pointedness increase; (2) wing load decreases; (3) slots on the wing tips become relatively shorter; (4) the alula tends to be shorter in relation to wing length; and (5) the tail is shorter in relation to wing length. Although body mass increases from southern to northern populations, changes in wing length and wing area are two to three times larger than expected for simple isometric relationships. Regarding the aerodynamic background of these changes, it can be stated that traits for energy-effective flight are more strongly developed and traits for maneuverability are less developed in birds traveling longer distances, presumably as a consequence of trade-offs. Nonmigratory blackcaps from Madeira and the Cape Verde islands do not always show the traits we would expect in view of their sedentary behavior. This can be seen as a result of recent colonization of these islands by migrants or of selection by factors other than migration behavior. In migratory populations, changes between the first and the second set of primaries during first complete molt show almost the same pattern as the changes from nonmigratory to migratory populations. During molt of the primaries, blackcaps of nonmigratory populations do not show these changes. Hybrids between migrating and nonmigrating blackcap populations (Moscow and Madeira) showed intermediate values between parent populations in wing length, wing shape, and wing area; in the other variables they resembled either parent population.

KEYWORDS: bird migration; bird flight; wing; tail; alula; emargination

INTRODUCTION

The flight apparatus of birds is shaped by the bird's requirements for uplift, propulsion, maneuverability, and energy budget.[1–3] However, these four factors reach their optima with different wing and tail structures. For instance, it has been shown that migrating bird species have more pointed wings than their sedentary relatives,[4,5] and to a lesser extent, the same pattern has been shown on an intraspecific level.[6] In

---

Address for correspondence: Wolfgang Fiedler, Schlossallee 2, D-78315 Radolfzell, Germany. Voice: +49-7732-150160; fax: +49-7732-150165.
fiedler@orn.mpg.de

many studies of wing shape differences, a trade-off between energy-efficient flight and maneuverability is assumed.

Here I test the prediction that on an intraspecific level, medium- and long-distance migrants, in contrast to short-distance or nonmigrants, show a flight apparatus that is more optimized for energy-efficient flight and constrained for maneuverability. My hypothesis is that birds from populations with higher migration activity have (1) more pointed, slender wings with a higher aspect ratio and a higher wing load, and (2) shorter tails, a shorter alula, and less slotted wingtips (see DISCUSSION for details and references). Data were obtained from European blackcaps: (*Sylvia atricapilla*), an abundant songbird species with high variability of migratory behavior.[7] I present a comparative analysis of eight traits of the flight apparatus in populations with different migratory behavior. In addition, individuals of different age classes and hybrids of parents originating from migratory and nonmigratory populations are compared to characterize traits of a migrant's flight apparatus and find out how it was shaped by factors other than migration distance.

MATERIALS AND METHODS

Blackcaps from eight different local populations (FIG. 1) were assigned to five main groups: (1) Santiago (Cape Verde Islands); (2) Madeira; (3) Gibraltar/Spain and Rome/Italy—"Mediterranean"; (4) southwestern Germany—"Central Europe";

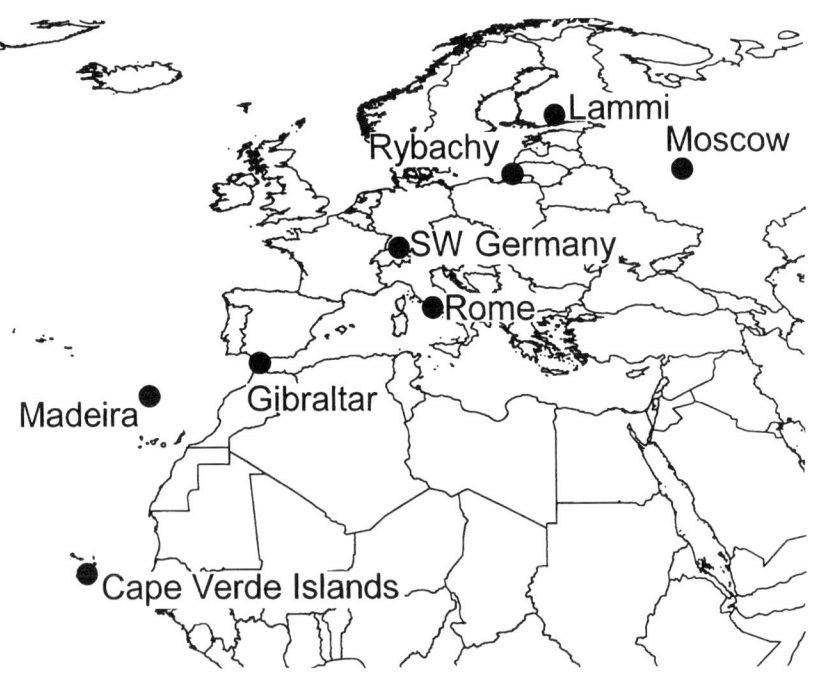

FIGURE 1. Sample localities.

and (5) Rybachy/Kaliningrad region in Russia, Moscow/Russia, and Lammi/Finland—"Northern Europe." All birds were either taken into captivity as nestlings (Madeira, Moscow, part of the birds from southwestern Germany) or were measured during summer, when the presence of migrants could be excluded. Only the birds in Gibraltar were measured in winter, but could easily be distinguished from local residents by morphology and plumage color.[8]

Currently the blackcap species is divided into five subspecies,[9] which are defined mainly by their breeding areas and show very large overlaps in morphological and plumage color traits. According to Cramp,[9] birds of group 1 belong to *S. a.* subsp. *gularis,* which was described by Bannermann and Bannermann[10] to be smaller than the nominate form, although this is clearly not the case (body mass, TABLES 1 and 2). Group 2 is formed by *S. a. heinecken*, and group 3, by *S. a. paulucci*. Groups 4 and 5 belong to the nominate *atricapilla*, but with respect to plumage color and wing length, group 5 would fall well within *S. a. dammholzi*,[11] which is described for Caucasia and Asia Minor.[9]

Blackcaps from Santiago, Madeira, and the "Mediterranean" group are nonmigrants or short-distance migrants (migration distances mainly 0–500 km); those of "Central Europe" migrate to the Mediterranean (600–1000 km), although the birds of the "Northern Europe" group also winter in the Mediterranean as well as in sub-Saharan Africa (1800–5000 km; Bird Ringing Database Vogelwarte Radolfzell and Ref. 9). Since blackcaps from Santiago and Madeira differed substantially in many traits from each other as well as from the Mediterranean birds, they are treated separately. All birds were measured by the author except most of the birds from Rybachy, which were measured by N. Zelenova and the staff of the Biological Station Rybachy. Blackcaps were measured either in nature or while kept in captivity for other experiments.[12,13]

All birds were assigned to age groups strictly based on the age of their primary feathers (age group 1: all birds with the first set of primaries ever grown; age group 2: older birds). For the analysis, geographic groups were regarded as nested in age groups.

To measure wing length, wing shape, and wing load, I used method W_{max}[11] for wing length and calculated a wing shape index from the lengths of all long primaries and the first secondary as proposed by Jenni and Winkler.[14] The index was derived by the following formula: (differences between longest primary and proxP − differences between longest primary and distP)/wing length. ProxP stands for all primaries proximal to the longest primary and the outermost secondary, and distP stands for all primaries distal to the longest primary except the rudimentary outermost one (modified Holinski Index[15]). A higher value indicates a more pointed wing. Wing area and length of spread wing were measured from digitized photographs of the opened right wing. Aspect ratio is defined here as the square of two times the length of the spread wing divided by two times the wing area (slightly modified from Pennycuick,[16] who used the square of the total wing span). A higher value indicates longer and narrower wings. Wing load was calculated by dividing the body mass by two times the wing area plus a constant of 13.5 cm² for body projection.

Body mass values were obtained with digital balances (0.1 g) from birds with fat deposition not more than class 3 on the Kaiser scale[17] for fat deposition ranging from 0 (no visible body fat) to 8 (maximum visible fat load). For estimation of wing load, too few birds with data on both body mass with low fat deposition and wing area

TABLE 1. Traits of the flight apparatus of blackcaps of four different groups in the first set of primaries (age group 1)

Age group = 1	Cape Verde	TK	Madeira	TK	Mediterranean	TK	Central Europe	TK	Northern Europe
	23		5		70		61		228
Body mass (g)	17.43		16.34		16.86		17.35	*	18.67
	±0.87		±0.36		±1.62		±1.18		±1.19
	25		11		69		67		238
Wing length (mm)	72.0	*	69.8		70.02	*	72.84	*	76.36
	±1.38		±1.19		±1.36		±1.43		±1.70
	25		11		67		66		232
Wing shape index	0.36		0.41		0.37	*	0.44	*	0.52
	±0.071		±0.04		±0.057		±0.063		±0.066
	12		9		65		51		4
Wing area (cm^2)	39.2	*	36.2	*	38.6	*	40.1	*	44.0
	±2.07		±1.52		±1.77		±1.83		±1.10
	12		9		65		51		4
Wing load (mPa)	1.90		1.91		1.86		1.86		1.73
	±0.09		±0.066		±0.120		±0.070		±0.04
	12		9		65		51		4
Aspect ratio	4.35		4.46		4.34		4.38		4.45
	±0.19		±0.237		±0.246		±0.252		±0.071
	25		11		68		66		4
Alula–wing ratio	0.26		0.26		0.26		0.25		0.25
	±0.01		±0.012		±0.011		±0.015		±0.012
	23		11		62		39		82
Wing–tail ratio	1.26		1.25	*	1.20	*	1.23	*	1.25
	±0.029		±0.035		±0.033		±0.038		±0.040
	25		10		68		66		81
Notch index	0.24		0.24	*	0.25	*	0.24		0.23
	±0.014		±0.011		±0.017		±0.015		±0.012

NOTE: In each cell the first row indicates sample size, the second row the mean value, and the third row the standard deviation. Columns labeled TK show significant differences ($P < .05$) between the two neighboring cells as revealed by the Tukey–Kramer comparison of means.

were available. Therefore, I used a mean body mass value that was calculated per age group and population. Although I am aware of the problems arising when body mass is used as a measure for body size, it was the only option in this study since another analysis (Fiedler, in preparation) showed that there are concerns that tarsus length really reflects body size in the mixed dataset of caged and free-living birds and other skeletal measures where not applicable in living birds in the field.

As traits indicating the degree of maneuverability,[2,3,18] besides wing shape I used the outermost (longest) tail feathers (measured with a pin ruler) as indicators for tail length, the length of the longest alula feather measured with caliper on the proximal side from insertion to feather tip, and the slots in the wingtips as described by the

TABLE 2. Traits of the flight apparatus of blackcaps of four different groups in the second or any later set of primaries (age group 2)

Age group = 2	Cape Verde	T K	Madeira	T K	Mediterra- nean	T K	Central Europe	T K	Northern Europe
Body mass (g)	23		33		36		49		71
	17.79	*	**16.53**		**17.09**		**17.80**		**18.46**
	±1.17		±1.21		±1.20		±1.70		±1.58
Wing length (mm)	27		54		38		77		96
	71.76		**71.54**	*	**70.34**	*	**75.0**	*	**76.4**
	±1.40		±1.43		±1.37		±1.75		±2.20
Wing shape index	26		54		32		77		83
	0.37	*	**0.44**	*	**0.37**	*	**0.50**		**0.52**
	±0.062		±0.065		±0.068		±0.071		±0.085
Wing area (cm^2)	15		41		28		58		27
	39.3		**37.1**		**38.0**		**40.7**	*	**42.6**
	±2.68		±1.51		±2.028		±1.56		±3.29
Wing load (mPa)	15		41		28		58		27
	1.94		**1.91**		**1.91**		**1.88**		**1.88**
	±0.11		±0.065		±0.083		±0.062		±0.1
Aspect ratio	15		41		29		58		27
	4.33	*	**4.52**	*	**4.34**	*	**4.58**	*	**4.74**
	±0.272		±0.147		±0.267		0.194		±0.20
Alula–wing ratio	26		54		37		72		44
	0.26		**0.25**		**0.25**		**0.24**		**0.25**
	±0.012		±0.01		±0.017		±0.011		±0.011
Wing–tail ratio	27		48		23		53		72
	1.26	*	**1.24**	*	**1.19**	*	**1.28**		**1.28**
	±0.034		±0.03		±0.028		±0.035		±0.039
Notch index	27		54		36		72		69
	0.23		**0.24**		**0.25**	*	**0.23**		**0.23**
	±0.018		±0.014		±0.019		±0.014		±0.015

NOTE: For further explanation, see TABLE 1.

emargination of the inner webs of primaries 9, 8, and 7.[11] An index value for the notches was calculated by dividing the sum of inner emargination lengths of the outermost three primaries by the sum of the corresponding feather lengths. A higher value indicates a more slotted wing tip. Length measurements of alula and tail are given in relation to wing length to correct for individual size variation.

I used JMP 5.0.1 (SAS Institute, Cary, NC) as the statistical software package. Differences among means in TABLES 1 and 2 were tested with the Tukey–Kramer honestly significant difference test.[19] This test is conservative when sample sizes differ.[20] Furthermore, we are expecting to deal with clines, so that significances may only stress the magnitude of changes and cannot necessarily be used here as a criterion regarding the existence or nonexistence of changes.

TABLE 3. Expected and found values for the allometric exponent a after Gould[21]

	$a_{expected}$ Under isometric growth	a_{found} Interspecifically[69]	a_{found} This study, age group 1	a_{found} This study, age group 2
Wing length	0.33	0.42*	0.63	0.64
Wing area	0.67	0.78	0.57	Not available
Wing load	0.33	0.22	0.17**	0.17**
Wing index	ca. 0.03	Not available	ca. 0.8	ca. 0.9
Tail length	0.33	Not available	0.3	0.24

* = Wingspan given; ** = estimated from mean body mass of population and age group.

RESULTS

Differences between populations with different migratory behavior are shown in TABLES 1 and 2. Trends are the same in young and older birds. Except for birds from the Atlantic Islands, blackcaps with increasing migratory behavior have longer and more pointed wings, and wing area and aspect ratio are increasing, whereas wing load decreases slightly. The alula is slightly shorter in relation to the leading edge of the wing, and the notch index is reduced. With increasing migration distances, the tail becomes shorter relative to the wings. Blackcaps from the Atlantic islands of Cape Verde and Madeira do not follow the trends expected from their sedentary life and show combinations of traits not recorded on the mainland.

To test whether these results are only allometric consequences of the fact that blackcaps from northern areas are larger than those from southern areas (TABLES 1 and 2), I used the exponential allometric growth formula:[21] $y = bx^a$ (value of variable y, constant b, body mass x) to calculate the exponent a for selected variables of empirical data. These values were compared with the predicted value for isometric growth and with interspecific data from Norberg[3] to estimate whether the results presented here may merely follow the normal pattern of changes between smaller and larger birds (TABLE 3). The expected value for isometric growth of the wing shape index was derived graphically by isogonal enlargement of a model wing and measuring "feather lengths" before and after the increase. As shown in TABLE 3, wing length and wing shape index increase more, and wing area and wing load increase less than expected for isometric growth, whereas mass increase in the tail follows the isometric expectation.

Between blackcaps in the first and in the second set of primaries (age groups 1 and 2) in the migratory southwestern German population, the differences in traits of the flight apparatus are almost the same as those between populations that cover shorter or longer migration distances (TABLE 4). Only wing load is the same when age groups are compared, but it decreases with migration intensity. In contrast, sedentary birds or birds of the least migrating population in Rome that belong to age group 2 do not have longer wings, more pointed wings, a higher aspect ratio, or a relatively longer tail than birds of these populations in age group 1. But, again in contrast, these birds in age group 2 have less slotted wingtips and a relatively shorter

TABLE 4. Changes of traits of the flight apparatus by changing from age group 1 to age group 2

	Southwestern Germany (same individuals)		Southwestern Germany (two groups)		Rome (two groups)	
	DF	Difference	DF	Difference	DF	Difference
Body mass (g)	6	+1.1*	80	+0.34 n.s.	93	+0.4 n.s.
Wing length (mm)	29	+3.03***	82	+1.87***	94	+0.22 n.s.
Wing shape index	29	+0.07***	81	+0.02***	86	< +0.001 n.s.
Wing area (cm^2)	16	+0.88 n.s.	62	−0.68 n.s.	80	−0.81 n.s.
Wing load (mPa)	16	+0.008 n.s.	61	+0.02 n.s.	80	+0.08***
Aspect ratio	16	+0.29***	61	+0.16***	81	−0.06 n.s.
Alula–wing ratio	28	< −0.001 n.s.	77	+0.005 n.s.	92	−0.08**
Wing–tail ratio	18	+0.05***	46	+0.043***	72	−0.006 n.s.
Notch index	28	−0.01*	76	+0.005 n.s.	93	−0.003**

NOTE: First column: changes within the same individual in captive birds from the southwestern German population; second column: differences between samples of both age groups of captive and free-living southwestern German blackcaps, each individual only present in one group; third column: like the second, but for birds from the central Italian population of Rome. DF: degrees of freedom, difference − mean difference and direction of change.
*** = $P < .001$; ** = $P < .01$; * = $P < .05$; n.s = not significant (Student's t-test for matched pairs in column 1; t-test for equal variances in columns 2 and 3).

alula than the ones in age group 1 (alula length decreasing by 0.5 mm on average; standard deviation of difference: 0.21; $P = .02$, t-test).

Hybrids between migrating and nonmigrating blackcap populations (Moscow and Madeira) produced in captivity[13] show intermediate values in some of the traits and resemble the migratory parent population in others (TABLE 5), thus exhibiting combinations of traits that were not found in natural populations.

DISCUSSION

Between-Population Differences

The results presented here are in accordance with earlier studies of wing length and wing shape on the same species, as well as ecomorphological interpretations of differences in skeletal traits.[22,23] These differences have been used to distinguish resident from migrant blackcaps in sympatric wintering areas.[24,25] This article, however, presents a more comprehensive analysis of the flight apparatus using more traits with potential for aerodynamic significance on a wider range of blackcap populations.

Regarding the results presented here, we can state that a "typical" blackcap with relatively long migration distance has (1) long, pointed wings with a high aspect ratio; (2) a lower wing load than its conspecifics that migrate less far; (3) a tendency toward relatively short flaps on the leading edge of the wing, as well as (4) relatively

TABLE 5. Traits of the flight apparatus of the nonmigratory population of Madeira, the migratory population of Moscow, and hybrids of mixed broods in captivity

	Madeira	Hybrids	Moscow	Differences (Tukey–Kramer test)
Wing length (mm)	15 **70.70** ±0.702	14 **73.75** ±1.341	25 **78.52** ±1.960	All different ($P < .05$)
Wing shape index	15 **0.42** ±0.079	14 **0.48** ±0.069	23 **0.53** ±0.128	Only parental pop. different from each other ($P < .05$)
Wing area (cm^2)	11 **36.85** ±1.27	12 **39.65** ±1.667	20 **42.64** ±3.542	All different ($P < .05$)
Wing load (mPa)	11 **1.91** ±0.055	12 **1.88** ±0.068	20 **1.85** ±0.138	No difference ($P > .05$)
Aspect ratio	11 **4.51** ±0.134	12 **4.63** ±0.106	20 **4.76** ±0.206	Moscow population different from rest ($P < .05$)
Alula–wing ratio	15 **0.25** ±0.008	14 **0.24** ±0.011	24 **0.24** ±0.009	Madeira population different from rest ($P < .05$)
Wing–tail ratio	12 **1.24** ±0.032	12 **1.29** ±0.026	22 **1.28** ±0.042	Madeira population different from rest ($P < .05$)
Notch index	15 **0.24** ±0.017	14 **0.23** ±0.018	25 **0.23** ±0.015	Madeira population different from rest ($P < .05$)

NOTE: All birds were measured in the adult state. In each cell the first row indicates sample size, the second row the mean value, and the third row the standard deviation.

short splits in the outermost part of the wing; and (5) a shorter tail relative to wing length. Thus, the hypothesis stated at the outset can be verified, except for the wing load, which apparently does not increase with migratory behavior. The size increase of the wing area is sufficient to compensate for the body mass increase from southern to northern populations. However, with high fat loads, blackcaps of the more northern populations during migration periods nevertheless could show higher wing loads than their less migratory conspecifics.

As results from studies of aerodynamics of bird flight[2,3] have shown, this combination of traits is suitable to produce a larger forward component in flight due to a more prominent distal part of the wing; furthermore, the induced drag at the wings is reduced by the more slender wings, smaller flaps, and a shorter tail relative to the wing. However, the decrease in Reynolds number due to a higher aspect ratio of the wing and a reduced ability of the tips of the long primaries to bend and generate lift due to relatively short notches at the wing tip result in a reduced capacity for very slow flights under high angles of attack.[3] Furthermore, we can expect that the relatively short tail generates relatively less lift in slow flights and reduces the ability of the tail to start or stop roll maneuvers[18] (nose up or down). This notion supports the idea that blackcaps with a "migratory type" flight apparatus may be more energy efficient in flight than their less migratory conspecifics, but are constrained in maneuverability. This concept is in accordance with the early predictions of Kipp[26] as well as many later studies on interspecific levels[15,27] and on intraspecific levels.[6]

Birds from Madeira and the Cape Verde islands do not show the traits we would have predicted from their sedentary behavior. For instance, blackcaps from Madeira have very short but pointed wings and an aspect ratio resembling that of typical migrants, whereas Cape Verde birds have wing lengths similar to those from southwestern Germany with a wing shape similar to that of Mediterranean birds. These findings might be explained as a result of the recent colonization of these islands or a constant genetic influx by way of migrants,[28,29] and perhaps also as a result of selection by factors other than migration behavior.

Although northern migratory blackcaps are larger than the southern nonmigrating birds, the differences discussed here cannot be explained by allometry alone, as shown by a comparison of some of the empirical data with predicted values. Because allometry of wing shape index is rather complicated,[30] this fact is important to note. An approach with which to isolate wing shape from body size has been proposed by Lockwood and colleagues.[15]

Morphological Changes with Age

Comparing blackcap age classes reveals more or less the same pattern of differences, as shown for a comparison of blackcaps with long migration versus shorter migration distance. On a different scale, a bird in age group 1 shows higher maneuverability performance, whereas the same individual in age group 2 shows traits for a more energy-efficient flight. Alatalo *et al.*[31] propose that inexperienced birds need to have higher maneuverability, and therefore can better escape from predators, than adult and more experienced birds. Therefore, it should be advantageous for older birds of populations with longer migration distances to reduce maneuverability for the benefit of flight-energetics efficiency. The opportunity to do so for a blackcap arises after 1 year, during the next molt of primary and tail feathers.

Data presented here support the results of Pérez-Tris and Tellería[25] found for Spanish blackcaps during winter, namely that age differences are more pronounced in migratory blackcap populations than in nonmigratory ones. The traits mainly responsible for energy-efficient flight are left largely unchanged in the population with little or no migration, giving further evidence that these traits are special adaptations in migrants, which presumably are forced to give up the greater maneuverability for the benefit of higher energy-efficiency of flight. However, it is remarkable that the associated advantages or constraints for migrants are effective only in the second year. The calculated survival probability for a blackcap in the United Kingdom of 36% in the first year[32] means that almost two-thirds of the birds never get the chance to receive any benefits from migrating in the second set of primaries. On the other hand, Naef-Daenzer et al.[33] have shown for great tits and coal tits (*Parus major, Parus ater*) that 47% of all birds die within the first 20 days after fledging. With predation as the major cause of mortality, this study gives evidence that young birds are in great need of escape ability, and hence flight maneuverability, during the first period after fledging. This evidence supports the proposals made by Alatalo et al.[31] concerning a greater importance of maneuverability in young birds.

Hybridization Experiment

Hybrids of migrants and nonmigrants show intermediate results regarding wing length, wing pointedness, and aspect ratio, but are similar to at least one parent in the other traits. These intermediate traits in hybrids support the expectation of a strong genetic component in the expression of the morphology of the flight apparatus. In traits of the flight apparatus of blackcaps, heritabilities well above 0.6 were found.[30]

Outlook

A major problem in the interpretation of the results is the lack of any valid test for the significance of the magnitude of the observed changes. It remains to be shown that the observed, small changes in feather lengths indeed have significant influence on maneuverability or energy-efficient flight performance. Rough estimates can be obtained by using models of aerodynamic theory,[34] but experimental tests of birds with different wing and tail traits are required, for example, by flying birds with manipulated traits in a wind tunnel and recording their energetic demands and maneuverability.

REFERENCES

1. PENNYCUICK, C.J. 1969. The mechanics of bird migration. Ibis **111:** 525–556.
2. RAYNER, J.M. 1988. Form and function in avian flight. Curr. Ornithol. **1:** 1–66.
3. NORBERG, U.M. 1990. Vertebrate Flight. Springer-Verlag. Berlin.
4. SEEBOHM, H. 1901. Birds of Siberia. Murray. London.
5. MARCHETTI, K., T. PRICE & A. RICHMAN. 1995. Correlates of wing morphology with foraging behaviour and migration distance in the genus *Phylloscopus*. J. Avian Biol. **26:** 177–181.
6. EGBERT, J.R. & J.R. BELTHOFF. 2003. Wing shape in house finches differs relative to migratory habit in eastern and western North America. Condor **105:** 825–829.
7. BERTHOLD, P. 2001. Bird Migration. 2nd ed. Oxford University Press. Oxford.
8. FINLAYSON, J.C. 1981. The morphology of Sardinian warblers *Sylvia melanocephala* and blackcaps *S. atricapilla* resident on Gibraltar. Bull. B.O.C. **101:** 299–304.
9. CRAMP, S., Ed. 1992. The Birds of the Western Palaearctic. Vol. VI. Oxford University Press. Oxford.
10. BANNERMANN, D.A. & W.W. BANNERMANN. 1965. A History of the Birds of Madeira, the Deserts, and the Porto Santo Islands. Vol. 2. Birds of the Atlantic Islands. Oliver & Boyd. Edinburgh.
11. SVENSSON, L. 1992. Identification Guide to European Passerines. Stockholm.

12. BERTHOLD, P. 1998. Bird migration: genetic programs with high adaptability. Zoology **101**: 235–245.
13. PULIDO, F. 2000. Evolutionary Quantitative Genetics of Migratory Restlessness in the Blackcap (*Sylvia atricapilla*). Vol. 224. Edition Wissenschaft, Reihe Biologie. Tecum Verlag. Marburg, Germany.
14. JENNI, L. & R. WINKLER. 1989. The feather-length of small passerines: a measurement for wing-length in live birds and museum skins. Bird Study **36**: 1–15.
15. LOCKWOOD, R., J.P. Swaddle & J.M. Rayner. 1998. Avian wingtip shape reconsidered: wingtip shape indices and morphological adaptations to migration. J. Avian Biol. **29**: 273–292.
16. PENNYCUICK, C.J. 1978. Fifteen testable predictions about bird flight. Oikos **30**: 165–176.
17. KAISER, A. 1993. A new multi-category classification of subcutaneous fat deposits of songbirds. J. Field Ornithol. **64**: 246–255.
18. THOMAS, A.L. 1996. The flight of birds that have wings and a tail—variable geometry expands the envelope of flight performance. J. Theor. Biol. **183**: 237–245.
19. KRAMER, C.Y. 1956. Extension of multiple range tests to group means with unequal numbers of replications. Biometrics **12**: 309–310.
20. HAYTER, A.J. 1984. A proof of the conjecture that the Tukey-Kramer multiple comparisons procedure is conservative. Ann. Mathemat. Stat. **12**: 61–75.
21. GOULD, S.J. 1966. Allometry and size in ontogeny and phylogeny. Biol. Rev. **41**: 587–640.
22. TELLERÍA, J.L. & R. CARBONELL. 1999. Morphometric variation of five Iberian blackcap *Sylvia atricapilla* populations. J. Avian Biol. **30**: 63–71.
23. TELLERÍA, J.L., J. PÉREZ-TRIS & R. CARBONELL. 2001. Seasonal changes in abundance and flight-related morphology reveal different migration patterns in Iberian forest passerines. Ardeola **48**: 27–46.
24. PÉREZ-TRIS, J., S. BENSCH, R. CARBONELL, *et al.* 1999. A method for differentiating between sedentary and migratory blackcaps *Sylvia atricapilla* in wintering areas of southern Iberia. Bird Study **46**: 299–304.
25. PÉREZ-TRIS, J. & J.L. TELLERÍA. 2001. Age-related variation in wing shape of migratory and sedentary blackcaps *Sylvia atricapilla*. J. Avian Biol. **32**: 207–213.
26. KIPP, F.A. 1959. Der Handflügel-Index als flugbiologisches Maß. Vogelwarte **20**: 77–86.
27. MÖNKKÖNEN, M. 1995. Do migrant birds have more pointed wings? A comparative study. Evol. Ecol. **9**: 520–528.
28. HELBIG, A.J. 1994. Genetic basis and evolutionary change of migratory directions in a European passerine migrant *Sylvia atricapilla*. Ostrich **65**: 151–159.
29. PÉREZ-TRIS, J., S. BENSCH, R. CARBONELL, *et al.* 2004. Historical diversification of migration patterns in a passerine bird. Evolution **58**: 1819–1832.
30. FIEDLER, W. 1998. The flight apparatus of the blackcap (*Sylvia atricapilla*): methods of measurement, intraspecific variability, pheno- and genotypic variance, selection and ecophysiological consequences [in German]. Ph.D. thesis. Eberhard-Karls-University. Tübingen, Germany.
31. ALATALO, R.V., L. GUSTAFSSON & A. LUNDBERG. 1984. Why do young passerine birds have shorter wings than older birds? Ibis **126**: 410–415.
32. SIRIWARDENA, G., S.R. BAILLIE & J.D. WILSON. 1998. Variation in the survival rates of some British passerines with respect to their population trends on farmland. Bird Study **45**: 276–292.
33. NAEF-DAENZER, B., F. WIDMER & M. NUBER. 2001. Differential post-fledging survival of great and coal tits in relation to their condition and fledging date. J. Anim. Ecol. **70**: 730–738.
34. HEDENSTRÖM, A. 1994. Ecology of avian flight. Ph.D. thesis. University of Lund. Lund, Sweden.

Melatonin and Nocturnal Migration

LEONIDA FUSANI[a] AND EBERHARD GWINNER[b,c]

[a]Università di Siena, Dipartimento di Fisiologia, via Aldo Moro, 53100 Siena, Italy

[b]Max-Planck Institute for Ornithology, 82346 Andechs, Germany

ABSTRACT: Many species of diurnal birds migrate nocturnally. Here, a series of studies of the blackcap (*Sylvia atricapilla*) on the relationship between nocturnal restlessness and melatonin, a hormone that in birds modulates day–night rhythms, are reviewed. Migratory populations from Sweden and Kenya were compared with resident populations from Cape Verde. In blackcaps of migratory populations, night levels of melatonin were lower during the migratory period, when birds showed nocturnal activity, than before and after this period, when birds did not show nocturnal activity. On the contrary, the occurrence of periodic or irregular phases of nocturnal activity in some nonmigratory birds from Cape Verde was not accompanied by a reduction in melatonin levels. In a second series of experiments, it was studied whether melatonin levels change when nocturnally active blackcaps are experimentally transferred from a migratory to a nonmigratory state. A long migratory flight and a refueling stopover were simulated by depriving birds of food for 2 days, subsequently readministering food. The experiments were done in autumn with birds collected in Sweden, and repeated in spring with birds collected in Kenya. In autumn, there was a suppression of nocturnal activity and an increase in melatonin in the night following food reintroduction. In spring, the effects were qualitatively similar, but their extent depended on the amount of body fat reserves. Taken together, the studies demonstrate the existence of a functional relationship between melatonin and nocturnal restlessness and of seasonal differences in the response of the migratory program to food availability.

KEYWORDS: bird migration; nocturnal migration; melatonin; nocturnal restlessness; Zugunruhe

INTRODUCTION

Many birds are nocturnal migrants; that is, they are normally active during the day, but perform migratory flights mainly or exclusively at night.[1] Thus, in these species migration is associated with a major switch in the circadian (24-h) pattern of

---

Address for correspondence: Leonida Fusani, Dipartimento di Fisiologia, Sez. Neuroscienze e Fisiologia Applicata, Università di Siena, via Aldo Moro, 53100 Siena, Italy. Voice: +39-0577-234106; fax: +39-0577-234037.

fusani@unisi.it

[c]This paper was written after the untimely death of Prof. E. Gwinner. However, the experiments and the ideas reported in this review originated from the collaboration and the many discussions between the two authors. Therefore, Prof. Gwinner made a fundamental contribution to the writing of this paper.

activity. The mechanisms controlling the activation of nocturnal migratory behavior are largely unknown; however, an involvement of the circadian system is likely. The pineal gland and its hormone melatonin are major components of the avian circadian system (reviewed in Refs. 2 and 3). In most bird species, the pineal gland contains an autonomous circadian oscillator that is responsible for the rhythmic release of melatonin.[4,5] Few studies have investigated whether and how melatonin is involved in the regulation of nocturnal migration. In the following sections, we will illustrate how our studies on *Sylvia* sp. warblers have provided some answers to these questions.

DOES THE MELATONIN PROFILE CHANGE WHEN DIURNAL BIRDS DEVELOP NOCTURNAL RESTLESSNESS?

There is good agreement in the literature that in birds and other organisms, melatonin levels are elevated only during nighttime, regardless of whether the species is diurnal or nocturnal.[6, 7] The onset of night activity in nocturnal migrants allows us to investigate whether a change in the activity pattern from diurnal to nocturnal is correlated with changes in the pattern of melatonin secretion. It has long been known that during the migratory periods, caged migrants may show a "Zugunruhe" or nocturnal migratory restlessness. Under constant photoperiodic conditions, garden warblers (*S. borin*) show a circannual rhythmicity of migratory disposition.[8] Garden warblers held in a constant 12-h photoperiod from September to June went through the autumn migratory season, winter quiescence, and spring migratory season during this 9-month experimental period.[9] During migratory periods, the pattern of melatonin secretion was not different from that recorded during the other times: Melatonin secretion remains high at night and low during the day.[9] However, peak night melatonin levels were lower during migration times than during the winter quiescent phase.[9] There are at least three hypotheses to explain this phenomenon. Reduction of night peak melatonin during migration could simply be a seasonal correlation; that is, melatonin and nocturnal activity are not functionally related but change seasonally in parallel with each other. Alternatively, the lower amplitude of melatonin during migratory periods could be a consequence of the fact that the birds are awake at night. Locomotor activity or increased light input through the open eyes may induce a decrease in the circulating concentration of melatonin (see discussion in Ref. 10). Finally, lowering melatonin release may facilitate nocturnal activity or serve some other migration-related purposes. In recent years, we conducted a series of experiments to test these hypotheses.

IS THE DECREASE IN THE NIGHT AMPLITUDE OF MELATONIN SPECIFICALLY RELATED TO NOCTURNAL MIGRATION?

We addressed this question with a within-species comparative study using blackcaps (*S. atricapilla*), a close relative of the garden warbler. Within this species, there are populations with very different migratory behaviors, ranging from the nonmigratory populations of Gibraltar and some Atlantic islands to the long-distance migrants of Northern and Eastern Europe that winter in Africa.[11] Thus, we compared seasonal

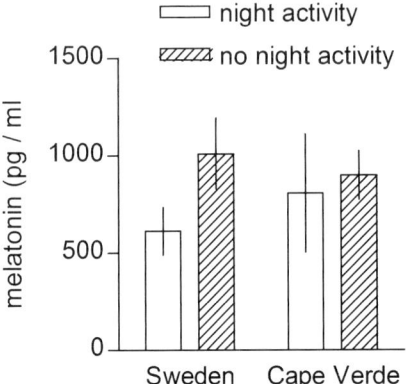

FIGURE 1. Peak night melatonin levels (mean ± SEM) in blackcaps kept in constant 12L:12D. In migratory blackcaps from Sweden, melatonin was lower when the birds showed migratory restlessness. In resident Cape Verde blackcaps, there were no differences in melatonin levels between night-active and -inactive birds sampled at the same time of the year. (Data from Fusani and Gwinner.[12])

changes in the night melatonin profile between resident (nonmigratory) blackcaps from the Cape Verde islands and migratory blackcaps from Sweden and Kenya.[12] Swedish birds were caught in late September, prior to the onset of southward autumnal migration. Kenyan birds were caught in late March, prior to the onset of northward spring migration. The birds of Cape Verde were caught in November. All birds were housed in the same room and were kept in a 12-h photoperiod, and food and water were given *ad libitum*. We found seasonal changes in the nocturnal peak of melatonin in the birds from migratory populations.[12] In these birds, the pattern of melatonin was similar to that previously described for the garden warblers.[9] Melatonin amplitude was lower during the migratory period, when birds showed nocturnal activity, than at other times of the year, when birds did not show nocturnal activity[12] (FIG. 1). On the contrary, in the birds from Cape Verde we found no significant changes in the amplitude of melatonin between seasons. In these birds, the pattern of nocturnal activity was less clear-cut than in those from the migratory populations. Some birds showed continuous activity with no clear night–day transition, whereas others showed short periods with nocturnal activity throughout the year.

Nonmigratory nocturnal restlessness ("Nachtunruhe") has been reported in several bird species,[13] but its causes are still unclear. To further investigate this point, we compared the melatonin peak between Cape Verde birds that either had or had not shown nocturnal activity within the previous 2 weeks. All the birds were sampled within a 5-day period to minimize possible seasonal effects. We found no differences in melatonin levels between night active and night inactive Cape Verde blackcaps (FIG. 1). Thus, the within-species comparison illustrated that only blackcaps from migratory populations show seasonal changes in melatonin amplitude, suggesting that such changes are directly related to migratory disposition.[12] In addition, the lack of significant differences in the melatonin peak between night-active and night-inactive birds of a nonmigratory population indicates that the presence of nocturnal ac-

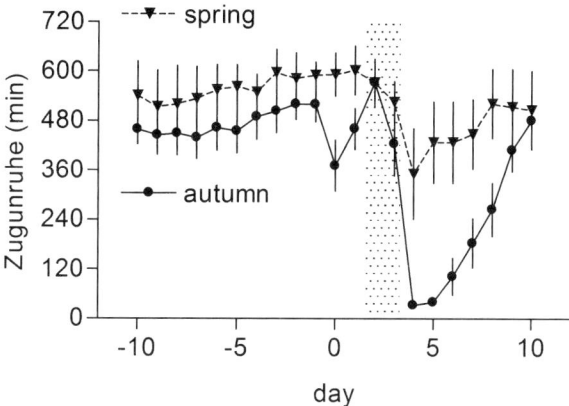

FIGURE 2. Effects of the "fasting and refeeding" protocol on Zugunruhe in blackcaps during autumn and spring migration. The hatched area represents the period of food deprivation. In autumn, Zugunruhe was dramatically reduced in the night following food reintroduction. In spring, the effects were similar, but there was a large variability in the individual response to the food treatment. (Data from Fusani and Gwinner.[17])

tivity *per se* does not induce changes in melatonin amplitude. In summary, these results indicate that the decrease in melatonin during Zugunruhe is specifically related to the activation of nocturnal migration and does not simply result from behavioral changes such as increased locomotor activity or higher light input through the open eyes.

IS THE REDUCTION IN MELATONIN AMPLITUDE DURING ZUGUNRUHE SPECIFICALLY RELATED TO NOCTURNAL MIGRATION?

Due to the rhythmic nature of melatonin release, it is difficult to test experimentally whether melatonin influences the expression or the amount of Zugunruhe. Removal of the pineal gland or melatonin implants that result in constantly low or high levels of melatonin disrupt circadian activity and may mask specific effects on Zugunruhe[3] (Gwinner, unpublished observations). We chose the opposite approach—that is, to manipulate Zugunruhe and look at changes in melatonin. It is known from previous studies on spotted flycatchers (*Muscicapa striata*)[14] and garden warblers[15,16] that food reintroduction after a short period of food deprivation temporarily suppresses Zugunruhe. This phenomenon presumably mimics the situation of a bird that has been fasting during a long migratory flight and interrupts migration upon reaching a suitable refueling site. The use of this experimental paradigm ("fasting and refeeding") does not allow us to tell whether changes in melatonin amplitude cause changes in Zugunruhe or vice versa; however, it allows us to test whether the behavioral and the hormonal effects are functionally associated.

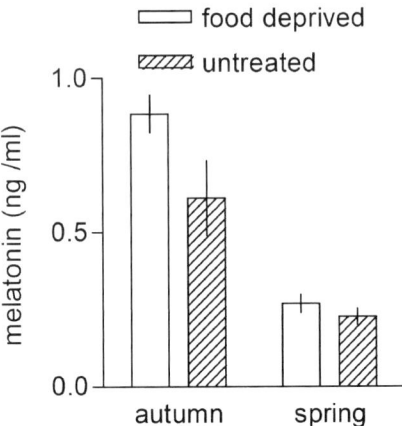

FIGURE 3. Peak night melatonin in food-deprived and control blackcaps in the night following food reintroduction. Melatonin levels were higher in autumn than in spring and in food deprived than in control birds. (Data from Fusani and Gwinner.[17])

In a preliminary study with garden warblers, melatonin levels increased after inducing suppression of autumnal Zugunruhe by means of the fasting and refeeding protocol (Gwinner, Schwabl-Benzinger, and Schwabl, unpublished results, reported in Ref. 10). Recently, we used the same protocol to alter Zugunruhe in blackcaps.[17] Blackcaps from long-distance migratory populations show robust, long-lasting Zugunruhe in captivity.[11,12] The experiments were done with birds taken in Sweden before autumn migration and in Kenya before spring migration. Experiments started at the end of October for autumnal migration and in mid-May for spring migration. The comparison between autumn and spring migration was intended to investigate the possible involvement of melatonin in differences reported between these two migratory periods, such as migration speed and duration.[1] Birds were held in a 12-h photoperiod (light on: 6:00 A.M.) with food and water *ad libitum*. Food was then removed for 60 h from 6:00 P.M. of Day 1 to 6:00 A.M. of Day 4, but fresh water was given as usual. On Day 4, at 6:00 A.M., fresh food was given to the birds. In autumn, the protocol induced a dramatic reduction of Zugunruhe in the first night following food reintroduction (Day 4; FIG. 2), which was accompanied by an increase in the night peak of melatonin (FIG. 3). In spring, the effects were similar, but there was a large variability in the individual response (FIG. 2). Interestingly, the levels of melatonin were lower in spring independently of the treatment (FIG. 3), an aspect of the results that is discussed below.

These experiments demonstrated that there is a functional correlation between Zugunruhe and melatonin: When we induce a reduction of Zugunruhe by simulating a stopover after a long migratory flight, melatonin increases to levels similar to those reported for nonmigratory periods.[12] Although these data strongly suggest that melatonin modulates nocturnal activity in migrants, a formal demonstration is still missing.

Besides the effects on Zugunruhe, it seems likely that a reduction in melatonin release may be functional to other migration-related physiological adaptations. For example, reduced melatonin amplitude may serve to adapt faster to a changing photoperiod as encountered during migration. Treatment of house sparrows with exogenous melatonin reduces resynchronization times by dampening the endogenous melatonin rhythm and thus making the internal clock more "passive."[18,19] A similar effect due to a reduction of melatonin amplitude during migration could facilitate a faster adjustment of the activity pattern to latitudinal changes in photoperiod.[10] This hypothesis is in line with the facts that melatonin levels are lower in spring,[17] when birds migrating north encounter more rapid changes in photoperiod than in autumn.[1]

Another open question is how refueling after a prolonged period of fasting might induce an increase in melatonin and suppress Zugunruhe. In mammals, the diet can affect the amplitude of melatonin release.[20] In the field, the effects of the fasting-and-refeeding protocol are stronger when the birds are brought down to a lean, nonmigratory body mass.[14] In our experiments with the blackcaps, we found that in spring the response to the fasting-and-refeeding protocol is highly variable and depends on the body mass and the amount of fat reserves reached at the end of the food-deprivation period.[17] Interestingly, in these birds melatonin levels are negatively correlated with the decrease in body mass,[17] suggesting that melatonin amplitude is regulated by nutrition-related factors.

CONCLUSIONS

Our studies on the blackcap have highlighted a role for the hormone melatonin in the control of nocturnal migration. The night levels of melatonin are reduced when blackcaps show migratory restlessness, that is, during the migratory periods, compared with other times of the year.[12] This seasonal correlation between night melatonin levels and migratory restlessness does not depend on external cues such as the photoperiod because it is shown by birds kept in constant 12D:12L. The seasonal changes in night melatonin amplitude seem to be strictly linked to the expression of nocturnal restlessness, because they are found only in birds of migratory populations, whereas nonmigratory birds do not show such seasonal changes. Birds from nonmigratory populations, which show nonmigratory nocturnal activity, do not have reduced melatonin levels compared with night-inactive birds of the same populations.[12] Thus, reduced night melatonin levels are specifically linked to the expression of nocturnal migratory restlessness, suggesting that this physiological phenomenon is a component of the "migratory adaptation syndrome," at least in nocturnal migrants. We do not yet know the causal link between nocturnal restlessness and melatonin. We do know, however, that the relationship between melatonin and restlessness is a functional one, because when an interruption in the migratory program is induced using a food deprivation protocol, night melatonin levels increase to nonmigratory levels.[17] In nature, this would correspond to a migration stopover, when birds temporarily interrupt migration for refueling. Interestingly, birds have lower melatonin levels during spring than during autumn migration, and their response to food deprivation, which depends on body mass and fat loss, differs between autumn and spring.[17] These results suggest that in blackcaps, the "migratory adaptation syndrome" might be designed to respond to different sets of external and internal cues, depending on whether it is autumn or spring migration. The blackcap studies have opened new perspectives in bird migration, particularly in the little-explored field of adaptations specific to nocturnal migration.

REFERENCES

1. BERTHOLD, P. 2001. Bird Migration: A General Survey. Oxford University Press. Oxford.
2. CASSONE, V.M. & M. MENAKER. 1984. Is the avian circadian system a neuroendocrine loop? J. Exp. Zool. **232:** 539–549.
3. GWINNER, E., M. HAU & S. HEIGL. 1997. Melatonin: Generation and modulation of avian circadian rhythms. Brain Res. Bull. **44:** 439–444.
4. TAKAHASHI, J.S., H. HAMM & M. MENAKER. 1980. Circadian rhythms of melatonin release from individual superfused chicken pineal glands in vitro. Proc. Natl. Acad. Sci. USA **77:** 2319–2322.
5. GWINNER, E. & R. BRANDSTAETTER. 2001. Complex bird clocks. Philos. Trans. R. Soc. Lond. B Biol. Sci. **356:** 1801–1810.
6. ARENDT, J. 1998. Melatonin and the pineal gland: influence on mammalian seasonal and circadian physiology. Rev. Reprod. **3:** 13–22.
7. KUMAR, V., E. GWINNER & T.J. VAN'T HOF. 2000. Circadian rhythms of melatonin in European starlings exposed to different lighting conditions: relationship with locomotor and feeding rhythms. J. Comp. Physiol. A Sens. Neural Behav. Physiol. **186:** 205–215.
8. GWINNER, E. 1986. Circannual Ryhthms. Springer-Verlag. Heidelberg.
9. GWINNER, E., I. SCHWABL-BENZINGER, H. SCHWABL & J. DITTAMI. 1993. 24-hour melatonin profiles in a nocturnally migrating bird during and between migratory seasons. Gen. Comp. Endocrinol. **90:** 119–124.
10. GWINNER, E. 1996. Circadian and circannual programmes in avian migration. J. Exp. Biol. **199:** 39–48.
11. BERTHOLD, P., U. QUERNER & R. SCHLENKER. 1990. Die Moenchsgrasmuecke. Ziemsen Verlag. Wittenberg-Lutherstadt.
12. FUSANI, L. & E. GWINNER. 2001. Reduced amplitude of melatonin secretion during migration in the blackcap (*Sylvia atricapilla*). In Perspectives in Comparative Endocrinology: Unity and Diversity. Proc. 14th Intl. Congr. Comp. Endocrinol. Sorrento, Italy, May 2001. H.J.T. Goos et al., Eds.: 295–300. Monduzzi. Bologna.
13. BERTHOLD, P. 1988. Unruhe-Aktivität bei Vögeln: eine Übersicht. Vogelwarte. **34:** 249–259.
14. BIEBACH, H. 1985. Sahara stopover in migratory flycatchers (*Muscicapa striata*): fat and food affect the time program. Experientia **41:** 695–697.
15. GWINNER, E., H. BIEBACH & I. VON KRIES. 1985. Food availability affects migratory restlessness in caged garden warblers (*Sylvia borin*). Naturwissenschaften **72:** 51–52.
16. GWINNER, E., H. SCHWABL & I. SCHWABL-BENZINGER. 1988. Effects of food-deprivation on migratory restlessness and diurnal activity in the garden warbler *Sylvia borin*. Oecologia **77:** 321–326.
17. FUSANI, L. & E. GWINNER. 2004. Simulation of migratory flight and stopover affects night levels of melatonin in a nocturnal migrant. Proc. R. Soc. Lond. B Biol. Sci. **271:** 205–211.
18. HAU, M. & E. GWINNER. 1995. Continuous melatonin administration accelerates resynchronization following phase-shifts of a light-dark cycle. Physiol. Behav. **58:** 89–95.
19. ABRAHAM, U., E. GWINNER & T.J. VAN'T HOF. 2000. Exogenous melatonin reduces the resynchronization time after phase shifts of a nonphotic zeitgeber in the house sparrow (*Passer domesticus*). J. Biol. Rhythms **15:** 48–56.
20. SELMAOUI, B., A. OGUINE & L. THIBAULT. 2001. Food access schedule and diet composition alter rhythmicity of serum melatonin and pineal NAT activity. Physiol. Behav. **74:** 449–455.

Phenotypic Flexibility of Skeletal Muscles during Long-Distance Migration of Garden Warblers: Muscle Changes Are Differentially Related to Body Mass

ULF BAUCHINGER[a,b] AND HERBERT BIEBACH[b]

[a]*Department of Biology II, University of Munich, D-82152 Planegg-Martinsried, Germany*

[b]*Max Planck Institute for Ornithology, D-82346 Andechs, Germany*

ABSTRACT: Mass changes of skeletal muscles occur in a variety of species during the migratory period. Phenotypic flexibility of flight muscle mass is considered to represent adaptations of the flight muscle to changing power requirements associated with changes in body mass. We analyzed the relationship between muscle masses and body mass for garden warblers (*Sylvia borin*) sampled during spring migration in Tanzania, Ethiopia, and Egypt, and during autumn migration in Turkey. Flight muscle mass was positively related to body mass of warblers at only one of the four sites, in Egypt where warblers had just arrived after a long migratory flight. Analysis of covariance (ANCOVA) revealed a significant interaction term between sampling site and body mass ($P < .001$), indicating that flight muscle and body mass differed from site to site. We therefore question the idea that changing power requirements associated with changes in body mass cause mass changes of the flight muscle. We further suggest that different migration strategies across different landscapes shape the relationship between flight muscle and body mass. Flights across major ecological barriers may cause substantial catabolism of flight muscle protein until the limit necessary for flight, while migration across "common landscape," which enables a bird to land and to feed and/or drink, may occur without the need to catabolize flight muscle protein. However, a differential relationship between flight muscle mass and body mass described here for a long-distance migrant seems as well to occur in a short-distance migrant and is therefore unlikely to be part of an adaptive syndrome typical for long-distance migration.

KEYWORDS: migratory strategy; endurance locomotion; differential; catabolism; passerine

Address for correspondence: Ulf Bauchinger, Department of Biology II, University of Munich, Grosshaderner Str. 2, D-82152 Planegg-Martinsried, Germany, Phone: +49-89-2180-74132; Fax: +49-89-2180-9974134.

bauchinger@lmu.de

INTRODUCTION

Phenotypic flexibility of muscles during the migratory period occurs in a variety of bird species and is usually associated with three phases of migration: (1) Increasing muscle mass during premigratory periods,[1-8] but see Refs 9 and 10; (2) decrease in muscle mass during flight;[11-17] and (3) increasing muscle mass during stopover periods.[18-23] In passerine birds, mass changes of flight and leg muscles account for up to 25% of changes in body mass, with most changes occurring after a flight across ecological barriers, such as the Sahara desert belt or the Mediterranean Sea.[15,24] Mass changes of flight muscles and leg muscles also may occur because of "use-hypertrophy" or "disuse-atrophy," as documented in grebes at migratory staging sites.[19] During the flightless molting period the flight muscle shrinks, but leg muscles increase at the same time.

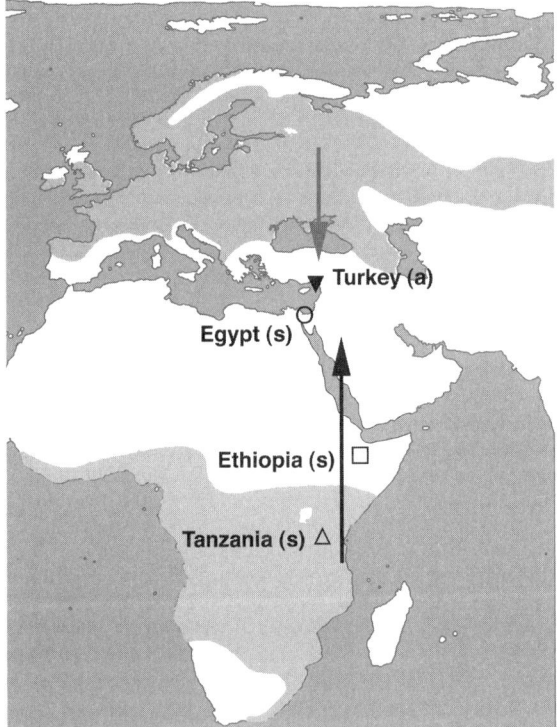

FIGURE 1. Map indicates the breeding range for garden warblers in Eurasia (*dark gray*) and the wintering range south to the Sahel zone in Africa (*light gray*).[45] [Symbols indicating the sampling sites correspond to the graphs in FIGS. 2 and 3. Spring migration (s): Tanzania (*open triangles*), Ethiopia (*open squares*), and Egypt (*open circles*). Autumn migration (a): Turkey (*filled triangles*). *Arrows* give overall migratory direction for spring and autumn migration.]

One contemporary paradigm postulates that flight muscle mass is adjusted directly in response to changes in body mass.[2,3,7,25–27] Body mass may double during migration,[28] a phenomenon, including mass changes of organs, besides fat metabolism.[9,22,29–31] According to scaling predictions, the parameters are expected to be linearly related.[32,33] For example, Lindström et al.[27] found a linear relationship between flight muscle mass and body mass when they quantified flight muscle thickness in great knots under different conditions, such as, during fasting, refuelling, and flight. Mass changes are the most obvious adjustments, but compositional changes may occur as well and may adjust muscle function to the changing power requirements.[3,15,34–36]

We measured the mass of flight muscle, leg muscle, and whole body of garden warblers (*Sylvia borin*) sampled during spring and autumn migration. Garden warblers breed throughout the Palearctic region and winter in Central and Southern Africa. During migration they encounter two major ecological barriers, the Mediterranean Sea and the Sahara desert belt, which together extend for about 2500 km (FIG. 1). Birds were collected during autumn migration in Turkey[15] and spring migration in Tanzania, Ethiopia, and Egypt.[24] Since the whole body mass changes during migration, these measurements allowed us to determine the relationship between mass of flight muscle, leg muscle, and, whole body mass in garden warblers.

METHODS

Autumn Migration

We used mist nets to capture 12 garden warblers betweem August 30 and September 5, 1996 during autumn migration in Turkey at the Mediterranean cost (36°40′N; 33°25′E). The catching site was an orchard with fig trees. Time of capture relative to phenology of migration that year suggests that birds were sampled during stopover and refueling prior to the flight across the Mediterranean Sea (for further details, dissections, and tissue analysis, see Ref. 15).

Spring Migration

We used mist nets to capture a total of 27 garden warblers at three sites during spring migration: between March 19 and March 27, 1997 in Tanzania ($n = 8$; 3°10′S; 37°05′E) while birds were in a premigratory state, between April 26 and May 2, 1998 in Ethiopia ($n = 9$; 9°21′N; 43°48′E) just prior to the bird's flight across the Sahara desert, and between April 29 and May 17, 1998 in Egypt ($n = 10$; 31°08′N; 33°25′E) immediately after the flight across the Sahara desert (for further details, dissections, and tissue analysis, see Ref 24).

Skeletal Muscles

The term "flight muscle" is used for the combined left and right *Musculus pectoralis* and *Musculus supracoracoideus* muscle. The term "leg muscle" is used for the combined leg muscles attached to the *Tibotarsus*. All muscle masses are presented as lean dry mass (for dissection details and tissue analysis, see Ref. 15).

Statistics

We performed statistical tests (Advanced Statistics 10.0, SPSS Inc., Chicago, IL) separately for the autumn and spring periods. All variables were tested for normality and homogeneity of variances. Analysis of covariance (ANCOVA) was performed on the autumn sample, with either flight or leg muscle as dependent variable and the covariates body mass and tarsus length. Tarsus was chosen as the measurement of structural size. While tarsus length was not measured for all birds sampled in spring, sternum length was measured in all birds by the same person and chosen as the measurement for structural size. For the spring data set, ANCOVA was performed with either flight or leg muscle as dependent variable, sampling site as the fixed factor, the covariates body mass, and sternum length, and the interactions "sternum length * sampling site" and "body mass * sampling site." In case the interaction(s) revealed a significant result ($P < .05$), the respective variable(s) (i.e., body mass, sternum length, or sampling site) was (were) removed and the analysis repeated. If interaction(s) did not reveal a significant result, the interaction term(s) was (were) removed and the analysis repeated without the interaction term(s). If the ANCOVA resulted in a nonsignificant result for the model (corrected model), analysis was stopped. In all cases, the entered factor body mass represented body mass minus either flight muscle or leg muscle mass.[37]

RESULTS

Spring Migration

ANCOVA results for the mass of the flight or leg muscle are presented in TABLE 1. During spring, no significant interactions (sampling site * body mass; sampling site * sternum length) were apparent for the leg muscle, and therefore the analysis was repeated with only sampling site, body mass, and sternum length in the model (FIG. 2). Significant effects were thereafter present for the relationship of leg muscle mass to body mass and sternum length, but leg muscle did not differ between sampling sites. In contrast, flight muscle masses showed a highly significant interaction term between sampling site and body mass (TABLE 1, FIG. 2), and the relationship between flight muscle mass and body mass differed significantly between the sampling sites. Sternum length had no significant effect on flight muscle mass. A significant relationship between flight muscle mass and body mass occurred only in garden warblers sampled in Egypt, but not in birds sampled in Tanzania and Ethiopia (FIG. 2).

Autumn Migration

Mean body mass of garden warblers sampled in Turkey was 20.7 ± 4.5 g [95% confidence interval (C.I.)], mean flight muscle and leg muscle mass amounted to 0.781 ± 0.134 g (95% C.I.) and 0.166 g ± 0.025 g (95% C.I.), respectively. ANCOVA for the birds sampled during autumn migration in Turkey only showed significance for the flight muscle, but not for the leg muscle (FIG. 2). Flight muscle mass is significantly related to tarsus length, however, not to body mass (TABLE 1).

TABLE 1. Results of ANCOVA for flight and leg muscles of garden warblers sampled during autumn migration and spring migration

Dependent variable	ANCOVA, entered factor	df	F value	P value
Autumn migration				
Flight muscle	Corrected model	2, 11	4.9	.036
	Body mass	1	3.3	.10
	Tarsus	1	7.5	.023
Leg muscle	Corrected model	2, 11	2.9	.11
	Body mass	1	1.4	
	Tarsus	1	4.8	
Spring migration				
Flight muscle	Corrected model	6, 26	19.6	.001
	Sampling site * body mass	3	26.0	.001
	Sternum	1	2.6	.12
Leg muscle	Corrected model	6, 22	8.1	.001
	Sampling site	2	2.3	.13
	Body mass	1	4.8	.044
	Sternum	1	6.1	.025

df = Degrees of freedom for the corrected model and total degrees of freedom.

DISCUSSION

Phenotypic Flexibility of Skeletal Muscle Mass

During migration, the muscle mass of garden warblers was highly flexible. Body mass contributes significantly to the mass changes of the leg muscle of garden warblers sampled during spring migration, but not in birds sampled during autumn migration. It is therefore suggested that phenotypic mass changes of the leg muscle during migration occur in relation to body mass changes, and do not depend on the sampling site. Flight muscle mass during spring was not significantly influenced by body size. However, the effect of body mass on flight muscle mass was significantly different between the sampling sites (FIG. 2). Phenotypic flexibility of the flight and leg muscles has been documented in many migratory birds, and especially in birds that migrate long distances. For example, skeletal muscle mass of migratory birds increased during the premigratory period[1-8] (but see Refs. 9 and 10) and during the stopover period [18-23] and decrease after migratory flight.[11-15,17,38]

Phenotypic flexibility of the flight and leg muscle masses in garden warblers during migration did not consistently track changes in body mass, as indicated by the significant interaction between sampling site and body mass for the flight muscle mass of birds sampled during the spring migration. Flight muscle mass of birds captured in Egypt was significantly related to body mass, whereas there was no such relationship in birds captured in Tanzania, Ethiopia, and Turkey. This argues against the prevailing hypothesis for coincident changes in mass of the flight muscle and body mass during migration[25] (for review, see Refs. 6, 9, 13, and 39).

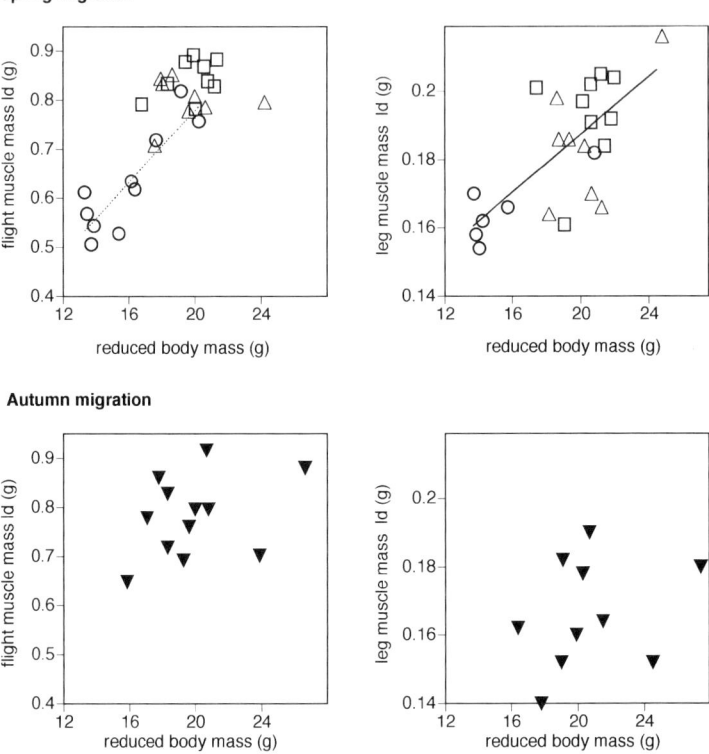

FIGURE 2. Lean dry masses (ld) of flight and leg muscle plotted against reduced body mass for garden warbler sampled during spring (Tanzania = *open triangle*; Ethiopia = *open square*; Egypt = *circles*) and autumn migration (Turkey = *filled triangle*). Reduced body mass represents body mass minus flight muscle mass, or body mass minus leg muscle mass, respectively. *Lines* represent linear regression lines. *Dotted lines* = only for garden warblers sampled in Egypt (= ANCOVA result with significant interaction between sampling site and body mass). *Full line* = for all birds from Tanzania, Ethiopia, and Egypt (= ANCOVA result with no significant interaction between sampling site and body mass, but significant relationship between muscle mass and body mass).

We propose three explanations for the lack of coincident changes in mass of flight muscle and body mass in garden warblers during migration: (1) a limited range of body masses within bird groups may limit the analysis; (2) changes in total body fat rather than total body mass may be related to flight muscle mass; and (3) the respective sampling site, or in other words, the stage of migration when the birds are at the respective sites.

(1) The body mass range for garden warblers sampled at the four sites was 18.3–25.0 g for Tanzania, 17.6–22.2 g for Ethiopia, 13.9–21.0 g for Egypt, and 16.5–27.5 g for Turkey. Although birds captured in Egypt had a relatively larger range in body mass (7.1 g) compared to other spring migration sites, birds captured in Turkey had

the largest range in body mass (11.0 g), yet did not show the predicted relationship between flight muscle mass and body mass.

(2) The major fuel for migratory flight is fat, and fat contributes the most to observed body mass changes.[29,30] Thus, changes in a bird's flight muscle might be more directly related to changes in body fat mass rather than to changes in whole body mass. For garden warblers during spring migration, flight muscle mass was differentially related to mass of body fat (ANCOVA; interaction sampling site * fat content; df 3, 25; $F = 10.8$; $P < .001$) and whole body mass (FIGS. 2 and 3). This is not surprising because the flight muscle has to move the whole body and not only the mass of stored fat. However, a higher fraction of birds from the Egyptian group were relatively lean (fat content of 5% of body mass or lower compared to the other sampling groups) even though we detected a strong relationship between mass of flight muscle and whole body mass in this group. During starvation, birds that run out of fat stores increase their protein metabolism to satisfy their energy and protein requirements.[40] For most species, including the garden warbler, migratory flight represents a period of in-flight starvation. Therefore, birds with low or depleted fat resources must power flight by protein catabolism, especially during flight across ecological barriers, which do not offer the possibility of forage on the ground. The 5% fat of total body mass just mentioned is generally seen as a critical value for the transition from phase II of fasting to phase III of fasting.[40,41] Interestingly, one bird captured in Tanzania and another captured in Ethiopia also had fat contents of 5%, yet still had relatively heavy flight muscle. This suggests, that the relationship between the flight muscle mass and body mass is not driven by increased protein catabolism in very lean birds. The strong relationship of flight muscle mass to body mass within the Egyptian group is a result shared by all birds, that is, very lean and fat birds, as well as by the ones with intermediate fat stores (FIG. 3).

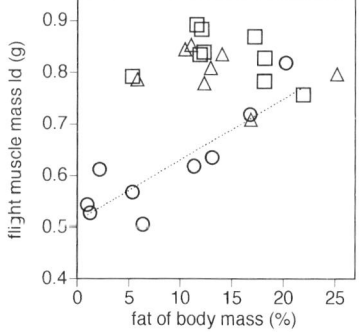 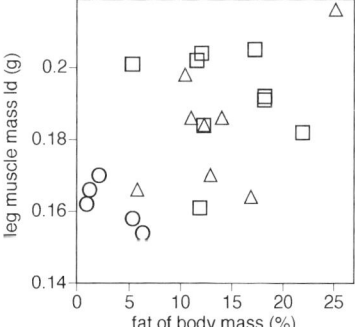

FIGURE 3. Lean dry mass (ld) of the flight muscle and the leg muscle plotted against percentage of total body fat of body mass for garden warblers sampled during spring migration in Tanzania (*triangles*), Ethiopia (*squares*), and Egypt (*circles*). A significant linear regression became apparent only for garden warblers sampled in Egypt (= ANCOVA result with significant interaction between sampling site and percentage of total body fat of body mass).

(3) Garden warblers sampled at the various sites obviously differed with respect to their stage of migration from birds captured in Ethiopia and Turkey about to fly across major ecological barriers of the Afro-Palearctic migration system, while birds captured in Egypt had just completed a flight across at least 1500 km of inhospital desert area (FIG.1). Thus, birds in Egypt had just completed a sustained flight period across an extended ecological barrier, whereas birds sampled at the other three sites had migrated across habitat that would have enabled food and water uptake. For birds flying across major ecological barriers, flight muscle may serve as a protein resource for periods when nitrogen intake is reduced or even zero.[6,9,13,31,39,42,43] If this functional hypothesis is correct, then birds would use the protein resources of their body (skeletal muscles and other organs) only when their protein budget is negative, such as when crossing large ecological barriers. This is clearly the case for birds sampled in Egypt, but this need not be the case for birds sampled in Tanzania, Ethiopia, and Turkey. Birds crossing large ecological barriers may selectively catabolize protein reserves for reasons other than satisfying their energy requirements. According to an alternative hypothesis, birds may break down protein in order to liberate the water bound to the protein[13,31,42,43] or to maintain blood glucose levels via gluconeogenesis.[44] While sustained periods of flight are necessary for overcoming major ecological barriers, the remaining part of migration can be performed in smaller steps. Migratory step length and the possibility of drink and/or feed could, according to the mentioned hypotheses, call into question the need to catabolize substantial amounts of flight muscle protein while migrating across "common landscape" (nonecological barriers). Uptake of water, nectar, which can be easily absorbed even with a reduced digestive tract, or other food could prevent a migratory bird from needing to catabolize protein, but uptake has to be seen as almost nonexistent during the flight across the Sahara and the Mediterranean Sea.

CONCLUSION

Mass changes of skeletal muscles during bird migration are a common phenomenon. It is suggested that changes in flight muscle mass occur in direct relationship to changes in body mass.[2,3,7,25–27] The data presented for garden warblers during migration show that flight muscle mass is not consistently related to whole body mass for birds during migration. A consistent relationship between flight muscle mass and body mass also seems questionable for the Eurasian golden plover (*Pluvialis apricara*) when gaining similar mass after arrival in its wintering area and prior to leaving this area in spring.[10] Protein tissue increase is associated with mass gain only in birds sampled in spring, while birds sampled in autumn deposit no additional protein, but only fat. A simple mass adaptation of the flight muscle to the changing power requirements associated with changing body mass seems questionable for both the garden warbler and other birds. We suggest that there are fundamentally different physiological challenges for birds that migrate across large ecological barriers compared to birds that migrate through "normal" habitat, and that these differences may explain why we found a significant relationship between flight muscle mass and body mass in garden warblers captured in Egypt but not in Tanzania, Turkey, or Ethiopia. Furthermore, the relationship between flight muscle mass and body mass (or fat mass) found in the Egyptian birds may represent the minimum requirements of the flight muscle to move the body, while the samples from the other three sites suggest that some of the birds have flight muscles that are bigger then necessary for pure flight requirements. Flight muscle is the main locomotor muscle, but has to be considered a store for proteins. Suggestions like the ones presented here are not entirely new. Flight muscle of the short-distance migratory robins (*Erithracus rubecula*) seemed to be differentially reduced after a flight across an ocean compared to flying across land.[11] Therefore, differential relationships between flight muscle and body mass as result of differential catabolism during migration across different landscapes may not be seen as a phenomenon typical of long-distance migration. The physiological requirements during migration are undoubtedly special, but the in-flight starvation may well be compared to other starvation periods among birds. Locomotor and resource function, which may shape the flight muscle and its relationship to body mass, exist as well in nonmigratory birds during breeding, molt, and inclement weather.

ACKNOWLEDGMENTS

This work was supported by German Science Foundation (DFG) grant no. BI 208/8-1 to H.B. We gratefully acknowledge the support of all the people contributing to successful data collection in the field and all the officials involved in this project. Special thanks are dedicated to Anke Wohlmann for her dedication during fieldtrips and in the laboratory, and to Lisa Trost, Matthias Starck, and Scott McWilliams for critical comments on the manuscript.

REFERENCES

1. DOLNIK, V.R. & T.K. BLYUMENTAL. 1967. Autumnal premigratory and migratory periods in chaffinch *(Fringilla coelebs coelebs)* and some other temperate-zone passerine birds. Condor **69**: 435–468.
2. FRY, C.H., I.J. FERGUSON-LEE & R.J. DOWSETT. 1972. Flight muscle hypertrophy and ecophysiological variation of yellow wagtail *Motacilla flava* races at Lake Chad. J. Zool. **167**: 293–306.
3. MARSH, R.L. 1984. Adaptations of the Gray Catbrid *Dumetella carolinensis* to long-distance migration: flight muscle hypertrophy associated with elevated body mass. Physiol. Zool. **57**: 105–117.
4. DAVIDSON, N.C. & P.R. EVANS. 1988. Prebreeding accumulation of fat and muscle protein by arctic-breeding shorebirds. *In* Acta XIX Congr. Int. Ornithologici, Ottawa. H. Ouellett, Ed.: 342–352. Universiy of Ottawa Press. Ottawa, Canada.
5. PIERSMA, T. 1990. Pre-migratory fattening usually involves more than the deposition of fat alone. Ringing & Migr. **11**: 113–115.
6. EVANS, P.R. 1992. The use of Balsfjord, north Norway, as a staging post by Knot during spring migration: fat depostion, muscle hypertrophy and flight strategies. Wader Study Group Bull. **63**(Suppl.): 126–128.
7. DRIEDZIC, W.R. *et al.* 1993. Adaptations in pectoralis muscle, heart mass, and energy metabolism during premigratory fattening in semipalmated sandpipers (*Calidris pusilla*). Can. J. Zool. **71**: 1602–1608.
8. BATTLEY, P.F. & T. PIERSMA. 1997. Body composition of Lesser Knots (*Calidris canutus rogersi*) preparing to take off on migration from northern New Zealand. Notornis **44**: 137–150.

9. LINDSTRÖM, Å. & T. PIERSMA. 1993. Mass changes in migrating birds—the evidence for fat and protein storage re-examined. Ibis **135:** 70–78.
10. PIERSMA, T & J. JUKEMA. 2002. Contrast in adaptive mass gains: Eurasion golden plovers store fat before midwinter and protein before prebreeding flight. Proc. R. Soc. Lond. B **269:** 1101–1105.
11. AKESSON, S. et al. 1992. Body composition and migration strategies: a comparison between robins (*Erithacus rubecula*) from two stop-over sites in Sweden. Vogelwarte **36:** 188–195.
12. BIEBACH, H. 1998. Phenotypic organ flexibility in garden warblers *Sylvia borin* during long-distance migration. J. Avian Biol. **29:** 529–535.
13. BAUCHINGER, U. & H. BIEBACH. 1998. The role of protein during migration in passerine birds. Biol. Cons. Fauna **102:** 299–305.
14. BATTLEY, P.F. et al. 2000. Empirical evidence for differential organ reductions during trans-oceanic bird flight. Proc. R. Soc. Lond. B **267:** 191–195.
15. BAUCHINGER, U. & H. BIEBACH. 2001. Differential catabolism of muscle protein in garden warblers (*Sylvia borin*): flight and leg muscle act as a protein source during long-distance migration. J. Comp. Physiol. B **171:** 293–301.
16. DEERENBERG, C., H. BIEBACH & U. BAUCHINGER. 2002. Spleen size variation during long-distance migration in a passerine. Avian Sci. **2:** 217–226.
17. SCHWILCH, R. et al. 2002. Protein loss during long-distance migratory flight in passerine birds: adaptation or constraint. J. Exp. Biol. **205:** 687–695.
18. GAUNT, A.S. et al. 1990. Rapid atrophy and hypertrophy of an avian flight muscle. Auk **107:** 649–659.
19. JEHL, J.R. 1997. Cyclical changes in body composition in the annual cycle and migration of the eared grebe *Podiceps nigricollis*. J. Avian Biol. **28:** 132–142.
20. PIERSMA, T. 1998. Phenotypic flexibility during migration: physiological optimization contingent on the risks and rewards of fuelling and flight. J. Avian Biol. **29:** 511–520.
21. PIERSMA, T. & R.E. GILL. 1998. Guts don't fly: small digestive organs in obese bar-tailed godwits. Auk **115:** 196–203.
22. PIERSMA, T., G.A. GUDMUNDSSON & K. LILLIENDAHL. 1999. Rapid changes in the size of different functional organ and muscle groups during refuelling in a long-distance migrating shorebird. Physiol. Biochem. Zool. **72:** 405–415.
23. PIERSMA, T. et al. 1999. Reversible size-changes in stomachs of shorebirds: when, to what extent, and why? Acta Ornithol. **34:** 176–181.
24. BAUCHINGER, U., A. WOHLMANN & H. BIEBACH. 2005. Flexible remodeling of organ size during spring migration of the garden warbler (*Sylvia borin*). Zoology. In press.
25. PENNYCUICK, C.J. 1975. Mechanisms of flight. *In* Avian Biology. D.S. Farner & J.R. King, Eds.: 1–75. Academic Press. New York.
26. PENNYCUICK, C.J. 1998. Computer simulation of fat and muscle burn in long-distance bird migration. J. Theor. Biol. **191:** 47–61.
27. LINDSTÖM, A. et al. 2000. Avian pectoral muscle size rapidly tracks body mass changes during flight, fasting and fuelling. J. Exp. Biol. **203:** 913–919.
28. BAIRLEIN, F. 1991. Body mass of garden warblers (*Sylvia borin*) on migration: a review of field data. Vogelwarte **36:** 48–61.
29. KLAASSEN, M. & H. BIEBACH. 1994. Energetics of fattening and starvation in the long-distance migratory garden warbler, *Sylvia borin*, during the migratory phase. J. Comp. Physiol. B **164:** 362–371.
30. KLAASSEN, M., Å. LINDSTRÖM & R. ZIJLSTRA. 1997. Composition of fuel stores and digestive limitations to fuel deposition rate in the long-distance migratory thrush nightingale, *Luscinia luscinia*. Physiol. Zool. **70:** 125–13.
31. KLAASSEN, M., A. KVIST & Å. LINDSTRÖM. 2000. Flight costs and fuel composition of a bird migrating in a wind tunnel. Condor **102:** 444–451.
32. HARTMAN, F.A. 1961. Locomotor mechanisms of birds. Smithson. Miscell. Coll. **143:** 1–91.
33. CALDER, W.A. 1984. Size, Function, and Life History. Havard University Press. Cambridge, MA.
34. PENNYCUICK, C.J. & M.A. REZENDE. 1984. The specific power output of aerobic muscle, related to the power density of mitochondria. J. Exp. Biol. **108:** 377–392.

35. LUNDGREN, B.O & K.H. KIESSLING. 1985. Seasonal variation in catabolic enzyme activities in breast muscle of some migratory birds. Oecologia **66:** 468–471.
36. PENNYCUICK, C.J. & P.F. BATTLEY. 2003. Burning the engine: a time marching computation of fat and protein consumption in a 5400-km non-stop flight by great knots, *Calidris tenuirostris*. Oikos **103:** 323–332.
37. CHRISTIANS, J. 1999. Controlling for Body mass effects: Is part-whole correlation important? Physiol. Bioch. Zool. **72:** 250–253.
38. BATTLEY, P.F. *et al.* 2001. Is long-distance bird flight equivalent to a high-energy fast? Body composition changes in freely migrating and captive fasting knots. Physiol. Biochem. Zool. **74:** 435–449.
39. BIEBACH, H. 1996. Energetics of winter and migratory fattening. *In* Avian Energetics and Nutritional Ecology. C. Carey, Ed.: 280–323 Chapman & Hall. New York.
40. GROSCOLAS, R. & J.-P. ROBIN. 2001. Long-term fasting and re-feeding in penguins. Comp. Biochem. Physiol. A **128:** 645–655.
41. JENNI, L. *et al.* 2000. Regulation of protein breakdown and adrenocortical response to stress in birds during migratory flight. Am. J. Physiol. Regul. Integr. Comp. Physiol. **278:** R1182–R1189.
42. KLAASSEN, M. 1996. Metabolic constraints on long-distance migration in birds. J. Exp. Biol. **199:** 57–64.
43. JENNI, L. & S. JENNI-EIERMANN. 1998. Fuel supply and metabolic constraints in migrating birds. J. Avian Biol. **29:** 521–528.
44. JENNI-EIERMANN, S. & L. JENNI. 1991. Metabolic responses to flight and fasting in night-migrating passerines. J. Comp Physiol. B **161:** 465–474.
45. GLUTZ VON BLOTZHEIM, U.N. 1991. Handbuch der Vögel Mitteleuropas. Handbuch der Vögel Mitteleuropas. Vol. 12, pp. 888–948.

Is There a "Migratory Syndrome" Common to All Migrant Birds?

THEUNIS PIERSMA,[a,b] JAVIER PÉREZ-TRIS,[c] HENRIK MOURITSEN,[d] ULF BAUCHINGER,[e] AND FRANZ BAIRLEIN[f]

[a]*Animal Ecology Group, Centre for Ecological and Evolutionary Studies, University of Groningen, Groningen, The Netherlands*

[b]*Department of Marine Ecology and Evolution, Royal Netherlands Institute for Sea Research (NIOZ), Den Burg (Texel), The Netherlands*

[c]*Department of Animal Ecology, Lund University, Lund, Sweden*

[d]*VW Nachwuchsgruppe "Animal Navigation," Department of Biology, University of Oldenburg, Oldenburg, Germany*

[e]*Department of Biology II, University of Munich, Munich, Germany*

[f]*Institute of Avian Research "Vogelwarte Helgoland," Wilhelmshaven, Germany*

ABSTRACT: Bird migration has been assumed, mostly implicitly, to represent a distinct class of animal behavior, with deep and strong homologies in the various phenotypic expressions of migratory behavior between different taxa. Here the evidence for the existence of what could be called a "migratory syndrome," a tightly integrated, old group of adaptive traits that enables birds to commit themselves to highly organized seasonal migrations, is assessed. A list of problems faced by migratory birds is listed first and the traits that migratory birds have evolved to deal with these problems are discussed. The usefulness of comparative approaches to investigate which traits are unique to migrants is then discussed. A provisional conclusion that, perhaps apart from a capacity for night-time compass orientation, there is little evidence for deeply rooted co-adapted trait complexes that could make up such a migratory syndrome, is suggested. Detailed analyses of the genetic and physiological architecture of potential adaptations to migration, combined with a comparative approach to further identify the phylogenetic levels at which different adaptive traits for migration have evolved, are recommended.

KEYWORDS: behavioral syndrome; bird migration; life history; phenotype; comparative method; trade off; coadapted trait complex; evolution of migration; exaptation

INTRODUCTION

Many people worldwide, including the community of professional biologists, continue to be impressed, inspired, and challenged by the seasonal movements of

Address for correspondence: Theunis Piersma, Animal Ecology Group, P.O. Box 14, 9750 Haren, the Netherlands. Voice: +31-50-3632040; fax: +31-50-3635205.
theunis@nioz.nl

migratory birds around the globe. Perhaps as a result, the phenomenon of bird migration has attracted much focused scientific attention in the past few decades, and this has generated several dozens of volumes dealing exclusively with bird migration.[1–6] All these efforts seem to have subsumed, or inspired, the idea that bird migration is a truly biologically distinct and unique phenomenon, that birds (and perhaps a few other groups of animals) possess an integrated group of special traits that enable these migrations. This is the migratory syndrome of birds. Here we summarize and develop the discussions at the ESF workshop "Are There Specific Adaptations for Long Distance Migration in Birds? The Search for Adaptive Syndromes" at the Max-Planck Institute for Ornithology in Andechs/Seewiesen, Germany, from January 6–8, 2005. In several different ways we actually develop the theme "evolution of bird migration" beyond the discussions provided by Zink and Rappole.[7,8]

Sih *et al.*[9] recently defined a *behavioral* syndrome as "a suite of correlated behaviors reflecting between-individual consistency in behavior across multiple (two or more) situations. A population or species can exhibit a behavioral syndrome. Within the syndrome, individuals have a behavioral type (e.g. more aggressive versus less aggressive behavioral types)." When referring to the *migratory* syndrome, we tend to mean something deeper and older than a set of traits that is distinct at the individual or population level (but see below). An important aspect of (life history) syndromes in general is that they are highly integrated at the morphological, sensory, physiological, and behavioral levels.[10] Although we restricted our search to birds, a clade of derived dinosaurs, Dingle[11] took it a step further by searching for a migratory syndrome in all animal groups. He defined five basic migration characteristics: (1) persistent movement between distant sites; (2) directional movement; (3) inert behavior to arresting stimuli; (4) zugdisposition (i.e., distinct behavior for departure and arrival); and (5) energy allocation (i.e., migratory fueling). These characteristics all hold for migrating birds.

To structure our discussion, we first briefly discuss what it takes to be a long-distance migrant in general, starting off with some clear exaptations to seasonal migration (i.e., traits that evolved as responses to diverse selection pressures not related to migration, but later turned out to be useful preadaptations for a migratory lifestyle);[12] we then compile a listing of the set of biological traits required for a migratory lifestyle. We would expect all these traits to be integrated in a single migratory syndrome at some phylogenetic level, if there is such a thing. We therefore proceed by putting some of these expected traits in a phylogenetic context (how old and phylogenetically deep are these traits), and round up by discussing whether the idea of an old and integrated migratory syndrome can be upheld.

EXAPTATIONS AND ADAPTATIONS TO LONG-DISTANCE MIGRATION

To start our search for a migratory syndrome, we first seek to define traits that, in isolation or combined, characterize birds with a migratory lifestyle. It seems particularly relevant to search for traits that are more elaborate in migrants compared with (most) residents, and which may therefore represent true adaptations to migration.

To be able to move efficiently over long distances, a land animal needs wings to reduce the cost of locomotion—that is, to fly.[13] It also needs modest body size to be able to become airborne in the first place, as well as pointed wings (and other morphological features) to make flight energy efficient.[14–16] To reduce the mass carried aloft, a flying animal would do well with a particularly light skeleton (e.g., two-layered skull bones or generalized reduction of bone structures), and migrants would also need efficient respiratory systems, such as the lungs of birds that recirculate air to ensure rapid and complete oxygen extraction.[17] Birds possess all these traits (flight capacity including light bones, modest size, aerodynamic morphology, and an efficiently extracting respiratory system), but as these traits are general to almost all birds, they should be regarded as preadaptations, or exaptations. Although such exaptive traits—or more specifically, trait values—may increase fitness of migrants compared with alternative trait values, they need to be distinguished from true adaptations, which are derived characters built by selection *for their current roles*.[12,18,19]

More to the point, perhaps, would be a listing of the traits directly associated with the specific issues of long-distance migration. During the ESF workshop we came to the following listing of migration-related problems and their phenotypic solutions. Our list is neither novel nor very original. It has been put forward in various disguises in the older literature.[20] However, so far no attempt has been made to distinguish between preadaptations and true adaptations to migration.

Long-distance migrant birds typically need to deal with the following issues and have come up with the following solutions.

(1) *Precise timing of seasonal physiological events*: taken care of by the development of sophisticated endogenous circannual clocks that function as evolutionary ecological memory systems.[21–26] Circannual clocks help to predictably time a very diverse range of aspects of a bird's life cycle, including molt,[2,24] migratory fattening,[21,23,27] migratory direction and distance,[25,28–30] gonadal growth,[21,24] and changes in the bird's internal organ physiology.[31–34]

(2) *Finding the way over large distances*: solved by evolution of sophisticated long-distance orientation systems, including long-distance compass orientation based on global cues such as the stars, the sun, and Earth's magnetic field[35–40] combined with endogenous information about migratory direction,[25,30,41,42] to which learned components are added with experience (e.g., detection of North based on celestial rotation[43] and identification of cues needed to relocate a bird's first breeding and wintering site[44]). Nighttime compass orientation in particular may require special physiological and molecular adaptations.[45,46] In some species, evolution of social behaviors facilitating successful orientation have evolved; some birds use a so-called guiding strategy, during which young birds follow parents or other adult conspecifics during first migratory journeys.[44,47]

(3) *Endurance performance (extended fasting and intense exercise)*: this is achieved by (a) quick adjustments of metabolism—for instance, extremely fast and efficient fat metabolism and storage;[48–52] (b) seasonally predictive fueling and molt;[26,53] (c) organ flexibility (e.g., reduction of digestive tract size and subsequent rebuilding of digestive system within a few days);[54–56] (d) endurance musculature;[57–60] and (e) specialized hemoglobin with par-

ticularly high oxygen affinity (left-displaced oxygen extraction curves) enabling high-altitude flight exercise.[61–63]

(4) *Contrasting environments (different food, competitors, predators, and parasites)* necessitate (a) flexible digestive systems (e.g., gizzard changes in shorebirds;[54] (b) nutritional flexibility;[51,52] (c) broad-spectrum immune defense systems;[64] (d) physiological flexibility;[65] and (e) specific cognitive abilities.[66]

(5) *Predation and potential for overheating during flight:* taken care of in some groups of birds by night-time migration,[67,68] which demanded the evolution of orientation mechanisms specialized for nighttime travel [see (2)].

(6) *Tracking or predicting food resources*: possible with the capacities listed under (2), but also needs orchestrated seasonal changes in physiology and metabolism (1, 3), as well as concomitant life-history adjustments, resulting from the combination of the reproductive benefits of exploiting seasonal habitats and the mortality costs associated with migration.[69]

(7) *Seasonal time pressures*, which can be solved by (a) multitasking (overlapping of physiological changes, e.g., testicular development during northward migration,[70] special abilities to speed up physiological processes); (b) accurate circannual clocks [see (1)]; (c) special adaptations allowing migrants to cope with sleep deprivation;[71] (d) selection for speeding up physiological capabilities; and (e) optimization of flight speed and efficiency (e.g., aerodynamic shape, long and pointed wings, ontogenetic variation in shape of the flight apparatus[72–75]); and (f) optimization of stopover and flight.[76,77]

(8) *Continuous variation and some degree of unpredictability of resource distribution.* Unpredictability in the environment has led to relatively large genetic variability in migratory traits within migrant populations. This genetic variation combined with the fact that phenotypic expression of migratory behavior seems to be determined by a genetic threshold, at which birds abruptly change from being migratory to being sedentary (the so-called Zugschwelle) means that changed selection pressures can result in bird populations switching between sedentary and migratory lifestyles within a few generations.[8,25,78,79]

EMPIRICAL FINDINGS

A review of the occurrence of the just-mentioned traits in various birds suggests that many specific traits are correlated with migration distance, but that no single trait seems to be unique to migrants. The single exception to this rule could be nighttime compass orientation and the consequential physiological and molecular adaptations, which may be unique to night migrants.[45,46] The reason seems to be that although long-distance migration results in several problems and constraints, most of these are of a more general nature such as endurance capabilities and coping with food-type variability, which are also facing several resident species living in special or extreme environments. Thus, while no single trait seems to be unique to long-distance seasonal migrant birds, migrants do seem to be found at one extreme end of several more or less continuous trait distributions.

PHYLOGENETIC PERSPECTIVES: MAPPING MIGRATORY TRAITS ON PHENOGRAMS

Arguably, the best way to decide whether the previously listed migration-related traits or trait complexes represent aspects of one or more highly integrated and old migratory syndromes is mapping of the traits onto phylogenetic trees.[7,8,18] By studying migration in a phylogenetically explicit context, several studies have shown that migratory habits of birds are evolutionarily labile (FIG. 1). The whole range from

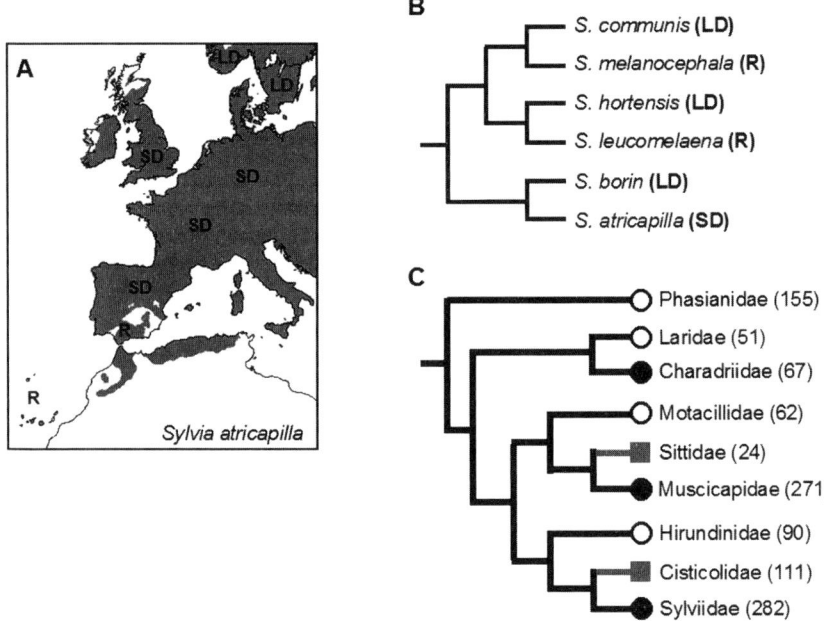

FIGURE 1. The complexity of the evolution of the "migratory syndrome" can be understood by analyzing its components in a phylogenetically explicit context. Variation in migratory behavior may be extensive within a single species, as shown in (**A**) by blackcaps *Sylvia atricapilla*, which include long-distance migrant (LD), short-distance migrant (SD), and resident populations (R) across their range (*shaded*). The same pattern is observed among closely related species, as shown in (**B**), by six *Sylvia* warblers (family Sylviidae[94]). However, evolutionary constraints are revealed at higher phylogenetic levels, as shown in (**C**), by nine bird families (the number of species in each family is shown in brackets). The tree (based on Refs. 95 and 96) shows six families that are primarily migratory (*circles*), contrasting with two families with conserved absence of migration (*gray branches* and *squares*). Within migratory families, the occurrence of nocturnal migration (*filled dots*), as opposed to diurnal migration (*open circles*), has been mapped as well. Although nocturnal migration seems to have repeatedly been gained and lost during the evolutionary history of birds, flying time is mostly constant within families.

long-distance migration to complete residency can be found within a single family or genus,[80] and such variation, together with the consequent changes in important components of the migratory syndrome like morphology, seasonal fattening, life history, and others, can evolve within a single species in just a few thousand years.[25,81] This circumstance makes us wonder whether or not the migratory syndrome is evolutionarily constrained.

However, looking at the occurrence of migration between bird taxa gives us little information on evolutionary constraints on each presumably adaptive trait. Migration is likely to be as old as the birds themselves (or even older), yet different bird taxa have evolved completely different solutions for moving long distances. For example, some species rely on endogenous programs to decide when and where to move, whereas others mostly follow social influences[47,82] or environmental cues.[83] Similar differences can be found for most of the traits common to all migratory birds (listed earlier), such as having orientation abilities (nocturnal and diurnal migrants use quite different clues) or putting on fat seasonally (warblers, waders, and geese may have quite different physiological mechanisms of fat deposition and use).[84]

Diverse solutions to the same problems posed by migration have probably evolved as independent responses to dissimilar selection pressures—not necessarily related to migration—in different bird taxa. As a consequence, if we are to identify evolutionary constraints on migratory syndromes, we first need to dissect such syndromes into traits and then determine whether such traits are homologous among the birds that share them or if they have evolved as independent adaptations in each group. For example, the occurrence of nocturnal migration is quite scattered along the phylogeny of birds, with nocturnal migrants being paraphyletic with respect to diurnal migrants (FIG. 1C). The presence of suspected physiological and molecular mechanisms for night orientation in nocturnally migrating garden warblers (*Sylvia borin*), but not in sedentary zebra finches (*Taeniopygia guttata*) or chickens (*Gallus gallus*) (which are included in mostly resident or otherwise diurnally migrating families[45,46]) suggests that diurnal migrants lack the machinery for nocturnal orientation (rather than keeping it unexpressed). Future studies should determine whether nocturnal migration evolved early and was repeatedly lost by different families or whether extant nocturnal migrants have independently evolved different physiological mechanisms of nighttime orientation (FIG. 1C).

The preceding example also illustrates the need to analyze the evolution of different potentially adaptive traits at different phylogenetic levels. The fact that all migratory families in FIGURE 1C include species with different behaviors (from resident to long-distance migrant) shows that some adaptations to migration can evolve very rapidly. Such traits can most elegantly be studied in species with diverse migratory behaviors, such as the blackcap (*Sylvia atricapilla*),[25] or in partial migrants and facultative migrants [e.g., redpoll (*Carduelis flammea*)[85]]. Considering the high susceptibility of migratory behavior to microevolution, different populations of the same species, or different species of the same genus, are likely to have the physiological and the molecular machinery needed to shift from migrant to nonmigrant in a relatively short time, as a response to ecological conditions.[79] However, such an evolutionary flexibility is not common to all species: although the adaptability of migratory patterns of some species seems little constrained,[27] other species retain apparently suboptimal migration patterns due to historical constraints.[86] Also very importantly, evolutionary flexibility is not common to all adaptations to migration.

Coming back to our example, all migratory families in FIGURE 1C are quite invariable with respect to time of migration (nocturnal or diurnal). This pattern indicates that some more fundamental physiological differences, such as the molecular or physiological machinery needed for nighttime orientation and/or magnetodetection, may be evolutionarily more constrained,[87] so their significance as true adaptations to migration can be evaluated only by comparing distantly related groups (families or higher taxa).

DISCUSSION

As we have seen, birds need many different adaptive traits to do a good job as a seasonal, long-distance migrant, even though any single species may possess only some of them. The phylogenetic perspective confirmed that there is no evidence for a single highly integrated and deep-rooted migratory syndrome. In fact, in various clades of birds, long-distance migrating species make up the tips of the trees, indicating that a migratory lifestyle, along with the necessary morphological, physiological, and behavioral adaptations, evolve and reevolve relatively quickly. At a deeper level, the ancestors to most birds (and certainly many of the dinosaurs[25]) have been migrants. Such an ancestry most likely provided them with the sensory acuity, clock- and-compass systems, and the basic performance machinery to become specialized long-distance migrants when the ecological need arose. However, with the evidence available today, we cannot discard the possibility that important traits associated with migration are true adaptations evolved during the radiation of birds, as innovative solutions to the same problems faced by their migratory ancestors (e.g., finding the way, putting on fat seasonally). To resolve this important issue, we need to disentangle the genetics, morphology, physiology, and ontogeny of the relevant traits. And such studies need to be combined with comparative analyses of the different trait features—from character architecture to current function—both within birds and between birds and other animals with variable migratory habits.

When considering adaptations to a migratory lifestyle, it is important to realize that migration requires a complex, highly integrated set of traits and that solutions to one problem may limit the possible solutions to another. For example, imprecision of the navigation system could limit the highest possible site fidelity and homing abilities of young birds.[44,88,89] In a similar way, remembrance of natal latitude[44] may limit otherwise adaptive breeding range change between years in individual birds.

Understanding the evolution of migratory adaptations is complicated by the different constraints (ecological, physiological, historical, and/or interactive) limiting adaptive changes in different traits. This also makes it difficult to predict how well migratory birds will be able to adapt to environmental change, either natural or induced by humans, which is an important conservation concern nowadays. Great advances in this direction have been achieved using some model species, such as the blackcap,[79] but an important implication of our review is that extending the conclusions obtained for one species to another may be problematic, particularly if the two species are distantly related.

One of the most critical challenges facing researchers in this field will be to understand the environmental triggers, cues, and regulators controlling seasonal and predictive changes in morphology, metabolism, and neuronal signaling, and how these changes are controlled on the molecular and physiological level in birds with different migratory strategies. Such approaches may enable us to identify truly integrated sets of traits (migratory syndromes) specific to one or more bird taxa with well-known evolutionary histories. Furthermore, understanding the architecture, integration, and evolution of particular traits or sets of migratory traits is not only important from a purely scientific perspective but will also help us to anticipate possible adaptive responses of migratory birds faced with natural or manmade environmental changes.[90–93]

CONCLUSION

Different evolutionary constraints on the various adaptations to migration make us discard the idea of the existence of an old and integrated migratory syndrome common to all migratory birds. Some traits may have been inherited from preavian common ancestors, whereas others have independently evolved at different times during the radiation of birds. We believe that different bird taxa, at various phylogenetic levels, share different sets of adaptations and preadaptations, which are expressed or suppressed depending on particular ecological circumstances. This hypothesis explains why birds like the blackcap can shift from resident to long-distance migrant in a few generations, changing endogenous programs, morphology, or life history adaptively. But it also explains why single species, or even whole families, rarely abandon an endogenous program in favor of socially influenced migrations or shift from nocturnal to diurnal migration.

ACKNOWLEDGMENTS

This contribution is dedicated to the memory of Eberhard Gwinner, who inspired so many of us and was such a master in asking the right questions in bird biology (and answering some of them!). We thank the European Science Foundation Bird Migration Program for financial support, the Max-Planck Institute for Ornithology for hosting the workshop, and the participants for their valuable and inspired input. Further financial support was received from the European Union (Marie Curie grant HPMF-CT-2002-02096 to JPT) and the Volkswagen Stiftung (to HM).

REFERENCES

1. ALERSTAM, T. 1990. Bird Migration. Cambridge University Press. Cambridge.
2. BERTHOLD, P. 1996. Control of Bird Migration. Chapman & Hall. London.
3. BERTHOLD, P. 2001. Bird Migration, a General Survey, 2nd ed. Oxford University Press. Oxford.
4. BERTHOLD, P., E. GWINNER & E. SONNENSCHEIN. 2003. Avian Migration. Springer-Verlag. Berlin.
5. GWINNER, E. 1990. Bird Migration. Physiology and Ecophysiology. Springer-Verlag. Berlin.
6. GREENBERG, R. & P.P. MARRA. 2005. Birds of Two Worlds: Ecology and Evolution of Migration. Johns Hopkins University Press. Baltimore, MD.

7. ZINK, R.M. 2002. Towards a framework for understanding the evolution of avian migration. J. Avian Biol. **33:** 433–436.
8. RAPPOLE, J.H., B. HELM & M.A. RAMOS. 2003. An integrative framework for understanding the origin and evolution of avian migration. J. Avian Biol. **34:** 124–128.
9. SIH, A., A. BELL & J.C. JOHNSON. 2004. Behavioral syndromes: an ecological and evolutionary overview. Trends Ecol. Evol. **19:** 373–378.
10. BADYAEV, A.V. 2004. Integration and modularity in the evolution of sexual ornaments. An overlooked perspective. *In* Phenotypic Integration. Studying the Ecology and Evolution of Complex Phenotypes. M. Pigliucci & K. Preston, Eds.: 50–79. Oxford University Press. Oxford.
11. DINGLE, H. 1996. Migration. The Biology of Life on the Move. Oxford University Press. New York.
12. GOULD, S.J. & E. VRBA. 1982. Exaptation—a missing term in the science of form. Paleobiology **8:** 4–15.
13. ALEXANDER, R. McN. 1998. When is migration worthwhile for animals that walk, swim or fly? J. Avian Biol. **29:** 387–394.
14. PENNYCUICK, C.J. 1978. Fifteen testable prediction about bird flight. Oikos **30:** 165–176.
15. PENNYCUICK, C.J. 1989. Bird Flight Performance. A Practical Calculation Manual. Oxford University Press. Oxford.
16. NORBERG, U.M. 1990. Vertebrate Flight. Mechanics, Physiology, Morphology, Ecology and Evolution. Springer-Verlag. Berlin.
17. POWELL, F.L. 2000 Respiration. *In* Sturkie's Avian Physiology, 5th ed. G.C. Whittow, Ed.: 233–264. Academic Press. San Diego.
18. BAUM, D.A. & A. LARSON. 1991 Adaptation reviewed—a phylogenetic methodology for studying character macroevolution. Syst. Zool. **40:** 1–18.
19. ROSE, M.R. & G.V. LAUDER, Eds. 1996. Adaptation. Academic Press. San Diego, CA.
20. DOLNIK, V.R. 1975. Migratory State of Birds [In Russian]. Nauka. Moscow.
21. GWINNER, E. 1977. Circannual rhythms in bird migration. Annu. Rev. Ecol. Syst. **8:** 381–405.
22. GWINNER, E. 1986. Circannual Rhythms. Springer-Verlag. Berlin.
23. GWINNER, E. 1996. Circadian and circannual programmes in avian migration. J. Exp. Biol. **199:** 39–48.
24. GWINNER, E. 1996. Circannual clocks in avian reproduction and migration. Ibis **138:** 47–63.
25. BERTHOLD, P. 1999. A comprehensive theory for the evolution, control and adaptability of avian migration. Ostrich **70:** 1–11.
26. GWINNER, E. & B. HELM. 2003. Circannual and circadian contributions to the timing of avian migration. *In* Avian Migration. P. Berthold, E. Gwinner & E. Sonnenschein, Eds.: 81–95. Springer-Verlag. Berlin.
27. BERTHOLD, P. *et al.* 1992. Rapid microevolution of migratory behaviour in a wild bird species. Nature **360:** 668–670.
28. BERTHOLD, P. 1973. Relationships between migratory restlessness and migratory distance in six *Sylvia* species. Ibis **115:** 594–599.
29. BERTHOLD, P. 1975. Migration: control and metabolic physiology. *In* Avian Biology. Vol. 5. D.S. Farner & J.R. King, Eds.: 77–128. Academic Press. New York.
30. GWINNER, E. & W. WILTSCHKO. 1978. Endogenously controlled changes in migratory direction of the garden warbler, *Sylvia borin.* J. Comp. Physiol. A **125:** 267–273.
31. PIERSMA, T. 1998. Phenotypic flexibility during migration: optimization of organ size contingent on the risks and rewards of fueling and flight? J. Avian Biol. **29:** 511–520.
32. PIERSMA, T., G.A. GUDMUNDSSON & K. LILLIENDAHL. 1999. Rapid changes in the size of different functional organ and muscle groups during refueling in a long-distance migrating shorebird. Physiol. Biochem. Zool. **72:** 405–415.
33. DIETZ, M.W., T. PIERSMA & A. DEKINGA. 1999. Body-building without power training: endogenously regulated pectoral muscle hypertrophy in confined shorebirds. J. Exp. Biol. **202:** 2831–2837.
34. LANDYS, M.M., T. PIERSMA & J. JUKEMA. 2003. Strategic size changes of internal organs and muscle tissue in the bar-tailed godwit during fat storage on a spring stopover site. Funct. Ecol. **17:** 151–159.

35. WILTSCHKO, W. & R. WILTSCHKO. 1972. Magnetic compass of European robins. Science **176:** 62–64.
36. WILTSCHKO, W. & R. WILTSCHKO. 1999. Das Orientierungssystem der Vögel—IV. Evolution. J. Ornithol. **140:** 393–418.
37. WILTSCHKO, W. & R. WILTSCHKO. 2002. Magnetic compass orientation in birds and its physiological basis. Naturwissenschaften **89:** 445–452.
38. EMLEN, S.T. 1975. The stellar-orientation system of a migratory bird. Sci. Am. **233:** 102–111.
39. COCHRAN, W.W., H. MOURITSEN & M. WIKELSKI. 2004. Migrating songbirds recalibrate their magnetic compass daily from twilight cues. Science **304:** 405–408.
40. MOURITSEN, H., G. FEENDERS, M. LIEDVOGEL & W. KROPP. 2004. Migratory birds use head scans to detect the direction of the earth's magnetic field. Curr. Biol. **14:** 1946–1949.
41. HELBIG, A.J. 1991. Inheritance of migratory direction in a bird species: a cross-breeding experiment with SE- and SW-migrating blackcaps (*Sylvia atricapilla*). Behav. Ecol. Sociobiol. **28:** 9–12.
42. HELBIG, A.J. 1996. Genetic basis, mode of inheritance and evolutionary changes of migratory directions in Palearctic warblers (Aves: *Sylviidae*). J. Exp. Biol. **199:** 49–55.
43. WILTSCHKO, W. *et al.* 1987. The development of the star compass in garden warblers, *Sylvia borin*. Ethology **74:** 285–292.
44. MOURITSEN, H. 2003. Spatiotemporal orientation strategies of long-distance migrants. *In* Avian Migration. P. Berthold, E. Gwinner & E. Sonnenschein, Eds.: 493–513. Springer-Verlag. Berlin.
45. MOURITSEN, H. *et al.* 2004. Cryptochromes and activity markers co-localize in bird retina during magnetic orientation. Proc. Natl. Acad. Sci. USA **101:** 14294–14299.
46. MOURITSEN, H. *et al.* 2005. A night vision brain area in migratory songbirds. Proc. Natl. Acad. Sci. USA, in press.
47. SCHÜZ, E. 1951. Überblick über die Orientierungsversuche der Vogelwarte Rossitten (jetzt: Vogelwarte Radolfzell). Proc. Xth Int. Ornithol. Congr., Uppsala, Sweden, pp. 249–268.
48. BAIRLEIN, F. & E. GWINNER. 1994. Nutritional mechanisms and temporal control of migratory energy accumulation in birds. Annu. Rev. Nutr. **14:** 187–215.
49. JENNI, L. & S. JENNI-EIERMANN. 1992. Metabolic patterns of feeding, overnight fasted and flying night migrants during autumn migration. Ornis Scand. **23:** 251–259.
50. JENNI, L. & S. JENNI-EIERMANN. 1998. Fuel supply and metabolic constraints in migrating birds. J. Avian Biol. **29:** 521–528.
51. BAIRLEIN, F. 2002. How to get fat: nutritional mechanisms of seasonal fat accumulation in migratory songbirds (Rev.). Naturwissenschaften **89:** 1–10.
52. BAIRLEIN, F. 2003. Nutritional strategies in migratory birds. *In* Avian Migration. P. Berthold, E. Gwinner & E. Sonnenschein, Eds.: 321–332. Springer-Verlag. Berlin.
53. PIERSMA, T. 2002. When a year takes 18 months: evidence for a strong circannual clock in a shorebird. Naturwissenschaften **89:** 278–279.
54. PIERSMA, T. & Å. LINDSTRÖM. 1997. Rapid reversible changes in organ size as a component of adaptive behaviour. Trends Ecol. Evol. **12:** 134–138.
55. MCWILLIAMS, S.R. & W.H. KARASOV. 2001. Phenotypic flexibility in digestive system structure and function in migratory birds and its ecological significance. Comp. Biochem. Physiol. A **128:** 579–593.
56. PIERSMA, T. & J. DRENT. 2003. Phenotypic flexibility and the evolution of organismal design. Trends Ecol. Evol. **18:** 228–233.
57. LUNDGREN, B.O. & K.H. KIESSLING. 1985. Seasonal variation in catabolic enzyme activities in breast muscle of some migratory birds. Oecologia **66:** 468–471.
58. BAIRLEIN, F. & U. TOTZKE. 1992. New aspects on migratory physiology of trans-Saharan passerine migrants. Ornis Scand. **23:** 244–250.
59. BAUCHINGER, U. & H. BIEBACH. 2001. Differential catabolism of muscle protein in garden warblers (*Sylvia borin*): flight and leg muscle act as a protein source during long-distance migration. J. Comp. Physiol. B **171:** 293–301.
60. GUGLIELMO, C.G., N.H. HAUNERLAND & T.D. WILLIAMS. 1998. Fatty acid binding protein in the flight muscle of migrating western sandpiper. Comp. Biochem. Physiol. B **119:** 549–555.

61. HIEBL, I. & G. BRAUNITZER. 1988. Anpassungen der Hämoglobine von Streifengans (*Anser indicus*), Andengans (*Chloephaga melanoptera*) und Sperbergeier (*Gyps rueppellii*) an hypoxische Bedingungen. J. Ornithol. **129:** 217–226.
62. LIANG, Y. et al. 2001. The crystal structure of bar-headed goose hemoglobin in deoxy form: the allosteric mechanism of a hemoglobin species with high oxygen affinity. J. Mol. Biol. **313:** 123–137.
63. KLAASSEN, M. 1996. Metabolic constraints on long-distance migration in birds. J. Exp. Biol. **199:** 57–64.
64. MØLLER, A.P. & J. ERRITZØE. 1998. Host immune defence and migration in birds. Evol. Ecol. **12:** 945–953.
65. BAIRLEIN, F. 1993. Ecophysiological problems of arctic migrants in the hot tropics. Proc. VIII Pan-Afr. Orn. Congr.: 571–578.
66. METTKE-HOFMANN, C. & E. GWINNER. 2003. Long-term memory for a life on the move. Proc. Natl. Acad. Sci. USA **100:** 5863–5866.
67. KERLINGER, P. & F.R. MOORE. 1989. Atmospheric structure and avian migration. Curr. Ornithol. **6:** 109-142.
68. PIERSMA, T., L. ZWARTS & J.H. BRUGGEMANN. 1990. Behavioural aspects of the departure of waders before long-distance flights: flocking, vocalizations, flight paths and diurnal timing. Ardea **78:** 157–184.
69. PÉREZ-TRIS, J. & J.L. TELLERÍA. 2002. Regional variation in seasonality affects migratory behaviour and life-history traits of two Mediterranean passerines. Acta Oecol. **23:** 13–21.
70. BAUCHINGER, U., A. WOHLMANN & H. BIEBACH. 2005. Flexible remodelling of organ size during spring migration in a long-distance migrant, the garden warbler (*Sylvia borin*). Zoology: in press.
71. RATTENBORG, N.C. et al. 2004. Migratory sleeplessness in the white-crowned sparrow (*Zonotrichia leucophrys gambelii*). PloS Biol. **2:** E212.
72. WINKLER, H. & B. LEISLER. 1992. On the ecomorphology of migrants. Ibis **134**(Suppl. 1): 21–28.
73. PÉREZ-TRIS, J. & J.L. TELLERÍA. 2001. Age-related variation in wing morphology of migratory and sedentary blackcaps, *Sylvia atricapilla*. J. Avian Biol. **32:** 207–213.
74. HEDENSTRÖM, A. 2003. Twenty-three testable predictions about bird flight. *In* Avian Migration. P. Berthold, E. Gwinner & E. Sonnenschein, Eds.: 563–582. Springer-Verlag. Berlin.
75. FIEDLER, W. 2005. Ecomorphology of the external flight apparatus of blackcaps (*Sylvia atricapilla*) with different migration behaviour. Ann. N.Y. Acad. Sci. **1046:** 253–263.
76. ALERSTAM, T. & A. LINDSTRÖM. 1990. Optimal bird migration. *In* Bird Migration. Physiology and Ecophysiology. E. Gwinner, Ed.: 331–351. Springer-Verlag. Berlin.
77. ALERSTAM T. & A. HEDENSTRÖM. 1998. The development of bird migration theory. J. Avian Biol. **29:** 343–369.
78. PULIDO, F., P. BERTHOLD & A.J. VAN NOORDWIJK. 1996. Frequency of migrants and migratory activity are genetically correlated in a bird population: evolutionary implications. Proc. Natl. Acad. Sci. USA **93:** 14642–14647.
79. PULIDO, F. & P. BERTHOLD. 2003. Quantitative genetic analysis of migratory behaviour. *In* Avian Migration. P. Berthold, E. Gwinner & E. Sonnenschein, Eds.: 53–77. Springer-Verlag. Berlin.
80. HELBIG, A.J. 2003 Evolution of bird migration: a phylogenetic and biogeographic perspective. *In* Avian Migration. P. Berthold, E. Gwinner & E. Sonnenschein, Eds.: 3–20. Springer-Verlag, Berlin.
81. PÉREZ-TRIS, J. et al. 2004. Historical diversification of migration patterns in a passerine bird. Evolution **58:** 1819–1832.
82. SUTHERLAND, W.J. 1998. Evidence for flexibility and constraint in migration systems. J. Avian Biol. **29:** 441–446.
83. ELKINS, N. 2004. Weather and Bird Behaviour. Christopher Helm. London.
84. LANDYS, M.M., T. PIERSMA, M. RAMENOFSKY & J.C. WINGFIELD. 2004. Role of the low-affinity glucocorticoid receptor in the regulation of behavior and energy metabolism in the migratory red knot *Calidris canutus islandica*. Physiol. Biochem. Zool. **77:** 658–668.

85. ZINK, G. & F. BAIRLEIN. 1996. Der Zug europäischer Singvögel: ein Atlas der Wiederfunde beringter Vögel, Vol. 3. AULA–Verlag. Wiesbaden.
86. RUEGG, K.C. & T.B. SMITH. 2002. Not as the crow flies: a historical explanation for circuitous migration in Swainson's thrush (*Catharus ustulatus*). Proc. Biol. Sci. **269**: 1375–1381.
87. MOURITSEN, H. & T. RITZ. Magnetoreception and its use in bird navigation. Curr. Opin. Neurol., submitted.
88. MOURITSEN, H. 1998. Modelling migration: the clock-and-compass model can explain the distribution of ringing recoveries. Anim. Behav. **56**: 899–907.
89. MOURITSEN, H. & O. MOURITSEN. 2000. A mathematical expectation model for bird navigation based on the clock-and-compass strategy. J. Theor. Biol. **60**: 283–291.
90. BOTH, C. & M.E. VISSER. 2001 Adjustment to climate change is constrained by arrival date in a long-distance migrant bird. Nature **411**: 296–298.
91. WALTHER, G.R. *et al.* 2002. Ecological responses to recent climate change. Nature **416**: 389–395.
92. BAIRLEIN, F. & O. HÜPPOP. 2004. Migratory fuelling and global climate change. Adv. Ecol. Res. **35**: 33–47.
93. PULIDO, F. & M. WIDMER. 2005. Are long-distance migrants constrained in their evolutionary response to environmental change? Causes of variation in the timing of autumn migration in a blackcap (*S. atricapilla*) and two garden warbler populations (*Sylvia borin*). Ann. N.Y. Acad. Sci. **1046**: 228–241.
94. SHIRIHAI, H., G. GARGALLO & A.J. HELBIG. 2001. Sylvia Warblers. Helm Identification Guide. Helm. London.
95. SIBLEY, C.G. & J.E. AHLQUIST. 1990. Phylogeny and classification of birds: a study in molecular evolution. Yale University Press. New Haven.
96. BARKER, F.K., G.F. BARROWCLOUGH & J.G. GROTH. 2001. A phylogenetic hypothesis for passerine birds: taxonomic and biogeographic implications of an analysis of nuclear DNA sequence data. Proc. Biol. Sci. **269**: 295–308.

Index of Contributors

Arlettaz, R., 81–95

Bairlein, F., 282–293
Baltic, M., 81–95
Bauchinger, U., ix–x, 214–215, 271–281, 282–293
Benowitz-Fredericks, Z M., 204–213
Biebach, H., 271–281
Both, C., 214–215

Chernetsov, N., 242–252

Fiedler, W., 253–263
Fusani, L., 264–270

Goymann, W., ix–x, 1–4, , 35–53
Groothuis, T.G.G., 168–180, 181–192
Gwinner, E., 216–227, 264–270

Hackl, R., 193–203
Helm, B., 216–227
Hirschenhauser, K., 138–153
Hunt, K.E., 109–137

Jenni-Eiermann, S., 1–4, 81–95, 96–108

Kitaysky, A.S., 204–213
Klasing, K.C., 5–16
Kotrschal, K., 138–153, 154–167
Kralj, S., 154–167

Möstl, E., 17–34, 138–153, 193–203
Mouritsen, H., 282–293

Palme, R., 17–34, 54–74, 75–80, 81–95, 96–108, 193–203
Pérez-Tris, J., 282–293
Piersma, T., 214–215, 282–293
Pulido, F., 228–241

Rettenbacher, S., 17–34, 193–203

Scheiber, I.B.R., 154–167

Thiel, D., 96–108
Touma, C., 54–74
Trost, L., 216–227

Von Engelhardt, N., 168–180, 181–192

Wasser, S.K., 109–137
Widmer, M., 228–241
Wingfield, J.C., 204–213